Molecular Aspects of Membrane Phenomena

International Symposium held
at the Battelle Seattle Research Center
Seattle, WA, USA, November 4–6, 1974

Edited by
H. R. Kaback H. Neurath G. K. Radda
R. Schwyzer W. R. Wiley

With 144 Figures

Springer-Verlag Berlin Heidelberg New York 1975

The Symposium was sponsored by
The Battelle Memorial Institute

In accord with that part of the charge of its founder, Gordon Battelle, to assist in the further education of men, it is the commitment of Battelle to encourage the distribution of information. This is done in part by supporting conferences and meetings and by encouraging the publication of reports and proceedings. Towards that objective, this publication, while protected by copyright from plagiarism or unfair infringement, is available for the making of single copies for scholarship and research, or to promote the progress of science and the useful arts.

ISBN-13:978-3-642-66226-3 e-ISBN-13:978-3-642-66224-9
DOI: 10.1007/978-3-642-66224-9

Library of Congress Cataloging in Publication Data. Main entry under title: Molecular aspects of membrane phenomena. Includes index. 1. Membranes (Biology)--Congresses. I. Kaback, H. Ronald, 1936-. II. Battelle Memorial Institute, Columbus, Ohio. Seattle Research Center. [DNLM: 1. Membranes--Metabolism--Congresses. 2. Membranes--Physiology--Congresses. 3. Energy transfer--Congresses. QH601 M718 1974] QH601.M63. 574.8'75. 75-25772.

Softcover reprint of the hardcover 1st edition 1975

Preface

This book is a compilation of formal presentations made during a three-day invitational conference at Battelle Research Center in Seattle, Washington. The purpose of organizing and publishing the proceedings of the conference is to provide a comprehensive survey of present knowledge on the determinants of membrane structure, the molecular specificity of membrane function, and the dynamic properties of membranes. Each presentation was followed by discussions which, because of their informal and spontaneous character, have not been included in this publication.

Molecular studies of biological membrane phenomena have progressed over the past decade to the point where it is now realistic to expect future resolution of the physico-chemical processes or forces governing the organization, function, and dynamic properties of membranes.

Drs. Radda, Kaback, and Schwyzer, each presenting a different approach to the biochemical and biophysical study of membranes, devoted four to six weeks as Battelle Visiting Fellows to developing the format and the roster of participants of the conference. The scientific "cluster" concept for planning the conference is reflected in the breadth of topics presented in this publication.

The overriding emphasis of the invited contributions is on the molecular aspects of three key membrane phenomena: *membrane dynamics*, *recognition*, and *energy coupling*. Presentations describe recent progress in defining the nature of the biochemical information which specifies membrane formation and the manner in which information, encoded in membranes, is functionally implemented at the molecular level. New research approaches are presented which delineate the dynamic properties of membranes and the significance of these properties to molecular membrane energy-coupling processes.

The organizing committee wishes to thank the Battelle Memorial Institute for the generous financial contribution which made the conference possible. The editorial staff of the Battelle Seattle Research Center has performed an invaluable service in the publication of this book. We also thank Shirley Lake, who was a perfect hostess for the conference. Finally, the Committee would like to express appreciation to the Session Chairmen: Drs. P. Siekevitz, B. Chance, V. T. Marchesi, and A. A. Eddy.

Seattle, Washington
April, 1975

H. R. Kaback (Nutley, NJ)
H. Neurath (Seattle, WA)
G. K. Radda (Oxford, England)
R. Schwyzer (Zürich, Switzerland)
W. R. Wiley (Richland, WA)

Contents

DYNAMIC PROPERTIES

RECOGNITION

ENERGY COUPLING

Participants

Dr. Nina M. Agabian
Department of Biochemistry
J567 Health Sciences SJ-70
University of Washington
Seattle, WA 98195, USA

Dr. Giovanna Ferro-Luzzi Ames
Department of Biochemistry
University of California
Berkeley, CA 94720, USA

Dr. Mordhay Avron
Department of Biochemistry
Weizmann Institute of Science
Rehovot, Israel

Dr. Anna Barker
Battelle
Columbus Laboratories
505 King Avenue
Columbus, OH 43201, USA

Dr. Eugene M. Barnes
Department of Biochemistry
Baylor College of Medicine
1200 Moursund Avenue
Houston, TX 77025, USA

Dr. William L. Bigbee
Battelle
Pacific Northwest Laboratories
P.O. Box 999
Richland, WA 99352, USA

Dr. Philip D. Bragg
Department of Biochemistry
University of British Columbia
Vancouver, B.C., Canada V6T 1W5

Dr. Max M. Burger
Department of Biochemistry
Biocenter of the University of Basel
4056 Basel
Klingelbergstrasse 70
Switzerland

Dr. James Callis
Department of Chemistry BG-10
University of Washington
Seattle, WA 98195, USA

Dr. Britton Chance
Johnson Research Foundation
Richards Building D-501
University of Pennsylvania
Philadelphia, PA 19174, USA

Dr. Dennis D. Cunningham
Department of Medical Microbiology
College of Medicine
University of California
Irvine, CA 92664, USA

Dr. Earl W. Davie
Department of Biochemistry
J317 Health Sciences SJ-70
University of Washington
Seattle, WA 98195, USA

Dr. Christoph de Haen
Division of Endocrinology
Department of Medicine RG-20
University of Washington
Seattle, WA 98195, USA

Dr. Harvey Drucker
Battelle
Pacific Northwest Laboratories
P.O. Box 999
Richland, WA 99352, USA

Dr. P. Leslie Dutton
Johnson Research Foundation
Medical School
University of Pennsylvania
Philadelphia, PA 19174, USA

Dr. Alan A. Eddy
University of Manchester
Institute of Science and Technology Medical School
Umist, Sackville Street
Manchester M60 1QD, England

Dr. Dave Fink
Battelle
Columbus Laboratories
505 King Avenue
Columbus, OH 43201, USA

Dr. Edmond H. Fischer
Department of Biochemistry
J367 Health Sciences SJ-70
University of Washington
Seattle, WA 98195, USA

Dr. Sidney Fleischer
Department of Molecular Biology
Vanderbilt University
Nashville, TN 37235, USA

Dr. C. Fred Fox
University of California
405 Hillgard Avenue
Los Angeles, CA 90024, USA

Dr. Carlos Gitler
Department of Chemical
Immunology
Weizmann Institute of Science
Rehovot, Israel

Dr. Milton P. Gordon
Department of Biochemistry
J391A Health Sciences SJ-70
University of Washington
Seattle, WA 98195, USA

Dr. David L. Gutnick
Roche Institute of Molecular
Biology
Kingsland Avenue
Nutley, NJ 07110, USA

Dr. Franklin M. Harold
National Jewish Hospital and
Research Center
3800 East Colfax Avenue
Denver, CO 80206, USA

Dr. Richard K. Haroz
Battelle
Geneva Research Centre
7, route de Drize
1227 Carouge-Geneva, Switzerland

Dr. Peter C. Hinkle
Section of Biochemistry, Molec-
ular and Cell Biology
Cornell University
Wing Hall
Ithaca, NY 14850, USA

Dr. Jen-shiang Hong
Brandeis University
415 South Street
Waltham, MA 02154, USA

Dr. H. Ronald Kaback
Roche Institute of Molecular
Biology
Nutley, NJ 07110, USA

Dr. Donald R. Kalkwarf
Battelle
Pacific Northwest Laboratories
P.O. Box 999
Richland, WA 99352, USA

Dr. Arthur Karlin
Columbia University
College of Physicians and
Surgeons
630 West 168th Street
New York, NY 10032, USA

Dr. John M. Keller
Department of Biochemistry
J529 Health Sciences SJ-70
University of Washington
Seattle, WA 98195, USA

Dr. George A. Kimmich
Department of Radiation Biology
and Biophysics
School of Medicine and Dentistry
University of Rochester
Rochester, NY 14624, USA

Dr. Leonard D. Kohn
National Institutes of Health
Section of Biochemistry of Cell
Reg., Lab. of Bioch. Pharm.
Building 4, Room B1-31
Bethesda, MD 20014, USA

Dr. Wilhelmus N. Konings
Laboratory of Microbiology
University of Groningen
Kerklaan 30, Haren (Gr)
The Netherlands

Dr. Jordan Konisky
Department of Microbiology
University of Illinois
323 Burrill Hall
Urbana, IL 61801, USA

Dr. William J. Lloyd
Department of Biochemistry
University of Oxford
South Parks Road
Oxford OX1 3QU, England

Dr. Vincent T. Marchesi
Yale School of Medicine
Department of Pathology
310 Cedar Street
New Haven, CT 06510, USA

Dr. Derek Marsh
Department of Biochemistry
University of Oxford
South Parks Road
Oxford OX1 3QU, England

Dr. Fritz Melchers
Basel Institute of Immunology
487 Grenzacherstrasse
4058 Basel, Switzerland

Dr. Mauricio Montal
Centro de Investigación y de
Estudios Avanzados del I.P.N.
Departamento de Bioquímica
Apartado Postal 14-740
México, D.F. 14

Dr. Hans Neurath
Department of Biochemistry
J405 Health Sciences SJ-70
University of Washington
Seattle, WA 98195, USA

Dr. William W. Parson
Department of Biochemistry
J535A Health Sciences SJ-70
University of Washington
Seattle, WA 98195, USA

Dr. Richard A. Pelroy
Battelle
Pacific Northwest Laboratories
P.O. Box 999
Richland, WA 99352, USA

Dr. Philip H. Petra
Departments of Obstetrics,
Gynecology, and Biochemistry
BB632 University Hospital RH-20
University of Washington
Seattle, WA 98195, USA

Dr. Robert L. Post
B-3307 Physiology Department
Vanderbilt Medical School
Nashville, TN 37235, USA

Dr. Herwig Puchinger
Battelle-Institut e.V.
6000 Frankfurt/Main 90
Postschliessfach 900160, Germany

Dr. George K. Radda
Department of Biochemistry
University of Oxford
South Parks Road
Oxford OX1 3QU, England

Dr. John P. Reeves
Department of Physiology
University of Texas
Southwest Medical School
5323 Harris Hines Boulevard
Dallas, TX 75234, USA

Dr. Roland Reiner
Battelle-Institut e.V.
6000 Frankfurt/Main 90
Postschliessfach 900160, Germany

Dr. Frederic M. Richards
Department of Molecular Bio-
physics and Biochemistry
Yale University
Box 1937 Yale Station
New Haven, CT 06520, USA

Dr. Russell Ross
Associate Dean for Scientific
Affairs
A315 D409 Health Sciences SJ-70
University of Washington
Seattle, WA 98195, USA

Dr. Gary Rudnick
Roche Institute of Molecular
Biology
Kingsland Avenue
Nutley, NJ 07110, USA

Dr. David F. Sargent
Biochemistry Department
University of Sydney
Sydney 2006
New South Wales, Australia

Dr. Gene A. Scarborough
Department of Biochemistry
University of Denver
Denver, CO 80220, USA

Dr. Michel Schneider
Battelle
Geneva Research Centre
7, route de Drize
1227 Carouge-Geneva, Switzerland

Dr. Richard P. Schneider
Battelle
Pacific Northwest Laboratories
P.O. Box 999
Richland, WA 99352, USA

Dr. Shimon Schuldiner
Roche Institute for Molecular
Biology
Kingsland Avenue
Nutley, NJ 07110, USA

Dr. Robert Schwyzer
Institute for Molecular Biology and Biophysics
Eidgenössiche Technische Hochschule
8049 Zürich, Switzerland

Dr. Bennett M. Shapiro
Department of Biochemistry
J591A Health Sciences SJ-70
University of Washington
Seattle, WA 98195, USA

Dr. Emanuel J. Shechter
Centre de Genetique Moleculaire
Centre National de la Recherche Scientifique
Gif-sur-Yvette 91, France

Dr. Steven A. Short
Roche Institute for Molecular Biology
Kingsland Avenue
Nutley, NJ 07110, USA

Dr. Philip Siekevitz
Rockefeller University
66th Street and York Avenue
New York, NY 10021, USA

Dr. Carolyn W. Slayman
Department of Physiology
Yale School of Medicine
New Haven, CT 06510, USA

Dr. Walther Stoeckenius
Cardiovascular Research Institute
University of California
San Francisco, CA 94143, USA

Dr. Christopher T. Walsh
Massachusetts Institute of Technology
Room 18-025
Cambridge, MA 02139, USA

Dr. William R. Wiley
Battelle
Pacific Northwest Laboratories
P.O. Box 999
Richland, WA 99352, USA

Dr. David F. Wilson
Johnson Research Foundation
University of Pennsylvania
Philadelphia, PA 19174, USA

Key to Photograph

1. Dr. Gary Rudnick
2. Dr. Dennis D. Cunningham
3. Dr. Anna Barker
4. Dr. Walther Stoeckenius
5. Dr. George K. Radda
6. Dr. H. Ronald Kaback
7. Dr. Robert Schwyzer
8. Dr. Nina M. Agabian
9. Dr. Jordan Konisky
10. Dr. Frederic M. Richards
11. Dr. Giovanna Ferro-Luzzi Ames
12. Dr. Robert L. Post
13. Dr. William J. Lloyd
14. Dr. Emanuel J. Shechter
15. Dr. Franklin M. Harold
16. Dr. Philip Siekevitz
17. Dr. John M. Keller
18. Dr. William L. Bigbee
19. Ms. Ellen Brandt
20. Dr. William R. Wiley
21. Dr. Hans Neurath
22. Dr. Christoph de Haen
23. Dr. John P. Reeves
24. Dr. Alan A. Eddy
25. Dr. Wilhelmus N. Konings
26. Dr. David L. Gutnick
27. Dr. Michel Schneider
28. Dr. Shimon Schuldiner
29. Dr. Carolyn W. Slayman
30. Dr. Britton Chance
31. Dr. Peter C. Hinkle
32. Dr. Sidney Fleischer
33. Dr. Vincent T. Marchesi
34. Dr. Mordhay Avron
35. Dr. Mauricio Montal
36. Dr. Carlos Gitler
37. Dr. Philip H. Petra
38. Dr. Philip D. Bragg
39. Dr. Jen-shiang Hong
40. Dr. Christopher T. Walsh
41. Dr. Richard K. Haroz
42. Dr. Arthur Karlin
43. Dr. Herwig Puchinger
44. Dr. Roland Reiner
45. Dr. Donald R. Kalkwarf
46. Dr. David F. Sargent
47. Dr. George A. Kimmich
48. Dr. Harvey Drucker
49. Dr. Dave Fink
50. Dr. Derek Marsh
51. Dr. Earl W. Davie
52. Dr. Fritz Melchers
53. Dr. C. Fred Fox
54. Dr. Steven A. Short
55. Dr. Eugene M. Barnes
56. Dr. Milton P. Gordon
57. Dr. David F. Wilson
58. Dr. Gene A. Scarborough
59. Dr. Max M. Burger
60. Dr. Russell Ross
61. Dr. Bennett M. Shapiro
62. Dr. Edmond H. Fischer
63. Dr. Richard P. Schneider
64. Dr. William W. Parson
65. Dr. Leonard D. Kohn

Not pictured:

Dr. James Callis
Dr. P. Leslie Dutton
Dr. Richard A. Pelroy

Dynamic Properties

Introduction

Philip Siekevitz
The Rockefeller University

The results of the many experiments being performed currently on membrane structure indicate that there can be little doubt that membranes have dynamic properties. What I think we all understand the term "dynamic" to mean in this case is that the structure is constantly undergoing change, that at any time the revealed structure, in a state of equilibrium, is the result of the interplay of forces operating within the structure and also impinging upon it from without. Thus, the dynamism occurs because there are forces holding the structure together in the face of forces tending to tear it apart. The endogenous forces--the static ones, if you wish--are contained in the configurations of the specific individual proteins and of the specific array of lipids, mostly phospholipids, that constitute any specific cellular membrane. If a membrane were merely a boundary enclosing metabolic segments of the cell, it would be enough to call it a static structure, held together by forces implicit in the structures of its constituent molecules. But no cellular membrane is merely a wall, for the very reason that most, if not all, of its proteins are enzymes. The mechanics of enzyme activity create movements in space, which of necessity, I think, jar the structure of which the enzymes are a part. Examples are the effects of light energy on chloroplast membranes and of substrate energy on mitochondrial membranes; such large-scale membrane movements can be easily observed when set in motion. The very existence of the transport properties of membranes also implies, in this case, a large-scale movement of membrane molecules relative to each other. Whether transport operates through the creation of "pores" or by the movement of "carriers" or by both, forces are involved that tend to disassemble a structure.

It should be remembered that there is also movement within the dimension of time. The exchanges of phospholipid molecules between various membranes of the cell, the turnover of membrane proteins, and dissimilar turnover rates among the proteins of the same membrane all point to the conclusion that, throughout a period of time, a membrane has a different constituency of its protein and lipid molecules; the protein and lipid specificities still exist, but the individual molecules are being constantly replaced. Again, the dynamic properties of a membrane are also revealed when we examine the natural, or unnatural, laboratory-induced development or differentiation of a membrane. The structure can be, and has been, modified in space by the addition to or the subtraction of some of its specific proteins or lipids.

I think, therefore, that the central problem in research in membrane structure is the elucidation of how a membrane continues to retain a specific structure, with specific properties, in the face of a seemingly unending series of blows tending to render it asunder. Thus, some of the recent research, and I think some of the chapters presented here, will deal with the elucidation of allowable limits: How much can a structure be modified, in space or in time, before it loses its specific identity? How much of the constituent molecules of a membrane can be eliminated without a loss of this identity? And of course, questions will be raised that perhaps are the primary ones: What do we mean by "identity"? What is the minimum structure that we can identify as being membranous? What is the minimum function that we can identify as being a necessity for a specific membrane? Once we get some answers to these questions, through the many techniques, biophysical and biochemical, that will be touched on in this book, then we can go on, I think with confidence, to the larger area--a description of the life of a biochemical structure.

Some New Approaches to ATPase Mechanisms in Energy Conversion*

Stuart J. Ferguson**, William J. Lloyd, and George K. Radda
Department of Biochemistry, University of Oxford, Oxford, England

INTRODUCTION

The plasma membranes of bacteria usually contain an ATPase which catalyzes oxidative phosphorylation or can couple the hydrolysis of ATP to transport. Since oxidative electron transfer can also drive active transport, it would appear that there is a common intermediate between the ATPase, electron transfer, and active transport. In higher organisms these functions are generally separated: ATP synthetic activity is located in the inner mitochondrial membrane while the plasma membrane, for instance, contains an ATPase that drives a Na^{+}/K^{+} pump. Other specialized ATPases include the Ca^{++} pump of sarcoplasmic reticulum and the enzyme which is found in the membranes of adrenaline storage vesicles (chromaffin granules).

In this chapter, we examine some novel features of three ATPases: those found in mitochondria, the bacterium *Micrococcus denitrificans*, and chromaffin granules. The fundamental problem is to understand how the chemical (hydrolytic) energy contained in ATP is linked to other energy forms (i.e., redox or chemical potential). In this presentation, we are not concerned with the nature of this linkage (this will be discussed in detail elsewhere in this symposium) but will concentrate on some new experimental approaches we have used to study events that occur at or around the membrane-bound ATPase. We summarize our objectives as follows:

- Can we discern any features of the isolated ATPases which can also be studied in the membrane-bound enzyme and which may give an indication of how they work in energy transducing processes?
- The various *energy fluxes must be controlled* in relation to one another in order to optimize the efficiency of energy coupling.

---

*Supported by the Science Research Council, London.
**SJF thanks the Medical Research Council for a training award.

It is therefore pertinent to ask, (a) to *what extent the phospholipid environment* modulates energy utilization, (b) whether *charge separation per se has a controlling influence* and whether this influence in some way is reflected in the properties of the membrane, and (c) if subunit interactions' *"conformational" control* and/or cooperativity are important in this type of control.

• There is clear indication that *vectorial movement of charge* (in or across the membrane) does take place. It is therefore important to try to point out how and why asymmetry arises in a membrane. This is particularly important in relation to the preparation of small vesicular systems for membranes (e.g., submitochondrial particles) where it is often accepted that "inversion" takes place. In such cases, it is important to ask, however, whether an apparent inversion might not represent some scrambling of the membrane components. We have, therefore, developed methods based on ^{31}P nuclear magnetic resonance that in the first instance allow us to define the factors that contribute to an asymmetric distribution of phospholipids in small bilayer vesicles.

MITOCHONDRIAL AND BACTERIAL ATPases: THE ROLE OF SUBUNITS

As a step toward identifying some of the functional groups and subunits in a membrane-bound ATPase, we have studied the chemical properties of mitochondrial ATPase (F_1), both in isolated and membrane-bound forms, and have compared them with those of the bacterial enzyme in particles prepared from *Micrococcus denitrificans*. We have shown in the last few years that the reagent NBD-chloride (7-chloro-4-nitrobenzo-2-oxa-1,3-diazole) is useful in investigating reactivities of SH-groups (BIRKETT et al., 1970); in conformational changes associated with ligand binding and enzyme regulation, by utilizing the fluorescence properties of the covalent label (BIRKETT et al., 1971); and in looking at half of the sites type reactivity in glyceraldehyde-3-phosphate dehydrogenase (PRICE, RADDA, 1972; 1974). The reaction of NBD chloride with isolated F_1 turned out to be rather unusual in that the spectrum of the reaction product differed from that of any other SH (or amino) derivative (Figure 1)--it had no fluorescence and the NBD-group could be readily removed from the enzyme by a variety of SH-compounds like N-acetyl cysteine (FERGUSON, LLOYD, RADDA, 1974). Figure 1 shows that after the addition of cysteine the spectrum shifts to the normal S-NBD derivative and that the material formed this way

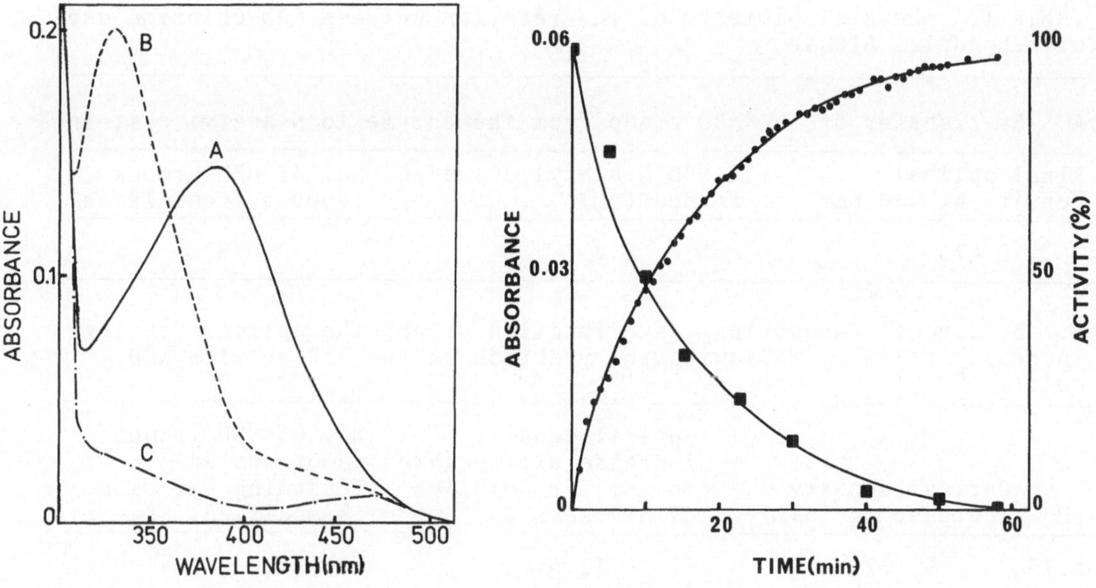

Fig. 1 Fig. 2

Fig. 1. Absorption spectra of F_1 after reaction with NBD-chloride. A (————) after ammonium sulfate precipitation and passage through Sephadex G-25; B (----------) "A" following addition of dithiothreitol (500 µm); C (-•-•-•-•-) "B" after ammonium sulfate precipitation and passage through Sephadex G-25

Fig. 2. Increase in absorption at 385 nm and decrease in ATPase activity with time on treatment with NBD-chloride. Enzyme activity = ■; absorption at 385 nm = ●

was removed by gel filtration. On the basis of a variety of spectroscopic and chemical criteria we have identified the reactive side chain in F_1 as tyrosine (FERGUSON et al., in press). We have also used two independent methods to establish that one mole of NBD (4-nitrobenzo-2-oxa-1,3-diazole-group) was incorporated per mole of F_1, Table 1, (mol wt 360,000). This incorporation led to the total loss of the ATPase activity which was recovered completely on removal of the NBD-group by SH-compounds. The stoichiometric relation between incorporation and activity was maintained throughout the reaction course (Figure 2). The rate constants for inactivation and NBD-incorporation are identical (FERGUSON, LLOYD, RADDA, 1974) and are at least 10^3 times larger than the rate constant for the reaction of N-acetyl-tyrosine ethyl ester with NBD-chloride.

On treatment of the O-NBD-tyrosyl ATPase with SDS, the label is transferred to SH-groups in the unfolded protein as seen by the spec-

Table 1. The stoichiometry of the reaction between NBD-chloride and mitochondrial ATPase

A. By transfer of the NBD group from the enzyme to N-acetyl cysteine[a]

Final optical density at 420 nm	S-NBD N-acetyl cysteine formed (μM)	No. of NBD groups removed from ATPase
0.076	5.8	0.97

B. By directly measuring, as a function of pH, the optical density increase at 385 nm following the reaction of the ATPase with NBD-chloride[b]

pH	Optical density increase $mg^{-1}ml^{-1}$*	Optical density increase extrapolated to a molar solution of ATPase x 10^{-3}	No. of NBD groups incorporated (assuming tyrosine is modified)
6.75	0.0320	11.6	1.0
7.00	0.0306	11.0	0.95
7.25	0.0320	11.6	1.0
7.5	0.0340	12.2	1.05
7.8	0.0295	10.6	0.91
8.0	0.0320	11.6	1.0
8.25	0.0340	12.2	1.05
8.75	0.0295	10.6	0.91
9.0	0.0300	10.8	0.93

[a] 6 μM NBD-modified soluble mitochondrial ATPase was incubated in a buffer containing 50 mM triethanolamine hydrochloride, 4 mM ATP, 2 mM EDTA, 200 mM sucrose pH 7.5 at 30°C. The reaction was initiated by addition of 1 mM N-acetyl cysteine and the production of S-NBD N-acetyl cysteine was followed spectrophotometrically at 420 nm. The results shown are the average of two experiments.

[b] 50 mM triethanolamine hydrochloride-KOH was used at 30°C at the pH values indicated in the table. Unneutralized tris was added to buffers above pH 8.25 in order to complement the buffering capacity of the triethanolamine. Buffers included 200 mM sucrose, 4 mM ATP, and 2 mM EDTA with 5 to 8 μM of mitochondrial ATPase and 100 μM NBD-chloride. Reactions were followed spectrophotometrically at 385 nm.

*Cuvettes had a 1 cm lightpath.

tral changes recorded in Figure 3. This prevents identification of the subunit that is modified, as SH-groups are present in several of the subunits (KNOWLES, PENEFSKY, 1972). When, however, the O-NBD-tyrosyl-ATPase is allowed to stand at pH 9.0, the NBD-group undergoes an intramolecular O $\rightarrow$ N shift as is shown in Figure 4. The N-NBD enzyme (which

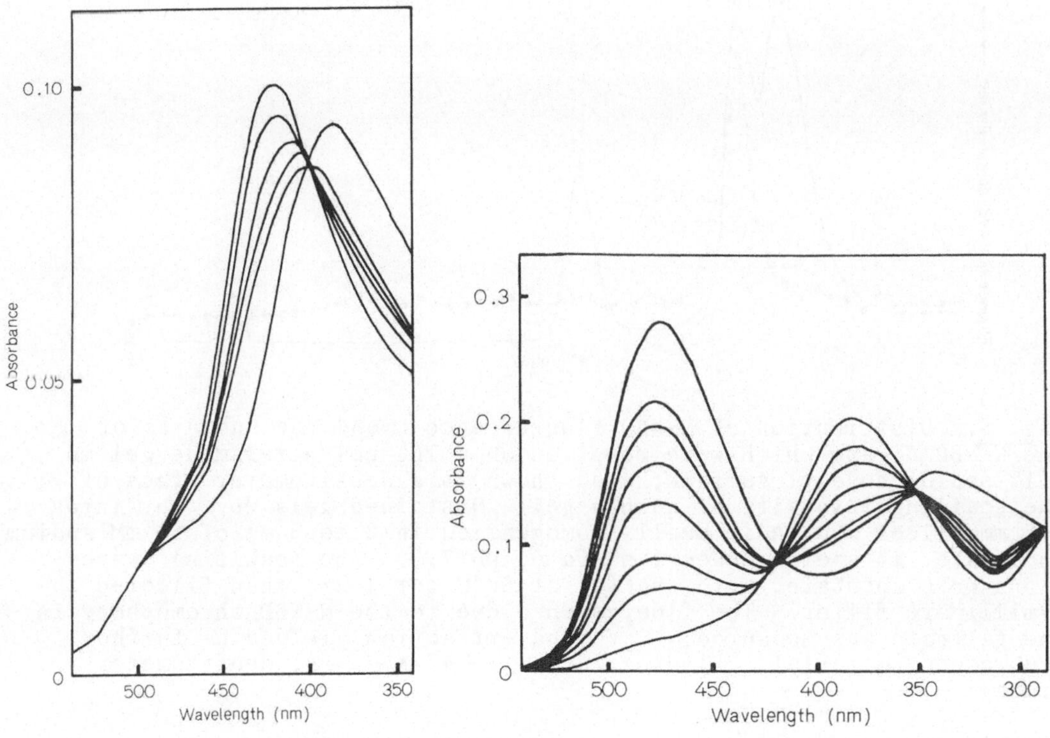

Fig. 3 Fig. 4

Fig. 3. Behavior of NBD-modified mitochondrial ATPase on addition of 0.1% sodium dodecyl sulfate. Reaction conditions were as described in the legend of Table 3 with the omission of a sulfhydryl reagent. Shift from absorbance at 385 nm to 420 nm was measured at the following times: 0, 2, 4, 5, 10, 30 minutes. NBD-modified ATPase, 6 μM. The reaction was initiated by addition of sodium dodecyl sulfate

Fig. 4. Transfer of the NBD-group from a tyrosyl oxygen to a nitrogen of mitochondrial ATPase followed by observing the shift in absorbance spectrum. Approximately 5 mg of an ammonium-sulfate precipitate of O^{tyr}-NBD-ATPase was dissolved in a buffer of 10 mM triethanolamine hydrochloride, 4 mM ATP, 2 mM EDTA, 200 mM sucrose and desalted on a column of Sephadex G-25 (medium), equilibrated with the same buffer. Subsequently sufficient 2M tris was added to bring the pH to 9.0. Incubation was at 30°C and spectra were taken at 0, 30, 80, 160, 250, 400, 600, and 1200 minutes

still contains nearly one mole of NBD-group per mole of enzyme and is now fluorescent) is still almost completely inactive but because of the stability of the N-NBD-group cannot be reactivated. This stability of the product also allows the examination of the subunit modified by gel electrophoresis (Figure 5). All the fluorescence is found in the β subunit, thus identifying one of these subunits as being "essential" for activity. Although it is possible that the O-NBD-group is also in

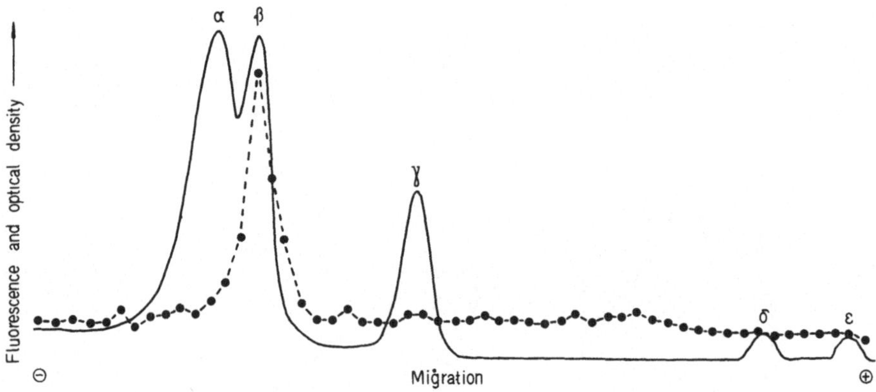

Fig. 5. Distribution of N-NBD fluorescence among the subunits of the N-NBD ATPase which were resolved on a 10% polyacrylamide gel in 0.1% sodium dodecyl sulfate; also shown is a densitometer trace of the staining intensity of such a gel. Unstained gels were cut into 1.5 mm slices and individually homogenized in 2 to 5 ml of 50 mM sodium phosphate, 1% sodium dodecyl sulfate, pH 7.5. The resulting suspensions were incubated with shaking at 30°C for 1 hr, then filtered on a millipore filter. The fluorescence due to the N-NBD chromophore in the filtrate was measured. A second extraction yielded no further fluorescent material. Fluorescence = ●—●—●—●—●; densitometer trace = ————————

the β subunit, it cannot be taken as proven, since we cannot exclude the possibility of *inter*-subunit O → N transfer of the label even though the reaction is clearly *intramolecular*.

Now a remarkable feature of this modification is the observation that it can also be done relatively cleanly with the membrane-bound ATPase. The rate constants for inactivation of the ATPase of submitochondrial particles and of particles prepared from *Micrococcus denitrificans* are also very similar to those of the isolated ATPase (Table 2) as is the case of reversibility by SH-compounds (Table 3).

The inactivation of the ATPase activity in submitochondrial particles is also accompanied by a concomitant decrease in the ATP-induced 1-anilino-naphthalene-8-sulfonate response (Figure 6), which again is recovered on reversal of the reaction.

We have examined the relation between loss of ATPase and ATP synthesis activities on modification of *Micrococcus denitrificans* particles (FERGUSON et al., 1974). Figure 7 shows that ADP-linked respiratory control is abolished in this system and is regenerated after addition of SH-compounds. The rates of the losses of the two activities at three pH values (Figure 8) are clearly the same. (It has also been demonstrated that ATP-synthesis follows the same pattern as ADP-linked respiratory control.) These observations are consistent

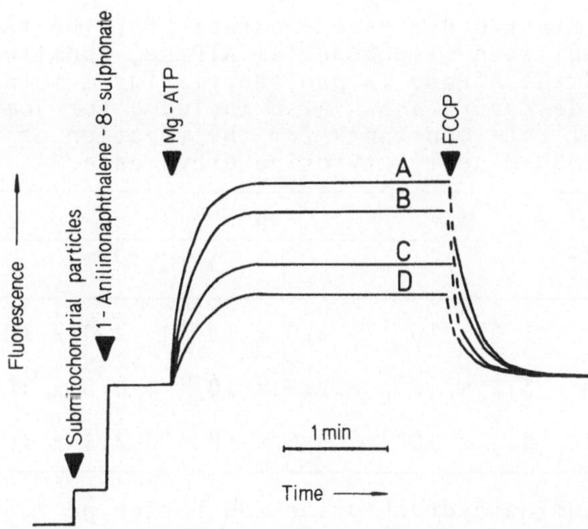

Fig. 6. The effect of preincubation of submitochondrial particles with NBD-chloride on 1-anilino-naphthalene-8-sulfonate fluorescence enhancements. Particles (20 mg/ml) were incubated at 0°C with 200 µM NBD-chloride in 50 mM tris acetate buffer pH 7.5. At the times indicated in the table, aliquots were added to 2 ml samples of a buffer containing 225 mM mannitol, 75 mM sucrose, and 20 mM tris pH 7.4 at 20°C. The particles, after addition of 5 µM 1-anilino-naphthalene-8-sulfonate, were energized with 1 mM $MgCl_2$ and 1 mM ATP. A = No incubation with NBD-chloride; ATPase activity 100%. C = 2 hr incubation with NBD-chloride; ATPase activity 50%. D = 3 hr incubation with NBD-chloride; ATPase activity 38%. B = "D" after incubation with 1 mM dithiothreitol; ATPase activity 95%. As indicated, 1 µM FCCP (carbonyl p-trifluoromethoxyphenylhydrazone), an uncoupler, returned the fluorescence to the unenergized level

Table 2. A comparison of the rates of inhibition of solubilized and submitochondrial particle ATPase and the ATPase from phosphorylating membrane fragments of *Micrococcus denitrificans*[a]

	Time required for 50% inhibition of ATPase activity (minutes)		
pH	Solubilized mitochondrial ATPase	Submitochondrial particles	*Micrococcus denitrificans*[b]
7.0	40	35	30
7.5	16	15	8
8.0	1.75	2	1.5

[a]The conditions for the inhibition of *Micrococcus denitrificans* ATPase can be found in FERGUSON et al. (1974). Solubilized mitochondrial ATPase (3 mg protein per milliliter) and submitochondrial particle ATPase (10 mg protein per milliliter) were incubated at 20°C with 200 µM NBD-chloride in a buffer containing 50 mM triethanolamine hydrochloride-KOH, 200 mM sucrose, 4 mM ATP, and 2 mM EDTA at pH 7.5.
[b]Calculated from the data of FERGUSON et al. (1974).

Table 3. Pseudo first order rate constants for the reactivation of NBD modified solubilized mitochondrial ATPase, submitochondrial particle ATPase, and the ATPase in phosphorylating membrane fragments from *Micrococcus denitrificans*. Also included for comparison, the pseudo first order rate constants for the reaction of sulfhydryl compounds with O^{tyr}-NBD-N-acetyl tyrosine ethyl ester[a]

Sulfhydryl reagent	A	B	C	D
	k(min^{-1})			
Dithiothreitol	1.6×10^{-1}	2.0×10^{-1}	2.0×10^{-1}	no value
Glutathione	3.2×10^{-2}	9.5×10^{-2}	6.0×10^{-2}	4.1×10^{-1}
N-acetyl cysteine	8.5×10^{-3}	5.0×10^{-2}	2.5×10^{-2}	1.7×10^{-1}

[a] 50 mM triethanolamine hydrochloride-KOH buffer pH 7.5 was used at 30°C, 200 mM sucrose, 4 mM ATP, and 2 mM EDTA with sulfhydryl compounds at 1.0 mM. A = NBD modified solubilized mitochondrial ATPase (6 μM); B = NBD modified submitochondrial particles ATPase (total particle protein 5 mg/ml); C = NBD modified ATPase of *Micrococcus denitrificans* membrane vesicles (total particle protein, 5 mg/ml); D = O^{tyr}-NBD-N-acetyl tyrosine ethyl ester (50 μm). A, B, and C were followed by the appearance of ATPase activity with time. D was followed spectrophotometrically at 420 nm.

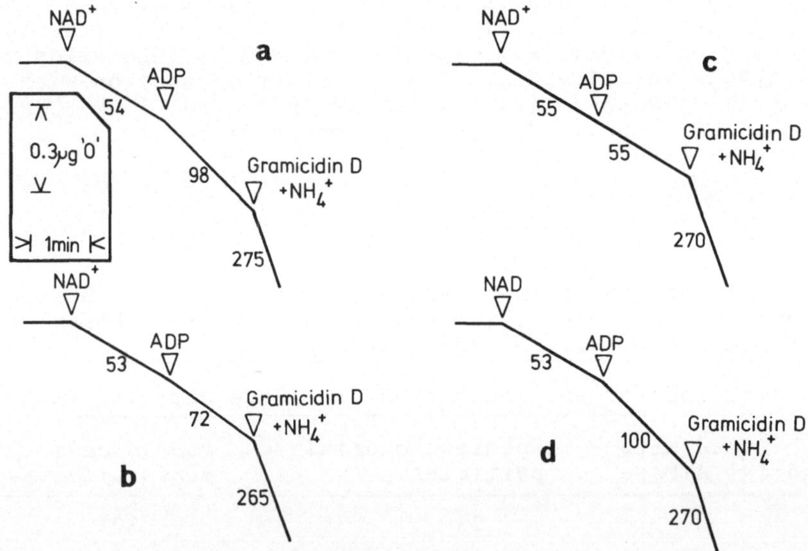

Fig. 7. Effect of NBD-chloride on respiratory control in phosphorylating particles from *Micrococcus denitrificans*. Particles were incubated at a concentration of 1.8 mg/ml in a medium containing 50 mM tris-acetate, pH 7.5, and 0.2 mM NBD-chloride at 20°C for (a) 0 min, (b) 10 min, (c) 30 min, and (d) 30 min followed by 5 min incubation with 1 mM dithiothreitol. Samples (0.1 ml) were taken for assay in a medium containing, in a total volume of 3 ml: 10 mM tris phosphate pH 7.3, 5 mM magnesium acetate, 0.2 mg yeast alcohol dehydrogenase, and 30 μl ethanol. The reaction was started by the addition of NAD^+ (0.6 mM), subsequently ADP (0.2 mM), and, as uncoupler, gramicidin D (1 μg) plus ammonium acetate (30 mM) were added as indicated. Oxygen uptake was measured at 30°C with a Clark-type oxygen electrode

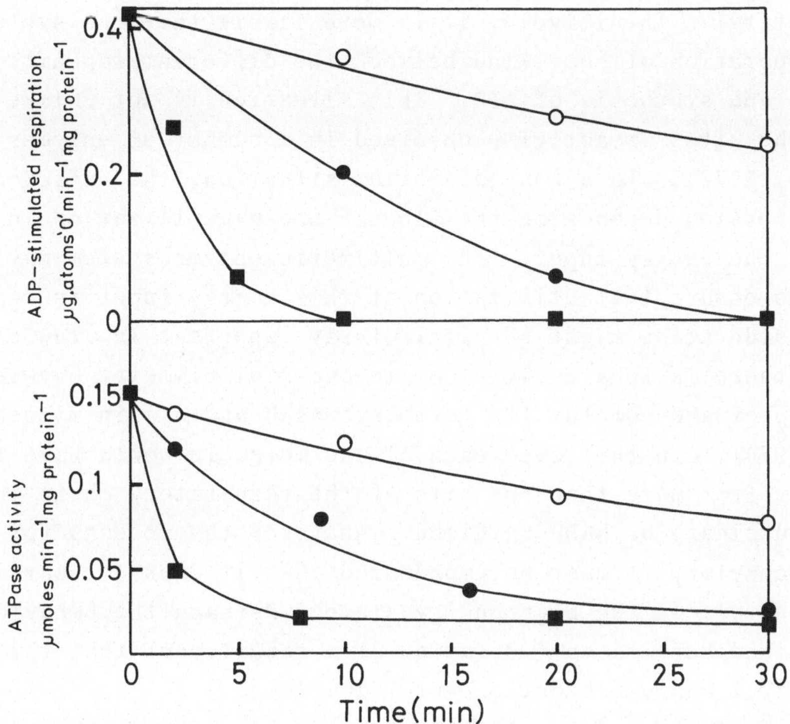

Fig. 8. The effect of the pH at which phosphorylating particles of *M. denitrificans* are incubated with NBD-chloride on the rate at which oxidative phosphorylation and ATPase activity are inhibited. o = pH 7.0; • = pH 7.5; ▪ = pH 8.0. Particles were incubated with NBD-chloride at the pH values shown and samples were assayed at the times indicated in the figure. Respiration rates were determined as described in Figure 7. ATPase activity was measured by monitoring phosphate release in an assay medium containing 5 mM ATP, 5 mM $MgCl_2$, 20 mM tris-acetate, 150 mM sodium bicarbonate pH 8.0

with a common mechanistic pathway for ATP synthesis and hydrolysis at the level of the ATPase.

The stoichiometry of the reaction between NBD-chloride and the membrane-bound ATPase and the site of modification has not yet been established. The reactions for the isolated and membrane-bound ATPases, however, are closely similar (in rate constants, pH-dependence, and reversibility). This suggests that the two processes are the same. It is therefore pertinent to ask what role the subunits play in a multimeric enzyme when modification of only one of them leads to total loss of activity. The simplest explanation is that there is only one catalytic site. This is not very likely, particularly since the three β subunits (SENIOR, 1973), one of which is presumably modified on an N-group, appear to be identical (although minor differences between them would not have been detected by the methods used to establish their identity). Intuitively, it is more likely that the system requires cooperation of some kind between the different subunits in the hydrolysis and synthesis of ATP. This situation is not unlike the "half-of-the-sites" reactivity observed in a number of enzymes (LAZDUNSKI, 1972). In a nonequilibrium situation, the efficiency of energy conversion depends on the flow of energy utilization in relation to that of the energy input. The multimeric enzyme system may be designed to ensure that utilization of this energy input is continuous. Such a consideration might be particularly important in view of the fact that there is apparently a one-to-one stoichiometry between the ATPase complex and complex III (cytochromes b and c_1) in mitochondria (SLATER, 1974). In this case each ATPase molecule would have to convert energy from more than one site of the respiratory chain into ATP for each succinate or NADH oxidized. Whatever the reasons for this one-site behavior, it must be emphasized that it does not say anything about the nature of the high energy state. Perhaps the behavior we have described here is an indication of a requirement that the ATPase should control and synchronize energy coupling.

It is worth pointing out that the reaction between NBD-chloride and ATPases might form the basis of a method for distinguishing ATPase activities in crude cell homogenates. The ATPases involved in oxidative phosphorylation, photophosphorylation, and bacterial transport are probably all sensitive to NBD-chloride in the manner described here, but are, in general, relatively insensitive to sulfhydryl reagents such as N-ethyl maleimide. In contrast, most other types of ATPase are sensitive to both N-ethyl maleimide and, we expect, in a different way, to NBD-chloride.

MODULATION OF ATPases BY PHOSPHOLIPIDS

There is good evidence on the basis of lipid substitution experiments that different phospholipids enhance or inhibit the activities of membrane-bound enzymes to different extents (WARREN et al., 1974). While such experiments do provide important and interesting information about the nature of protein-lipid interactions, it is worthwhile to consider another possibility. If, for example, the activity of a membrane-bound ATPase is controlled by an "energy pressure" which is present in the form of a charge or potential gradient of some kind, it is conceivable that in the mobile phase of a membrane some reorganization takes place. Rearrangements of this type could alter the immediate lipid environment of the ATPase and hence its activity, thus providing a mechanism for expressing the "energy pressure" as a molecular control element in the membrane. For instance, the dissociation of the mitochondrial ATPase inhibitor under the influence of an "energy pressure" might depend on such a mechanism.

The ATPase in adrenaline storage vesicles is thought to be involved in the active accumulation of adrenaline (TAUGNER, 1971) which in vivo is 0.5 M in the vesicles. The temperature dependence of the ATPase activity is rather unusual and shows anomalous behavior close to physiological temperatures (RADDA, in press). This temperature profile can be compared with the temperature dependence of phospholipid mobility as observed either by fluorescence polarization or by ^{31}P nuclear magnetic resonance. The motional characteristics (apparent rotational relaxation time) of two fluorescent probes, (12-[9-anthroyl stearate] AS, and 2-[9-anthroyl palmitate] AP) as a function of temperature (1/T) showed a deviation from linearity with a small break in the plots at the same temperature as the change seen in the Arrhenius plot of ATPase activity (RADDA, in press). Again, measurement of phospholipid ^{31}P nuclear magnetic resonance relaxation times as a function of temperature revealed a discontinuity in behavior at the same temperature (BARKER, RADDA, unpublished observations). Although these observations do not establish a causal relationship between the phenomena, they indicate the possibility that small changes in lipid structure (clustering, phase separation, or chain motion) can exert significant control on enzymatic activity. It is interesting in this case that the temperature-dependent change is close to the usual operating temperature of the system. One could imagine that ionic or charge perturbations at this temperature are sufficient to cause changes of the magnitude required for this control.

CHROMAFFIN GRANULE ATPase AND CHARGE SEPARATION

Control of ATPase activities and uncoupler effects on them in mitochondrial and bacterial systems is well known, but much less is known about the control of ATPases which are solely involved in transport. However, this phenomenon has recently been studied with the chromaffin granule system (BASHFORD, RADDA, RITCHIE, 1975). The fluorescent probe 1-anilino-naphthalene-8-sulfonate (ANS), previously extensively used to study the nature of the energized state in mitochondria and submitochondrial particles (AZZI et al., 1969), once again proved to be very useful. Addition of ATP and Mg^{2+} to chromaffin granules (and membrane vesicles prepared from them) in the presence of ANS results in a slow fluorescence increase which is abolished by the addition of conventional uncouplers. Associated with these uncoupling effects is a significant enhancement in the ATPase activity. The ATP/Mg^{2+}-induced ANS response and the uncoupler-stimulated ATPase activity both show pH optima at pH 6.4, while the "coupled" ATPase shows no such pH dependence. On the other hand, the uncoupling of the ANS response by a given concentration of uncoupler has a very different pH dependence. Below about pH 5.5 the effect of adding uncoupler is to further enhance the fluorescence of ANS, but the ATPase activity ceases to be uncoupler-sensitive and indeed the activity may be inhibited by the uncoupler. The observations suggest that if the role of uncoupler is to equilibrate protons across the membrane, then the pH of the chromaffin granule matrix is about 5.5.

At pH 6.6 metal ions other than Mg^{2+} produce the same activity from the "coupled" ATPase as Mg^{2+} itself. In contrast, in the uncoupled system, the different ions (Mn^{2+}, Co^{2+}) show different efficiencies (BASHFORD, RADDA, RITCHIE, 1975). It seems, therefore, that in the "coupled" system the rate-limiting step is not the turnover rate of ATPase, and that this step is independent of both the medium pH and the metal ion cofactor.

It is known that the "energy linked" ATPase activities of mitochondria, submitochondrial particles, and the plasma membranes of some microorganisms are sensitive to uncoupling agents and inhibitors of oxidative phosphorylation. Changes in ANS fluorescence have also been correlated with "energization" in submitochondrial particles (AZZI, et al., 1969). We believe that the chromaffin granule system is the first example of an uncoupler-sensitive ATPase activity not of this general type. One interpretation of these observations is that the ATPase activity of chromaffin granule membranes is associated with the movement of cations, probably protons, across or into the membrane. Hence, the sensitivity to uncoupling agents can be understood. Such a movement of protons would lead to the development of both electrical and chemical potential within or across the chromaffin granule membrane. The ANS response to "energization" may be due to either or both of these components. However, the uncoupling observed under conditions where the membrane is leaky with respect to charged species (e.g., in the presence of a permeant anion or of ionophores that move cations other than protons) indicates that the element due to electrical potential is important both in terms of ATPase activity and in the ANS response to "energization."

LIPID ASYMMETRY

It has been demonstrated by KAGAWA, KANDRACH, and RACKER (1973) that in order to reconstitute the mitochondrial ATPase into phospholipid vesicles with $^{32}P_i$-ATP exchange activity, a strict stoichiometry of lecithin to phosphatidylethanolamine is required, and that the optimum ratio depends on the presence or absence of cardiolipin. Clearly, the asymmetric assembly of these phospholipids may be very important. New ways of looking at phospholipid asymmetry in mixed phospholipid vesicles have been developed and used to try to define the factors that contribute to asymmetry (BERDEN, BARKER, RADDA, 1975).

The method relies on the selective perturbation of the magnetic resonance signals of phospholipid head-groups (either from the protons of the $\overset{+}{N}Me_3$ group or from the phosphorus) on the inside or outside of a phospholipid bilayer vesicle by a nonpenetrating paramagnetic shift or broadening reagent. For example, the $\overset{+}{N}Me_3$ resonance of phosphatidylcholine is split into two signals by the addition of ferricyanide (BERDEN, BARKER, RADDA, 1975). The ratio of the intensities gives the number of lecithin molecules on the inside and outside surfaces of the vesicle. In vesicles containing mixed lipids, the individual components can be observed by ^{31}P in favorable cases. Addition of Co^{2+} to the outside of such a vesicle broadens out the resonances of the groups on that side while inclusion of Co^{2+} into the vesicles has the opposite effect (BERDEN, BARKER, RADDA, 1975).

From such experiments, it can be shown that at pH 7.0 the ratio of phosphatidylcholine outside/inside is about 1:8, while the same ratio for phosphatidylserine is only 1:2 in a mixed vesicle of these two phospholipids, indicating the preference of phosphatidylserine for the inside. This preference is accentuated with decreasing vesicle size (BERDEN, BARKER, RADDA, 1975). Also as the pH of the medium is lowered, the asymmetry is increased. The same final distribution is obtained both when the lipid vesicles are prepared at a given pH or when the external pH is adjusted to the desired value. In the latter case, however, it takes about 2 hr for the lipid redistribution to take place. Similar measurements on mixtures of phosphatidylcholine with phosphatidylethanolamine, phosphatidylinositol, and phosphatidic acid show that all these lipids have a preference for the inside, while in contrast sphingomyelin is more favorably placed at the outside. To what extent this kind of asymmetry is important in reconstitution experiments is not known, but there is clear chemical evidence for lipid asymmetry in natural membranes.

SUMMARY

The properties of three ATPases--those found in mitochondria, *Micrococcus denitrificans*, and chromaffin granules--are examined. The role of specific functional groups in the first two were examined by their reactivity toward 7-chloro-4-nitrobenzo-2-oxa-1,3-diazole. This reagent modifies a single essential tyrosine group per enzyme molecule. The total loss of ATPase activity (and the ability to synthesize ATP by the membrane-bound enzyme) may relate to the possible role of subunit interactions in energy utilization.

The mobility of membrane phospholipids (studied by fluorescence techniques) modulates the activity of the chromaffin granule ATPase, and the activity of this enzyme is also enhanced by the addition of uncouplers.

Membrane asymmetry may be important in the sectional aspects of energy coupling. The asymmetric distribution of phospholipids in small vesicles was studied by ^{31}P nuclear magnetic resonance.

REFERENCES

AZZI, A., CHANCE, B., RADDA, G. K., LEE, C. P.: Proc. Nat. Acad. Sci. U.S.A. 62, 612-619 (1969).

BASHFORD, C. L., RADDA, G. K., RITCHIE, G. A.: FEBS Letters 50, 21-24 (1975).

BERDEN, J. A., BARKER, R. W., RADDA, G. K.: Biochim. Biophys. Acta 375, 186-208 (1975).

BIRKETT, D. J., DWEK, R. A., RADDA, G. K., RICHARDS, R. E., SALMON, A. G.: Eur. J. Biochem. 20, 494-508 (1971).

BIRKETT, D. J., PRICE, N. C., RADDA, G. K., SALMON, A. G.: FEBS Letters 6, 346-348 (1970).

FERGUSON, S. J., JOHN, P., LLOYD, W. J., RADDA, G. K., WHATLEY, F. R.: Biochim. Biophys. Acta 357, 457-461 (1974).

FERGUSON, S. J., LLOYD, W. J., LYONS, M. H., RADDA, G. K.: Eur. J. Biochem. (in press).

FERGUSON, S. J., LLOYD, W. J., RADDA, G. K.: FEBS Letters 38, 234-236 (1974).

KAGAWA, Y., KANDRACH, A., RACKER, E.: J. Biol. Chem. 248, 676-684 (1973).

KNOWLES, A. F., PENEFSKY, H. S.: J. Biol. Chem. 247, 6624-6630 (1972).

LAZDUNSKI, M.: In: Current Topics in Cellular Regulation (ed. HORECKER, STADTMAN) Vol. VI, pp. 267-310. London and New York: Academic Press 1972.

PRICE, N. C., RADDA, G. K.: In: Structure and Function of Oxidation Reduction Enzymes (ed. A. AKESON, A. EHRENBERG), pp. 161-169. Oxford and New York: Pergamon Press 1972.

PRICE, N. C., RADDA, G. K.: Biochim. Biophys. Acta 371, 102-116 (1974).

RADDA, G. K.: Phil. Trans. R. Soc. Lond. B. (in press).

SENIOR, A. E.: Biochim. Biophys. Acta 301, 249-277 (1973).

SLATER, E. C.: In: Dynamics of Energy Transducing Membranes (ed. ERNSTER, ESTABROOK, SLATER), pp. 1-20. Amsterdam: Elsevier 1974.

TAUGNER, G.: Biochem. J. 123, 219-225 (1971).

WARREN, G. B., TOON, P. A., BIRDSALL, N. J. M., LEE, A. G., METCALFE, J. C.: Proc. Nat. Acad. Sci. U.S.A. 71, 622-626 (1974).

Chemical Probes for the Geometry of Membrane Proteins*

Frederic M. Richards, James V. Staros, Kuan Wang, and H. Heitzmann
Department of Molecular Biophysics and Biochemistry, Yale University, New Haven, Connecticut 06520

INTRODUCTION

The position of the protein components of biological membranes both perpendicular to and in the plane of the permeability barrier has been the subject of much recent interest. The "sidedness" of the identifiable peptide chains has been examined with low-molecular-weight reagents similar to those used in modification studies on soluble proteins, and with such macromolecular reagents as enzymes or binding proteins. With the former category, use is made of nucleophilic groups on the exposed side chains of the proteins as the required reaction sites (BERG, 1969; BRETSCHER, 1971a, b, c; STECK, 1972; WHITELEY, BERG, 1974), while enzymes need exposed peptide bonds for proteolysis (BENDER et al., 1971; STECK et al., 1971; STECK, 1972; TRIPLETT, CARRAWAY, 1972) or aromatic side chains for iodination reactions (PHILLIPS, MORRISON, 1971a, b; 1973); lectins act as specific sugar binding reagents (e.g., NICOLSON, SINGER, 1974). These studies have recently been reviewed by WALLACH (1972), BRETSCHER (1973), ZWALL et al. (1973), STECK (1974), and SINGER (1974).

To be useful as a vectorial probe, the reagent must not penetrate the permeability barrier, so that its action, if any, may be said to be confined to the exterior or interior face of the membrane. For intact cells, normally only the exterior components are accessible to such reagents. Direct experiments on the interior components are then done either on special vesicular preparations where there may be good evidence for inversion of the membrane, or by difference between the intact cell results and those obtained on fragmented preparations. In both approaches one must assume conservation of molecular structure during a rather drastic change at the cellular level.

*This work was supported by Grant GM-12006 from the Institute of General Medical Sciences of the National Institutes of Health.

The two-dimensional arrangement of the proteins in the plane of the membrane has been examined extensively at the cytochemical level by a variety of different labeling procedures and where position specification is made with respect to cellular dimensions. Distributions at the molecular level represent a much less developed area of research. Some specialized systems with high concentrations of single protein species such as the disks of retinal rod outer segments (BLASIE et al., 1969; BLASIE, 1972) or the purple membrane of halobacteria (HENDERSON, 1975) have permitted X-ray diffraction studies revealing distributions in the range of molecular dimensions. Chemical approaches have been limited to a few cross-linking studies whose interpretation, until recently, has been very difficult.

GENERAL LABELING WITH A PHOTOACTIVATABLE REAGENT

For problems requiring general probes, the uncertain nature of the protein functional groups available (if any) makes results from the use of normal chemical reagents subject to some uncertainty because of reaction specificity. With enzymes as reagents, the reactive group problem is added to possible difficulties arising from steric restrictions because of the size of such macromolecular probes. Small size will minimize the latter problem, while omnivorous reactivity will in principle solve the former problem. Reagents making use of carbenes and nitrenes seem to fill the bill, at least on paper. Both classes of compounds have been proposed as affinity labels: carbenes some years ago by WESTHEIMER (SINGH et al., 1962), and, more recently, nitrenes by KNOWLES (see FLEET et al., 1969; review by KNOWLES, 1972).

Aryl nitrenes may suffer less than aliphatic carbenes from internal rearrangements. We thus picked this class of compounds for some initial attempts at general photolabeling in a membrane system. The difficulties that these procedures may encounter for specific affinity labeling have been well described by RUOHO et al. (1973), but we hoped to minimize these by seeking a general label where specificity would be avoided as much as possible.

The principal reagent that we have used so far is the nitrene generated by photolysis of N-(4-azido-2-nitrophenyl)-2-amino-ethylsulfonate (abbreviated NAP-taurine).

$$^{-}SO_3CH_2CH_2NH\text{-}C_6H_3(NO_2)\text{-}N_3 \xrightarrow{h\nu} {}^{-}SO_3CH_2CH_2NH\text{-}C_6H_3(NO_2)\text{-}N$$

Precursor — Active Reagent

The azide is easily prepared containing either tritium (STAROS et al., 1974) or ^{35}S (STAROS, RICHARDS, 1974). In experiments on intact erythrocytes, no detectable labeling of hemoglobin occurred, indicating little or no penetration of the reagent through the membrane. Gel electrophoresis patterns of the membrane protein components obtained with both whole cell and leaky ghost preparations are shown in Figure 1. While bands 1, 2, 4, 5, and 6 are not labeled in the whole-cell preparations (nomenclature of STECK, 1972), detectable labeling is observed in the regions of the minor bands 2.1 to 2.6, in band 3, and in the periodic acid-Schiff (PAS) staining region following band 3. The latter two are expected on the basis of the earlier work of a number of investigators. Bands 2.1 to 2.6, however, have not previously been reported as accessible to external membrane probes. It is not yet known whether steric accessibility or the absence of appropriate functional groups at the external surface is a proper explanation of the previous results.

The nitrene produced by photolysis is highly reactive, and labeling can be carried out at any reasonable temperature. While the cells do appear to be impermeable to the precursor reagent near 0°C, slow penetration of the azide does occur at 37°C. Intact cells can thus be loaded with the reagent in the dark at 37°C. The temperature is then lowered and the cells thoroughly washed, also in the dark. Photolysis and labeling of the interior components can subsequently be carried out. Most of the reagent is "lost" by reaction with hemoglobin, but enough is left to get some labeling of the internal membrane components. Many of the components expected to be on the cytoplasmic side of the membrane do get labeled. However, some peculiar effects are observed. Band 4.1, for instance, is labeled very poorly or not at all from either the inside or the outside with NAP-taurine. However, as will be seen below, there is good evidence that this component is complexed to spectrin, can be cross-linked to hemoglobin in intact erythrocytes, and thus must extend at least some distance out from the cytoplasmic surface. The data discussed above show some of the possibilities and also some of the difficulties connected with these reagents. The high reactivity of the active species assures the labeling of all compounds within physical range of the reagent, lipids, and sugars, as well as proteins. The inert azide can be placed where one wants before the labeling is started by illumination of the sample. This triggering capability is unique to photochemical probes as a general class. Unfortunately, while the probe is inert in the dark, the reagent has other properties that may distort the simple picture. In

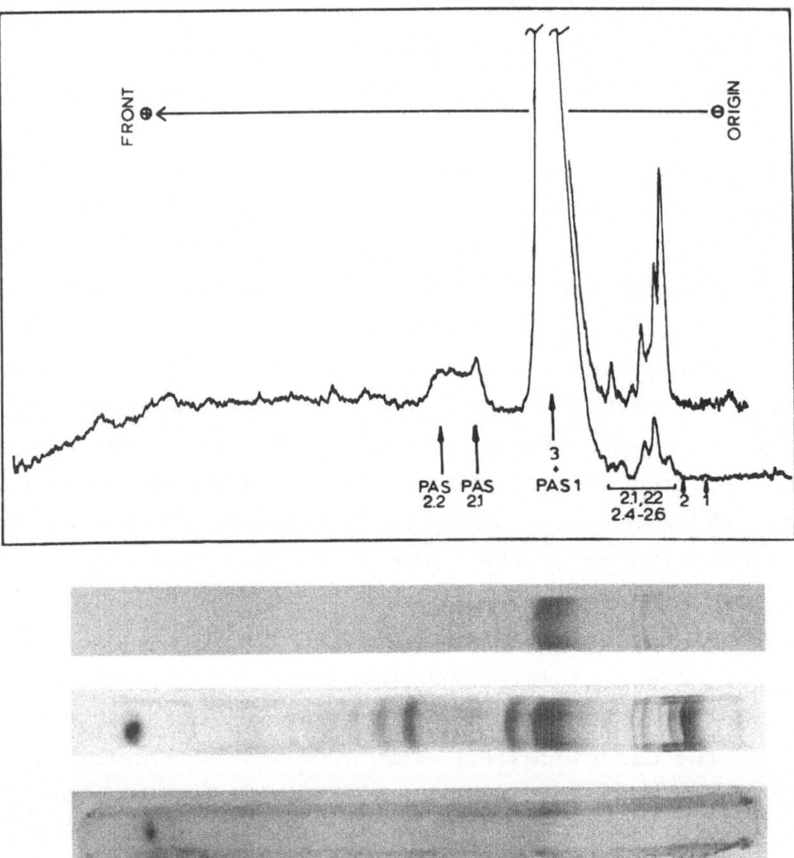

Fig. 1a

Fig. 1. Gel patterns from labeled erythrocytes and ghosts. (a) Intact erythrocytes labeled with [^{35}S]NAP-taurine. The panel shows a densitometer trace of an autoradiograph of the 4.5% sodium dodecyl sulfate gel. The upper trace on the right of the diagram was made from another autoradiograph with longer exposure. Below the panel and in register with it are shown (top) the autoradiograph corresponding to the full trace above; (center) the dried, Coomassie blue-stained gel slice from which the autoradiograph was taken; and (bottom) a dried slice from a periodic acid-Schiff (PAS) stained parallel gel. (b) Erythrocyte ghosts labeled and separated as above. Labeling of all visibly stain-

Figure 1, it can be noted that band 3 is very heavily labeled compared to all other components. Although this band does represent a large amount of the membrane protein, the very heavy labeling probably reflects the anion binding function of this component(s) (CABANTCHIK, ROTHSTEIN, 1972, 1974). Thus the quantitative aspects of the observed pattern cannot be used with confidence to compare, for example, the accessible surface area of the various proteins except in an all-or-none sense. In the internal labeling pattern, the small amount of

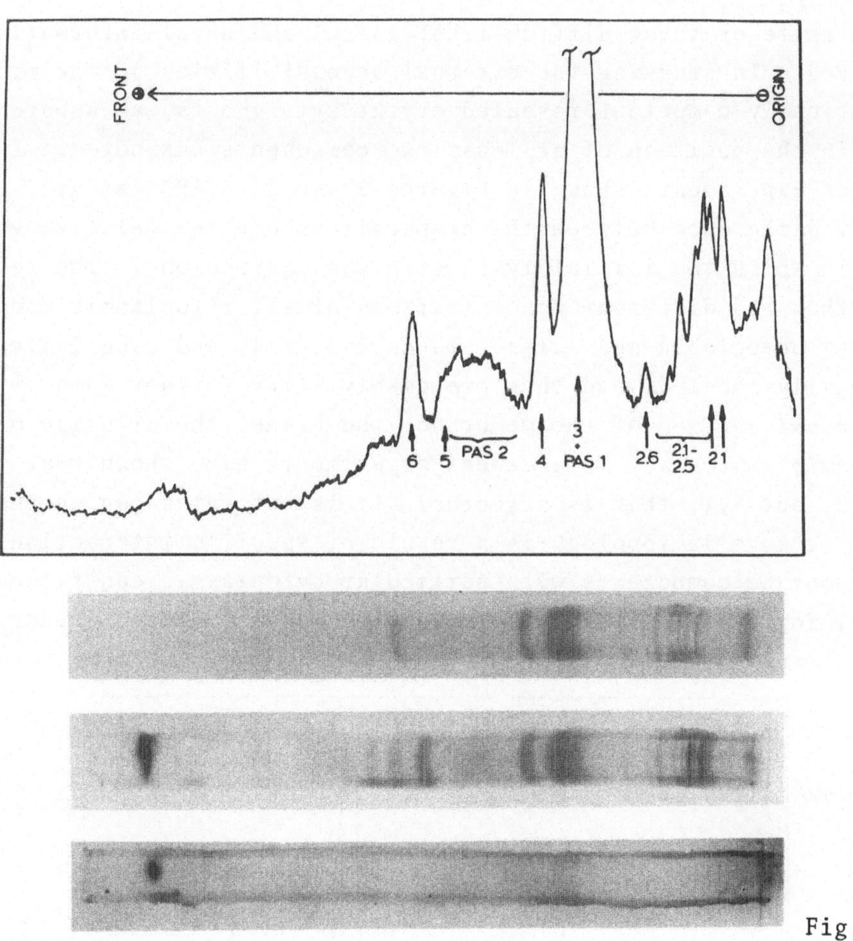

Fig. 1b

ing bands can be seen. The peak at the extreme right in the densitometer trace is highly polymerized protein which barely entered the gel, presumably formed by oxidation and incompletely reduced in this sample. Note: Untreated erythrocyte ghosts that were purified, electrophoresed, and stained as described gave visible sodium dodecyl sulfate gel patterns identical with those shown in this figure for NAP-taurine treated intact erythrocytes and ghosts. The numbering of the protein and glycoprotein bands is adapted from that of FAIRBANKS, STECK, and WALLACH (1971). (Reprinted from STAROS and RICHARDS, 1974.)

hemoglobin always isolated in a membrane preparation has a very high specific activity and almost overwhelms the low-molecular-weight part of the pattern. Dimers, trimers, and perhaps tetrameters of globin chains are seen in the gel patterns as well as the major monomer peak. These cross-linked complexes are not normally seen with monofunctional reagents and may reflect a type of complication to be found with these radical-forming reagents.

In spite of these difficulties, useful and novel information can be derived. In studying the external accessibilities of the proteins in osmotically competent resealed erythrocyte ghosts, an apparent change in the position of at least two components was noted. In the series of experiments shown in Figures 2 and 3 (STAROS et al., 1974), the only difference between the preparations was the relative volume of buffer in which the initial lysis step was carried out. The resealed ghosts thus had different concentrations of all cytoplasmic components, including hemoglobin and salts. Bands 2.3, 2.4, and especially 4 are more heavily labeled, and thus presumably stick farther into or through the *external surface* of the membrane, the higher the dilution of the *cytoplasmic contents*. More recent experiments have shown that it is band 4.2, not 4.1, that is affected. It is not yet known whether this apparent change in topology is a result of specific interactions of these membrane components with particular cytoplasmic constituents or whether, for example, it may be a general ionic strength effect.

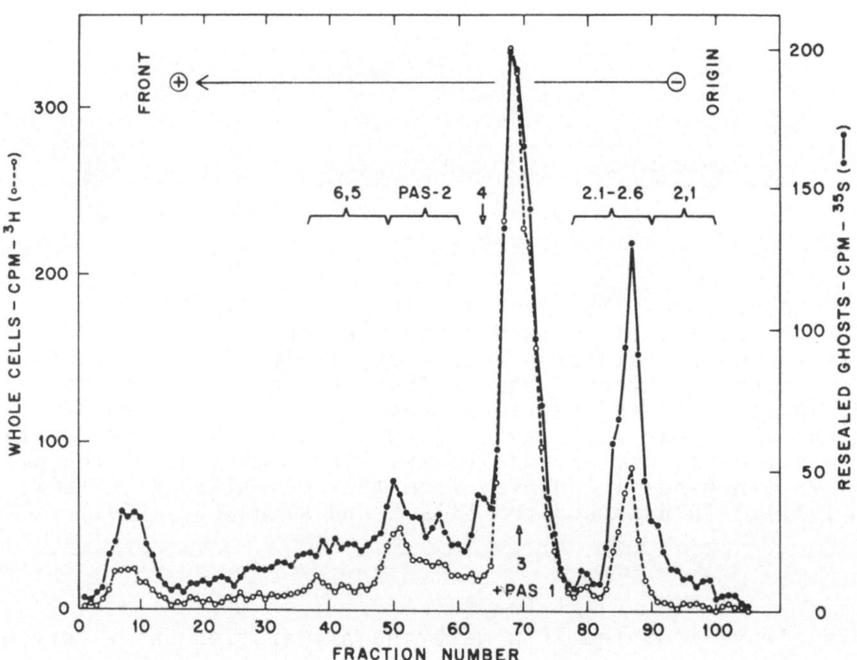

Fig. 2. Gel electrophoretic pattern of double label experiment on whole cells and resealed ghosts. Intact erythrocytes were labeled with NAP-[^{3}H]taurine; resealed ghosts were prepared from cells lysed at a hemolytic ratio of 20 and were labeled with NAP-[^{35}S]taurine. Aliquots of the membrane solutions were then mixed and electrophoresed together in one gel, which was then fractionated and counted. The isotope patterns were arbitrarily normalized at fraction 68, the top of band 3. The numbering system for the protein and glycoprotein bands is adapted from that of FAIRBANKS, STECK, and WALLACH (1971). (Reprinted from STAROS, HALEY, and RICHARDS, 1974.)

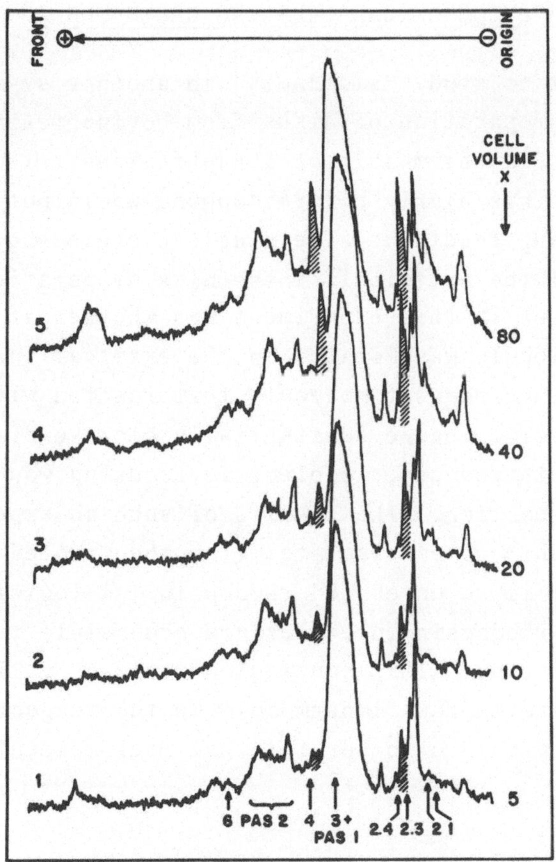

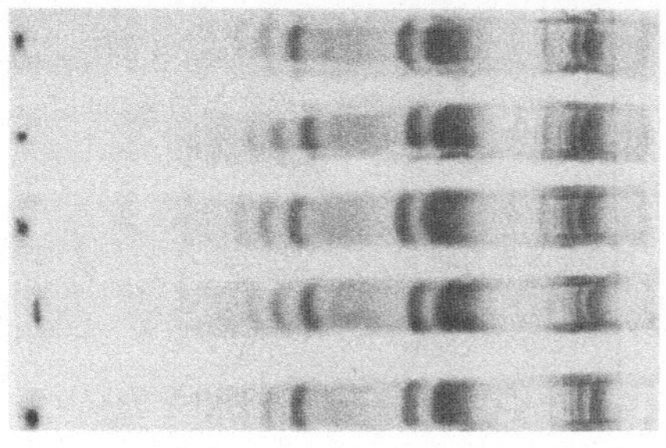

Fig. 3. Densitometer tracings of autoradiographs from SDS-polyacrylamide gels of NAP-[^{35}S]taurine labeled resealed ghost preparations. Traces 1 through 5 correspond to experiments in which the ghosts were prepared from cells lysed at the hemolytic ratios shown to the right of each curve. The unlabeled peak nearest the origin in each of the traces is the result of incompletely reduced oligomeric photooxidation products. The variable, unlabeled peak at the front of each of the traces results from material running with the SDS micelle front. The fixing and staining procedure used to prepare these gels results in an uneven retention of radioactivity in this frontal region, making the apparent differences among the traces inaccurate in this part of the gel. Variations in the PAS-2 area correspond to a component of that region that migrates at a slightly higher apparent molecular weight than band 5. Below is a photograph of the Coomassie blue-stained dried gel slices from which the autoradiographs were obtained. (Reprinted from STAROS, HALEY, and RICHARDS, 1974.)

A CYTOCHEMICAL LABELING PROCEDURE

In an attempt to study "sidedness" in another system, HEITZMANN (1974) treated a preparation of disks from bovine retinal rod outer segments with the diazonium salt of a naphthalene trisulfonic acid. It was expected that this highly polar compound would not penetrate the disk membrane during reaction. The single protein rhodopsin comprises 90% or more of all the protein in these disk preparations. The question being addressed in this experiment was whether at least part of each rhodopsin molecule was located on the external surface of the disk membrane. Since each reagent molecule that reacted with the protein increased the negative charge substantially because of the three sulfonic acid groups introduced, isoelectric focusing was a useful method of following the reaction. The results of such an experiment are shown in Figure 4. In the protein isolated from the treated membrane it can be seen that there is no unreacted rhodopsin. A logical conclusion might be that all rhodopsin molecules are accessible to the external surface of the membrane. This conclusion would be valid, however, only if the membrane were in fact impermeable to the reagent. There is no equivalent of hemoglobin or other internal macromolecule whose labeling

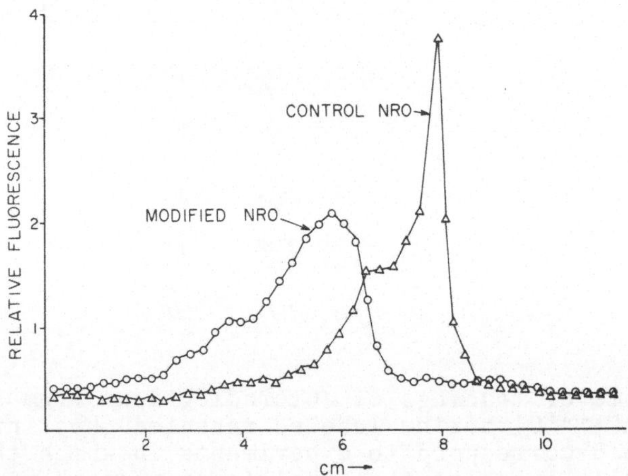

Fig. 4. Isoelectric focusing of samples from bovine retinal rod outer segments before and after reaction with naphthalene-1,3,6-trisulfonate-8-diazonium fluoroborate. Just prior to focusing, all rhodopsin in the samples was converted to the fluorescent derivative N-retinyl-opsin (NRO). The detergent-solubilized protein was run in focusing gels containing 2% Triton X-100. The N-retinyl opsin was detected by fluorescence in the gels. The pH gradient of the gels runs from about pH 3.7 at the left of the figure to pH 8.7 at the right. The principal peak of the unmodified protein corresponds to a pH of about 5.7. (Reprinted from HEITZMANN, 1974.)

or lack of it could be used as a test for reagent penetration. The only approach that we could think of was to try to locate the reagent molecule cytochemically in intact disk preparations. Attempts to prepare an antibody to the reagent as a haptene failed. The following procedure was developed as an alternative.

The vitamin biotin combines with avidin, a protein isolated from egg white, to form a noncovalent complex with an association constant of the order of 10^{15} M^{-1} (see review by GREEN, 1974). This binding is orders of magnitude tighter than that observed in most antigen-antibody complexes. The carboxyl group of biotin does not seem critical to formation of the avidin-biotin complex and provides a point of attachment for other groups. Use of this interaction, including protein modification with a biotin reagent, is described in an interesting study by BECKER and WILCHEK (1972) on the bacteriophage T4.

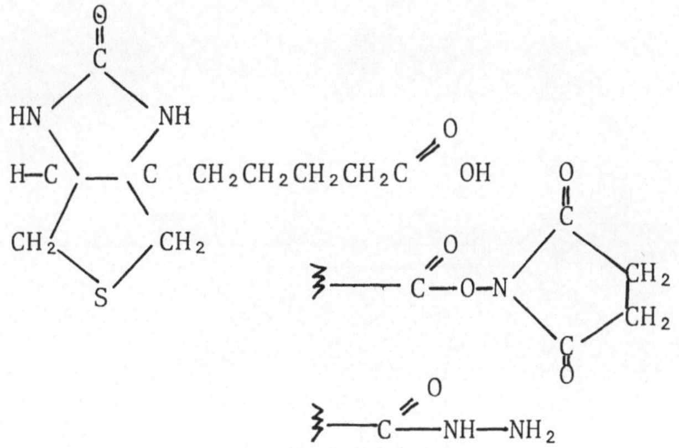

O
C
HN NH
O
H–C C $CH_2CH_2CH_2CH_2C$ OH
O
C
CH_2 CH_2
O
S
–C–O–N CH_2 CH_2
C
O
O
–C–NH–NH_2

In the proposed procedure (HEITZMANN, RICHARDS, 1974), the reagent to be used as a membrane probe is modified by the addition of biotin using whatever standard organic synthetic procedures are required. The reaction with the membrane is carried out with this altered reagent. Reagent molecules that have become attached to the membrane are located by treatment of the sample with avidin that has previously been coupled to ferritin as an electron dense marker. The position of the ferritin is then examined in the electron microscope according to standard procedures.

As a test of the various steps, two general reagents were prepared, the N-hydroxysuccinimide ester of biotin and biotinyl hydrazide. The active ester will react with many nucleophiles but especially with the ubiquitous amino groups. The hydrazide is known to react with aldehyde functions that can easily be produced by oxidation of the sugar moieties of membrane components. Both of these compounds prob-

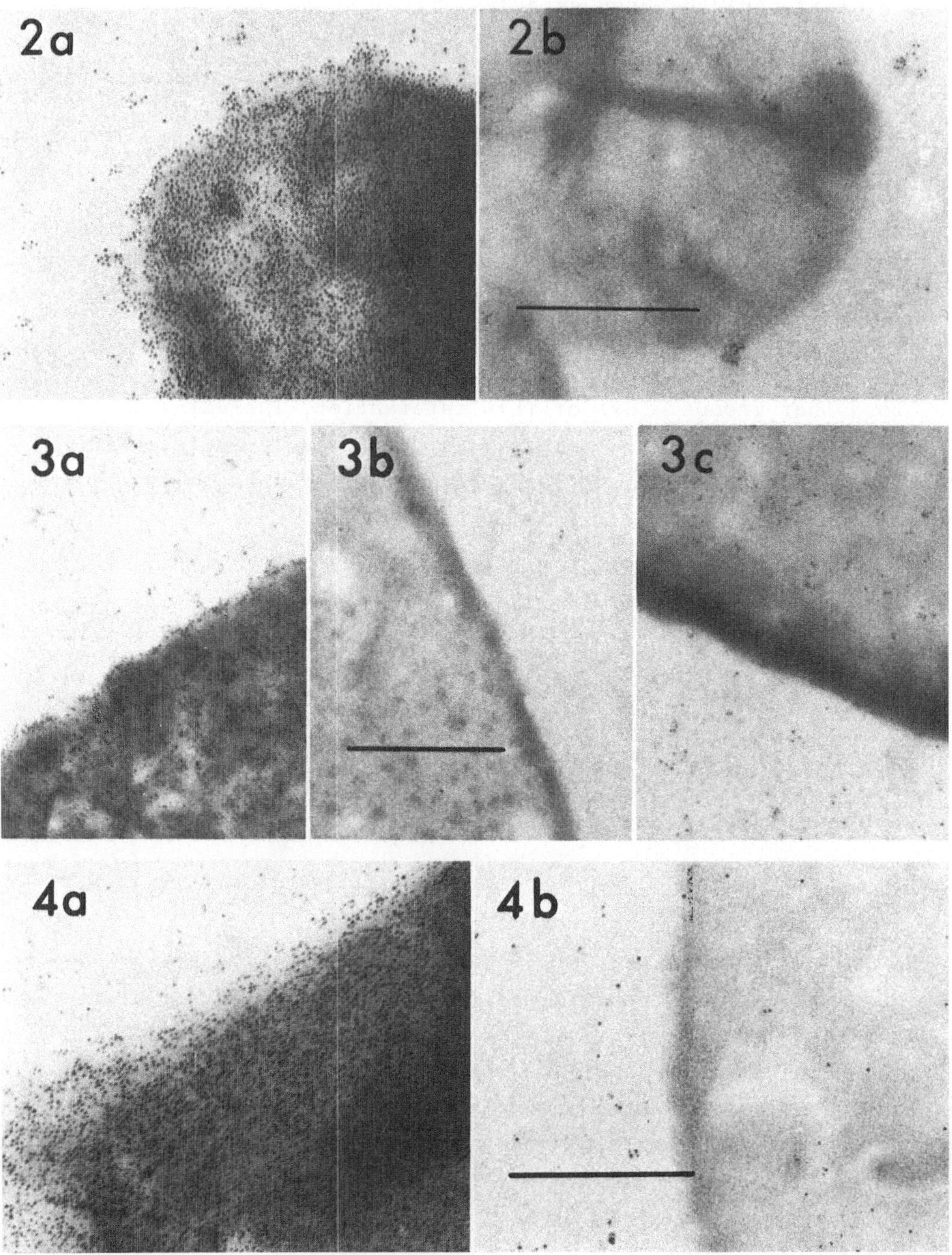

Fig. 5. The staining of some membrane preparations with ferritin-linked avidin (FAv). (2a) Ghosts of *Acholeplasma laidlawii* after reaction with biotinyl-N-hydroxy succinimide ester and labeling with FAv. (2b) Same ghost preparation as 2a but without reaction with the biotin reagent before labeling. (3a) Erythrocyte ghosts labeled with FAv after treatment with galactose oxidase followed by biotin hydrazide. (3b) Same as 3a but galactose oxidase step omitted. (3c) Same

ably penetrate the membrane easily and are not themselves very useful as vectorial probes, although advantage can be taken of the macromolecular size of the stain which clearly will not penetrate most membranes. Some tests on membranes from both *Acholeplasma laidlawii* and human erythrocytes are shown in Figure 5. The phospholipid components of the latter contain no amino groups and thus reaction of the biotin-active ester is restricted to protein. The sugars in the erythrocyte membrane were oxidized either with periodic acid or galactose oxidase to generate aldehydes. The differences between the control samples and the stained preparations give some feeling for the signal-to-noise ratio and indicate that the procedure should be useful. The original problem of the disks in the retinal rods has not yet been satisfactorily resolved due in part to the large size of the ferritin component and steric problems resulting from failure to get a suspension of well-separated disks.

The general approach can easily be modified for specific problems. The biotin reagents available depend only on imagination and the degree of synthetic skill available. Specific affinity probes for a variety of systems should be possible. Other and hopefully smaller electron dense particles might be used to tag the avidin. The "off" dissociation step of the biotin-avidin complex is so slow that sequential staining (avidin complexes with different metals) of a single specimen should be possible, following reaction with different biotinyl reagents. This may be particularly useful when energy discrimination detectors become available on high-resolution scanning or transmission electron microscopes.

---

as 3a but FAv sample pretreated with excess biotin before application to the membranes. (4a) Erythrocyte ghosts labeled with FAv after treatment with periodate followed by biotin hydrazide. (4b) Same as 4a but FAv was pretreated with excess biotin before use. In each case the calibration bar = 0.4 µm. (Reprinted from HEITZMANN and RICHARDS, 1974.)

NEAREST-NEIGHBOR ANALYSIS IN THE PLANE OF THE MEMBRANE

DAVIES and STARK (1970) showed how chemical cross-linking could be used to great advantage in establishing the quaternary structure of soluble oligomeric proteins. The technique has since been used for nearest neighbor analysis in more complex systems such as ribosomes (ACHARYA et al., 1973; BICKLE et al., 1972; LUTTER et al., 1974), membranes (STECK, 1972; HULLA, GRATZER, 1972; LOUIS, SHOOTER, 1972; JI, 1974), and nuclei (OLINS, WRIGHT, 1973).

In most of these studies, the reagents used have produced stable, covalently linked products. Further analysis has required that these cross-linked complexes be isolated as such and the observed molecular weight interpreted on the basis of prior knowledge of the system or with the help of prior labeling of specific components. In general, in complex systems, the mixtures produced by partial cross-linking can only severely complicate already difficult separation problems.

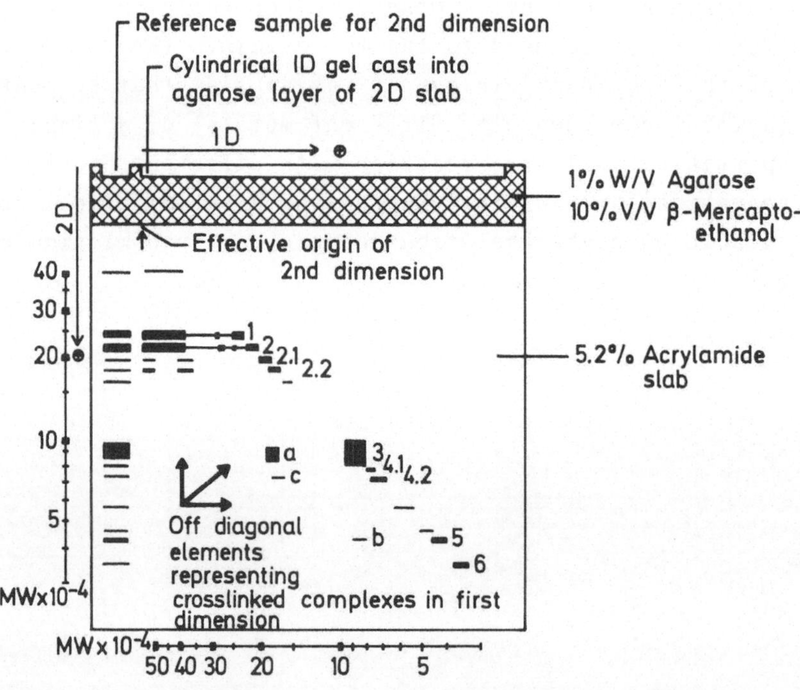

Fig. 6. Schematic diagram of the two-dimensional diagonal procedure employing SDS polyacrylamide gels. The sample is run in a cylindrical gel for the first dimension. This gel is then cast into the agarose layer of the two-dimensional slab gel. The reference sample shown for the second dimension is omitted in all present runs and in the subsequent figures below. (Adapted from WANG and RICHARDS, 1974a.)

In an attempt to overcome this analytical difficulty, we have recently reported the synthesis and use of several bifunctional reagents that contain the disulfide group and are thus cleavable (WANG, RICHARDS, 1974a, b). The principal compound that we have used so far is dimethyl-3,3'-dithiobispropionimidate dihydrochloride. Disulfide cross-links have also been produced by the copper (II) (orthophen-

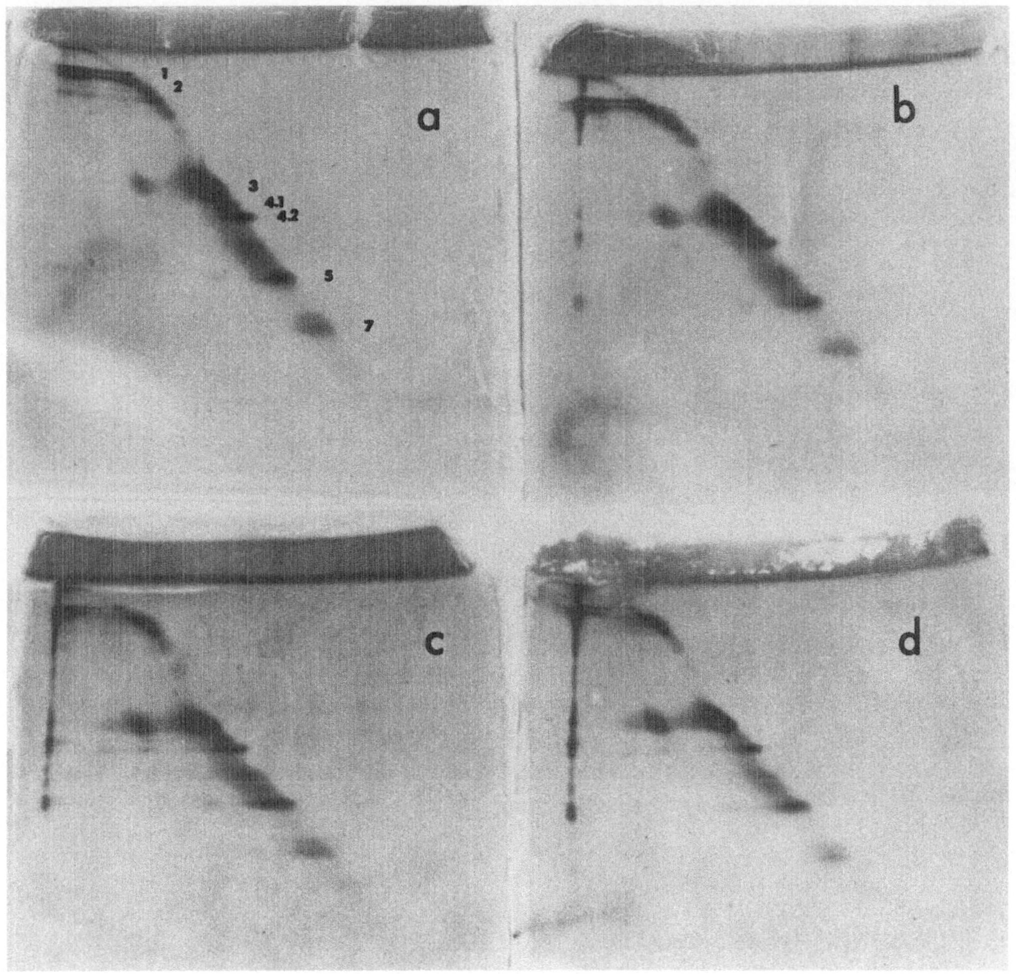

Fig. 7. The cross-linking of proteins in isolated erythrocyte membranes by dimethyl dithiobispropionimidate (DTBP). Ghosts (1 mg of protein per milliliter) were treated with DTBP at 0.08 mg/ml (a), 0.1 mg/ml (b), 0.3 mg/ml (c), 0.5 mg/ml (d), respectively, for 1 hr at room temperature. The reaction was quenched by 0.05 M NH_4Ac for 10 min. The suspensions were washed with 20 times their volume of ice-cold 5 mM phosphate buffer at pH 8.0 by centrifugation, and the pellet was analyzed after SDS solubilization. Band 6 is absent in this series of samples. (Reprinted from WANG and RICHARDS, 1974b.)

anthroline)$_2$-complex-catalyzed oxidation of intrinsic sulfhydryl groups in the membrane proteins. It should be noted that in all these studies, only limited cross-linking is useful. The products must be soluble in SDS-containing buffers for subsequent fractionation. It is easy to let the cross-linking proceed too far and end up with preparations that are totally insoluble in any nondegradative solvent.

A simple gel electrophoresis "diagonal mapping" technique has been developed based on the procedures introduced by BROWN and HARTLEY (1966) for locating disulfides in peptide sequence analysis. The

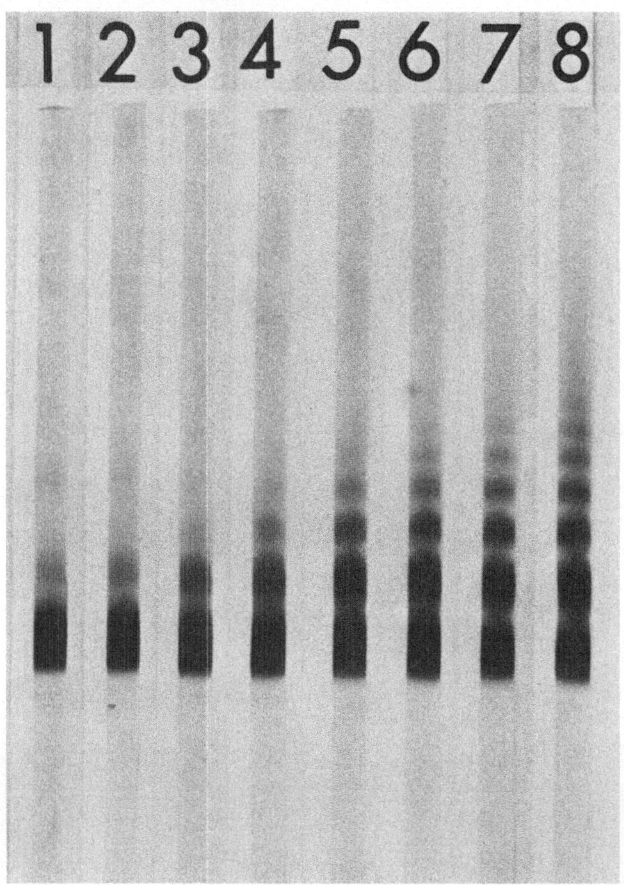

Fig. 8. Lysate fractions from DTBP-cross-linked intact erythrocytes. Erythrocytes in isotonic phosphate buffer (pH 8.0) were treated with DTBP at (1) 0 mg/ml, (2) 0.08 mg/ml, (3) 0.2 mg/ml, (4) 0.4 mg/ml, (5) 0.8 mg/ml, (6) 1.2 mg/ml, (7) 1.6 mg/ml, and (8) 2.0 mg/ml for 30 min at room temperature. The cross-linked erythrocytes were washed and lysed in the presence of N-ethyl maleimide. The membrane-free lysate fractions were analyzed on 3.2% polyacrylamide gels containing 0.2% SDS. Oligomers of globin chains as high as octomers are easily seen. (Reprinted from WANG and RICHARDS, 1975.)

analytical procedure involves two-dimensional SDS polyacrylamide gel electrophoresis. An SDS solution of the cross-linked preparation is run in the first dimension with no reducing agent. In the second dimension, the materials pass through an agarose layer containing a high concentration of reducing agent. Cross-linked complexes are thus split into their components. These products move faster in the second dimension than in the first and appear as off-diagonal spots in the developed electrophoretogram. A schematic diagram of the procedure is shown in Figure 6. Photographs of actual gels showing the patterns

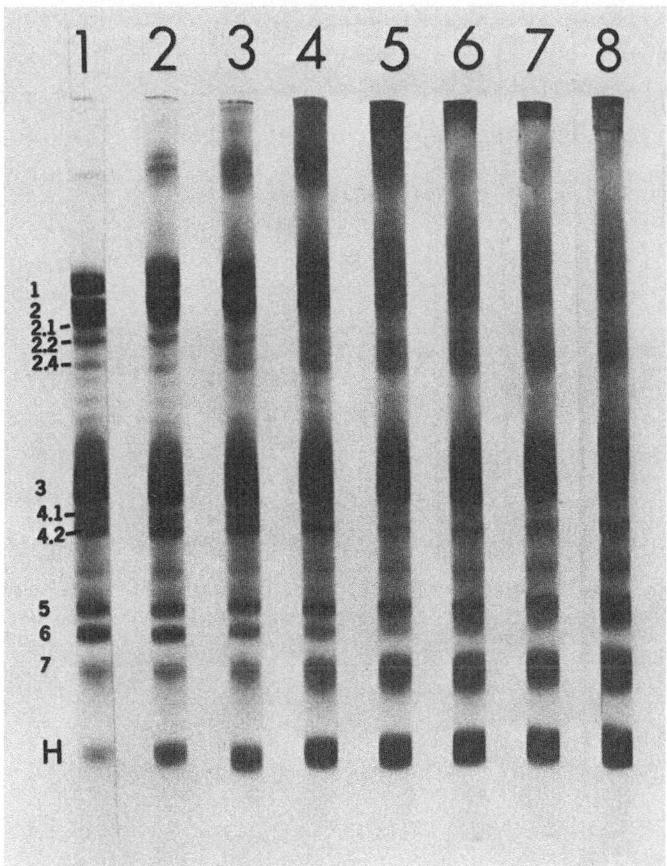

Fig. 9. Membrane fractions from DTBP-cross-linked intact erythrocytes. Membrane fractions from the corresponding samples in Figure 8 were analyzed on 3.2% polyacrylamide gels containing 0.2% SDS. Band numbers following STECK (1972) are shown in the left margin. H is the globin chain. The sharp, narrowly spaced bands representing globin chains attached to bands 1 and 2 are easily seen. (Reprinted from WANG and RICHARDS, 1975.)

derived from cross-linked erythrocyte ghosts and the membrane fraction of whole erythrocytes are shown in Figures 7 through 10. Summary diagrams are shown in Figures 11 through 13. The detailed conclusions from these studies are given in two papers by WANG and RICHARDS (1974a, b, 1975).

In overview, the cross-linking patterns of the ghost preparations strongly indicate that a number of the proteins represented by the bands in the standard SDS gel pattern exist in oligomeric form in the native membrane. This statement would be true whether or not a given gel band represents a unique chemical species. There are also some clear examples of heterologous complexes in which the partners have significantly different molecular weight; for example, band 2 + band 4.5 and bands 2.1 to 2.5 + band 3.

A striking characteristic of all but the most lightly cross-linked patterns is the very high-molecular-weight material that does not even

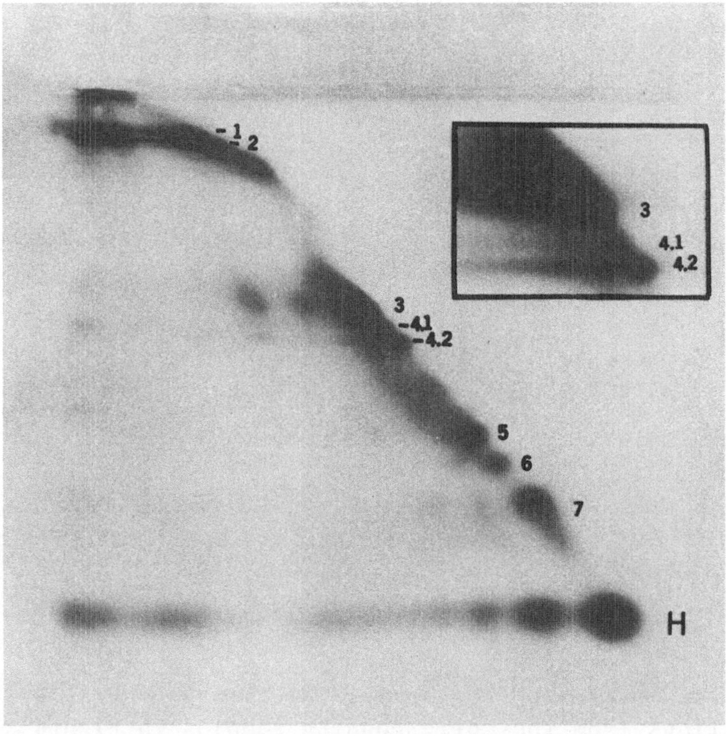

Fig. 10. Diagonal electrophoresis of a membrane fraction from DTBP-cross-linked intact erythrocytes. The sample corresponds to the membrane fraction used for gel 5 in Figure 9. The insert shows an expanded view of the band 3 to band 4.2 region of the diagonal map, where complexes of 4.2 and 4.1 with globin chains are seen. (Reprinted from WANG and RICHARDS, 1975.)

enter the first dimension gel. Following reduction, this material yields a line of spots at the left of the figures. Almost all of the original components are present in this line. Other experiments have shown that the line is not an artifact. The most straightforward explanation is that all the components appearing in this position are bound to spectrin in the intact membrane and thus become cross-linked to it during the reaction. Spectrin, represented by bands 1 and 2 in the gel pattern, starts out as the highest molecular-weight material present. It is very rapidly cross-linked to yield aggregates of 1 million or more in molecular weight. The smaller chains are attached to

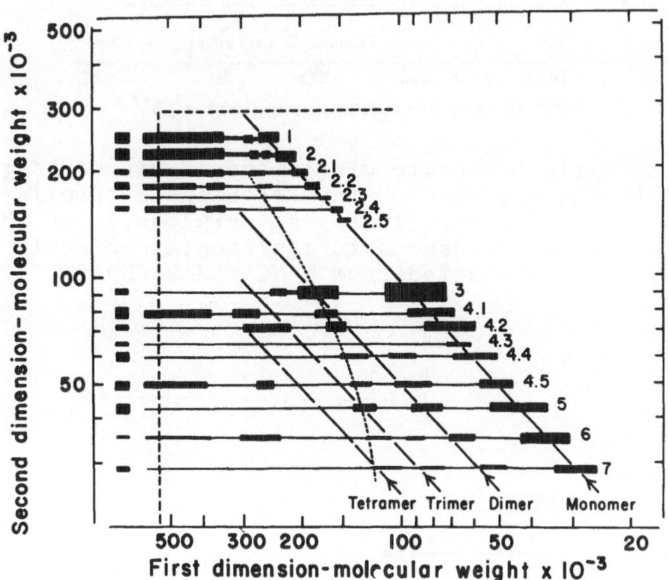

Fig. 11. A schematic composite diagram of DTBP cross-linking patterns of erythrocyte ghosts. Off-diagonal spots were plotted according to their estimated molecular weights in both the first and the second dimensions. Only those off-diagonal spots that could be reproducibly detected in samples from at least two different preparations of ghosts were included in this diagram. The width of each spot represents the range of molecular weight covered by the spot, and does *not* represent error bars for the distributions of the centers of a given spot in different gels. The height of each spot is roughly proportional to the staining intensity. *Dashed lines* corresponding to the expected positions for dimers, trimers, and tetramers are shown. The *curved dotted line* corresponds to the locus of spots representing complexes of a given band and a molecular weight increment of 88,000. The *small dash* near the 500,000-molecular-weight region represents the limit of the resolution of our gel electrophoresis system. The line of spots on the extreme left outside these dashes on the diagram represents the spots that were derived from material not entering the one-dimensional gel when ghosts were highly cross-linked. (They do *not* represent a reference sample.) The heights of these spots approximate the relative staining intensity. (Adapted from WANG and RICHARDS, 1974.)

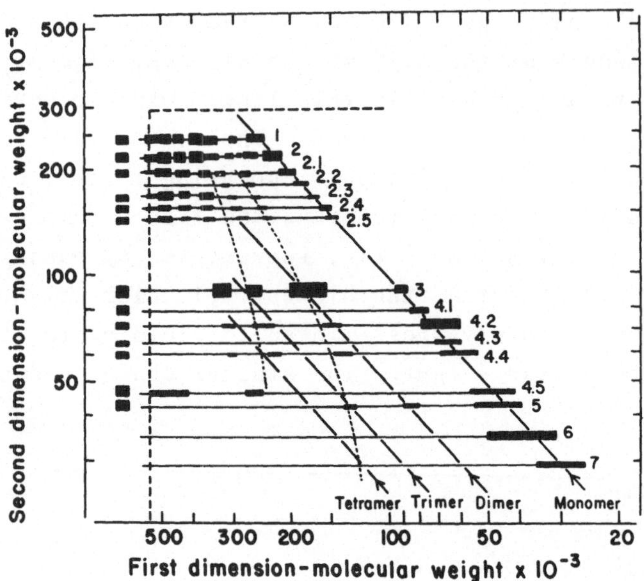

Fig. 12. The schematic composite diagram for erythrocyte ghosts of cross-linking patterns produced by the (orthophenanthroline)$_2$ Cu(II) reagent. (See legend of Figure 11 for description.) In this case the two *curved dotted lines* correspond to incremental molecular weights of 88,000 and 176,000. (Adapted from WANG and RICHARDS, 1974.)

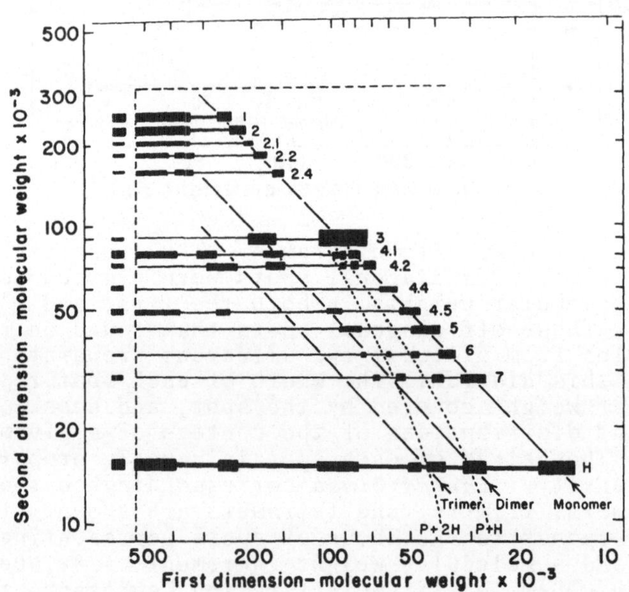

Fig. 13. The schematic composite diagram for intact erythrocytes of cross-linking patterns produced with DTBP. (See legend of Figure 11 for a description.) In this case the two dotted curved lines correspond to complexes with incremental molecular weights of 16,500 (P+H) and 33,000 (P+2H). (Reprinted from WANG and RICHARDS, 1975.)

these aggregates. It is hard to imagine how these smaller chains would appear in such high-molecular-weight material through some other route. The picture thus emerges of homo or hetero oligomers of most of the membrane proteins held in position in the lipid matrix through association with the spectrin meshwork that exists just under the cytoplasmic surface. Such a conclusion agrees well with current electron microscopic evidence (ELGSAETER et al., 1973; NICHOLSON, PAINTER, 1973).

The cross-linking reagent penetrates the erythrocyte membrane. Thus when whole cells are carried through the procedure, the cross-linking pattern discussed above is observed, and superimposed on it are additional complexes involving the chains of hemoglobin. This major cytoplasmic component thus serves as a macromolecular probe of the inner membrane surface. Complexes of hemoglobin with many of the components known to be present can easily be seen. One curious anomaly is the failure to observe any hemoglobin complexed to band 5. There is abundant evidence that this protein is present on the cytoplasmic surface. The reagent also cross-links both hemoglobin and band 5 separately, so the necessary functional groups are present on both molecules. In this case the negative cross-linking evidence (no complex observed), which is usually not useful evidence, may be significant. Either band 5 (erythrocyte actin, SHEETZ et al., 1974) is protected sterically from contact with hemoglobin or it forms a very specific, tight complex in which the functional groups of the two partners are not properly positioned for cross-linking by this reagent. This type of uncertainty can in principle be solved by using a series of cross-linking reagents of different lengths and group specificities.

REFERENCES

ACHARYA, A. S., MOORE, P. B., RICHARDS, F. M.: Biochemistry 12, 3108-3114 (1973).
BECKER, J. M., WILCHEK, M.: Biochim. Biophys. Acta 264, 165 (1972).
BENDER, W. W., GARAN, H., BERG, H. C.: J. Mol. Biol. 58, 783 (1971).
BERG, H. C.: Biochim. Biophys. Acta 183, 65 (1969).
BICKLE, T. A., HERSHEY, J. W. B., TRAUT, R. R.: Proc. Nat. Acad. Sci. U.S.A. 69, 1327-1331 (1972).
BLASIE, J. K.: Biophys. J. 12, 191 (1972).
BLASIE, J. K., WORTHINGTON, C. R., DEWEY, M. M.: J. Mol. Biol. 39, 407 (1969).
BRETSCHER, M. S.: J. Mol. Biol. 58, 775 (1971a).
BRETSCHER, M. S.: J. Mol. Biol. 59, 351 (1971b).
BRETSCHER, M. S.: Nature (London) New Biol. 231, 229 (1971c).
BRETSCHER, M. S.: Science 181, 622 (1973).
BROWN, J. R., HARTLEY, B. S.: Biochem. J. 101, 214-228 (1966).
CABANTCHIK, Z. I., ROTHSTEIN, A.: J. Membrane Biol. 10, 311 (1972).
CABANTCHIK, Z. I., ROTHSTEIN, A.: J. Membrane Biol. 15, 207 (1974).

DAVIES, G. E., STARK, G. R.: Proc. Nat. Acad. Sci. U.S.A. 66, 651-656 (1970).
ELGSAETER, A., SHOTTON, D., BRANTON, D.: J. Cell Biol. 59, 89a (1973).
FAIRBANKS, G., STECK, T. L., WALLACH, D. F. H.: Biochemistry 10, 2606 (1971).
FLEET, G. W., PORTER, R. R., KNOWLES, J. R.: Nature (London) 224, 511 (1969).
GREEN, N. M.: Advances in Protein Chemistry 28 (in press, 1974).
HEITZMANN, H.: Ph.D. Thesis, Yale University, New Haven, Conn. (1974).
HEITZMANN, H., RICHARDS, F. M.: Proc. Nat. Acad. Sci. U.S.A. 71, 3537 (1974).
HENDERSON, R.: Manuscript in preparation (1975).
HULLA, F. W., GRATZER, W. B.: FEBS Lett. 25, 275-278 (1972).
JI, T. H.: Proc. Nat. Acad. Sci. U.S.A. 71, 93-95 (1974).
KNOWLES, J. R.: Accounts Chem. Res. 5, 155 (1972).
LOUIS, C., SHOOTER, E. M.: Arch. Biochem. Biophys. 153, 641-655 (1972).
LUTTER, L. C., BODE, U., KURLAND, C. G., STÖFFLER, G.: Mol. Gen. Genet. 129, 167-176 (1974).
NICOLSON, G. L., PAINTER, R. G.: J. Cell Biol. 59, 345 (1973).
NICOLSON, G. L., SINGER, S. J.: J. Cell Biol. 60, 236 (1974).
OLINS, D. E., WRIGHT, E. B.: J. Cell Biol. 59, 304-317 (1973).
PHILLIPS, D. R., MORRISON, M.: Biochemistry 10, 1766 (1971a).
PHILLIPS, D. R., MORRISON, M.: FEBS Lett. 18, 95 (1971b).
RUOHO, A. E., KIEFER, H., ROEDER, P. E., SINGER, S. J.: Proc. Nat. Acad. Sci. U.S.A. 70, 2567 (1973).
SHEETZ, M., PAINTER, R., SINGER, S. J.: J. Cell Biol. (in press, 1974).
SINGER, S. J.: Ann. Rev. Biochem. 43, 805 (1974).
SINGH, A., THORNTON, E. R., WESTHEIMER, F. H.: J. Biol. Chem. 237, PC 3006 (1962).
STAROS, J. V., HALEY, B. E., RICHARDS, F. M.: J. Biol. Chem. 249, 5004 (1974).
STAROS, J. V., RICHARDS, F. M.: Biochemistry 13, 2720 (1974).
STECK, T. L.: In: Membrane Research (ed. C. F. FOX), p. 71. New York, N.Y.: Academic Press 1972.
STECK, T. L.: J. Cell Biol. 62, 1-19 (1974).
STECK, T. L., FAIRBANKS, G., WALLACH, D. F. H.: Biochemistry 10, 2617 (1971).
TRIPLETT, R. B., CARRAWAY, K. L.: Biochemistry 11, 2897 (1972).
WALLACH, D. F. H.: Biochim. Biophys. Acta 265, 61 (1972).
WANG, K., RICHARDS, F. M.: Israel J. Chem. 12, 375 (1974a).
WANG, K., RICHARDS, F. M.: J. Biol. Chem. (in press, 1974b).
WANG, K., RICHARDS, F. M.: J. Biol. Chem. (submitted, 1975).
WHITELEY, N. M., BERG, C. H.: J. Mol. Biol. 87, 541 (1974).
ZWALL, R. F. A., ROELOFSEN, B., COLLEY, C. M.: Biochim. Biophys. Acta 300, 159 (1973).

Structure-Function Relation in *Escherichia coli* Membranes

Emanuel Shechter, Lucienne Letellier, and Tadeusz Gulik-Krzywicki
Laboratoire des Biomembranes, Université de Paris-Sud, 91405 Orsay, and
Centre de Génétique Moléculaire du C.N.R.S., 91190 Gif-Sur-Yvette, France

INTRODUCTION

To study the relations between membrane function and membrane structure, one must often first establish the relationship between membrane function and one or more actual membrane structural parameters.

One of the most analyzed and best understood membrane structural parameters is the conformation of membrane lipid hydrocarbon chains. Extensive X-ray diffraction studies of lipid-containing systems (LUZZATI et al., 1972; DUPONT et al., 1972; RANCK et al., 1974), and particularly of lipid-water systems (RANCK et al., 1974), have shown that the lipid hydrocarbon chains are able to undergo conformational transitions whose nature have been thoroughly analyzed. It has been shown in particular that order-disorder conformational transitions of the lipid hydrocarbon chains involve only two types of well-defined conformations: a highly disordered or "liquid-like" conformation (α-type conformation); and an ordered conformation in which the chains are stiff and parallel, with rotational disorder in a two-dimensional hexagonal lattice (β-type conformation). Transitions between these two conformations involve complex phenomena, strongly dependent on the heterogeneity of the hydrocarbon chains.

Similar order-disorder transitions have by now been observed in a variety of biological membranes, and considerable efforts have been made to determine whether they influence some membrane-associated functions. Although clear-cut cases have been described, mainly in bacterial membranes (see references in OVERATH et al., in press) and mycoplasma membranes (see references in STEIM, 1972), the field appears rather confused and controversial. Some of the reasons for this confusion may be trivial: use of indirect techniques to determine the order-disorder conformational transitions, or determination of these transitions by direct technique but in a qualitative way; or use of

Arrhenius plots with all their possible misinterpretations to relate physiological to structural transitions. Other reasons, however, might be of more importance: in particular the observation that many membrane-associated functions display physiological transitions at temperatures well above the order-disorder transitions. Although some of these physiological transitions can be accounted for by structural changes of the membrane proteins, the exact nature of some of the others is not understood. Attempts have been made to relate these last transitions to the lipid moiety of the membranes, and concepts such as clusters of "solid-like liquid lipids" have been advanced (LEE et al., 1974). Although such concepts could, in principle, explain the existence of physiological transitions related to the lipids at temperatures above their order-disorder transitions, they are based uniquely on electron spin resonance experiments using tempo partitioning between the membrane and the water phases. We find it hard to understand how such clusters can still exist some 50°C above the order-disorder transitions; further experiments using different techniques are needed in order to clarify this matter.

We present and discuss here the relations that exist between the order-disorder conformational transitions of lipid hydrocarbon chains and membrane-associated functions in *E. coli* membranes.

In our experiments, an *E. coli* unsaturated fatty acid auxotroph was grown on various unsaturated fatty acids (elaidic, oleic, linoleic, linolenic acids). Cytoplasmic membranes were prepared according to KABACK (1971); outer membranes were prepared according to OSBORN (OSBORN et al., 1972; OVERATH et al., in press). The order-disorder transitions were determined quantitatively and in detail using high-angle X-ray diffraction. The morphological changes induced by these transitions in the hydrophobic core of the membranes were followed by freeze-etch electron microscopy. Two membrane functions were followed: active transport of proline and vectorial phosphorylation of glucose across the cytoplasmic membranes.

Our results show the following: (1) the characteristics of the order-disorder transitions (shape of the transition, cocrystallization, or segregation of various lipid species, and amount of lipid taking part in the transition) depend on the fatty acid composition and the protein concentration of the membranes; (2) the ordering of the membrane lipid hydrocarbon chains results in the exclusion of membrane-penetrating proteins from the ordered domains leading to the appearance on freeze-fracture electron micrographs of smooth surfaces devoid of particles and particulated surfaces highly concentrated in membrane-penetrating proteins; (3) some membrane functions (active transport) respond to the order-disorder transitions, while others (vectorial phosphorylation) do not; (4) the relation between order-disorder transition and membrane-associated functions is an indirect one (the membrane-associated function responds primarily to the immobilization [in our case, through the aggregation] of the membrane-penetrating proteins concomitant with the transition); and (5) the use of Arrhenius plots to relate order-disorder transitions and membrane-associated functions can easily result in erroneous interpretation and should be handled with extreme care.

RESULTS AND DISCUSSION

Membrane Composition

The fatty acid composition of the membranes studied is shown in Table 1. The composition is very similar for a given cytoplasmic membrane and the corresponding outer membrane.

Table 1. Fatty acid composition (%) of *E. coli* membranes[a]

	Supplemented fatty acid					
	Trans Δ^9 $C_{18:1}$ (elaidic)		Cis Δ^9 $C_{18:1}$ (oleic)		Cis $\Delta^{9,12}$ $C_{18:2}$ (linoleic)	Cis $\Delta^{9,12,15}$ $C_{18:3}$ (linolenic)
Fatty acid	cytopl.	outer	cytopl.	outer	cytoplasmic	cytoplasmic
$C_{12:0}$	2	6	1	1	12	
$C_{14:0}$	16	6	16	9	12	9
$C_{16:0}$	10	5	36	36	51	63
Trans $C_{18:1}$	72	83				
Cis $C_{18:1}$			47	54		
Cis $C_{18:2}$					37	
Cis $C_{18:3}$						28
Unsaturated	72	83	47	54	37	28
Saturated	28	17	53	46	63	72
Unsaturated/ saturated ratio	2.6	4.9	0.9	1.2	0.6	0.4

[a]Data for cytoplasmic membranes taken from SHECHTER et al. (in press). Data for outer membranes taken from OVERATH et al. (in press).

The only unsaturated fatty acid in the membrane is the one supplemented during the growth of the cell. The ratio of unsaturated to saturated fatty acid decreases in the order elaidic, oleic, linoleic, and linolenic membranes and parallels the increasing fluidizing effect of these fatty acids on the membrane: at one extreme, elaidic acid, a monounsaturated transfatty acid, resembles more a saturated than an unsaturated fatty acid and has a poor fluidizing effect; at the other extreme, linolenic acid with its three double bonds has a great fluidizing effect.

The protein-to-phospholipid ratio of the membranes is shown in Table 2. It is greater for an outer membrane than for the corresponding cytoplasmic membrane. Elaidic cytoplasmic membranes display, under our growth conditions, a relatively low ratio of 1; oleic, linoleic, and linolenic cytoplasmic membranes display a protein-to-phospholipid ratio of 2, which is comparable to that observed in wild-type cytoplasmic membranes.

Table 2. Composition (%) of the membranes[a]

Membrane	Protein (w/w)	P-lipid (w/w)	LPS (w/w)	Protein/ P-lipid (ratio)
Elaidic				
Cytoplasmic	53	47	-	1.1
Outer	66	17	17	3.9
Oleic				
Cytoplasmic	70	30	-	2.3
Outer	60	17	23	3.9
Linoleic				
Cytoplasmic	67	33	-	2.0
Linolenic				
Cytoplasmic	65	35	-	1.9

[a]Data for cytoplasmic membranes taken from SHECHTER et al. (in press). Data for outer membranes taken from OVERATH et al. (in press).

High-Angle X-Ray Diffraction

At sufficiently high temperature, the high-angle X-ray diffraction of all preparations (cytoplasmic membranes, outer membranes, total lipid extracts) display a broad band centered at around 4.5 Å (Figures 1 and 2). This band is characteristic of disordered lipid hydro-

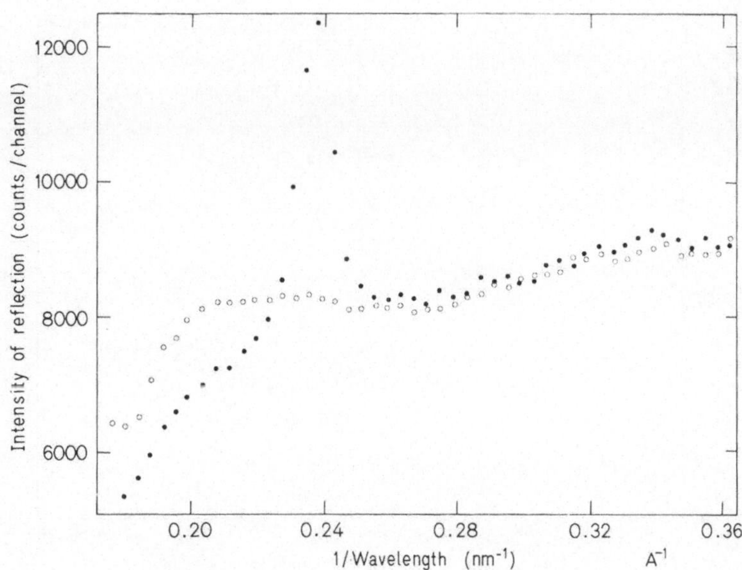

Fig. 1. High-angle X-ray diffraction of the cytoplasmic membranes isolated from cells grown in the presence of elaidic acid (• = membranes at 2°C; o = membranes at 45°C). The membranes are centrifuged overnight at 100,000 x *g*. The pellet (approximately 70% water) is placed in a temperature-controlled sample holder. An anode-rotating Elliot GX6 generator is used as the X-ray source. The diffracted X-ray beams are detected using a linear-position, sensitive, proportional counter (DUPONT et al., 1972). The high-angle X-ray spectra are recorded directly on a multichannel analyzer. The data in the figure represent part of these spectra. The sharp reflection centered at 0.24 $Å^{-1}$ (4.2 Å) characterizes the ordered hydrocarbon chains of the membrane lipids, while the broad band centered at around 0.22 $Å^{-1}$ (4.5 Å) is characteristic of the disordered hydrocarbon chains. Part of the background in this region is the result of the diffusion of water and of the proteins present in the sample; as a consequence, the broad lipid band at 4.5 Å is deformed. The same types of spectra are observed for the other cytoplasmic membranes

carbon chains (RANCK et al., 1974). As the temperature is decreased, a sharp reflection centered at 4.2 Å appears, indicating that some of the hydrocarbon chains have become ordered (Figures 1 and 2)(RANCK et al., 1974). The integrated intensity of this reflection increases with decreasing temperature and eventually reaches a constant value. This implies the existence of disorder-order transitions, and thus the existence of a segregation of lipids in different types of domains: those containing lipids with disordered chains and those containing lipids with ordered chains.

The ratios of ordered-to-total hydrocarbon chains for the various membranes and total lipid extracts, as a function of temperature and as determined from the integrated intensity of the 4.2 Å reflection (see

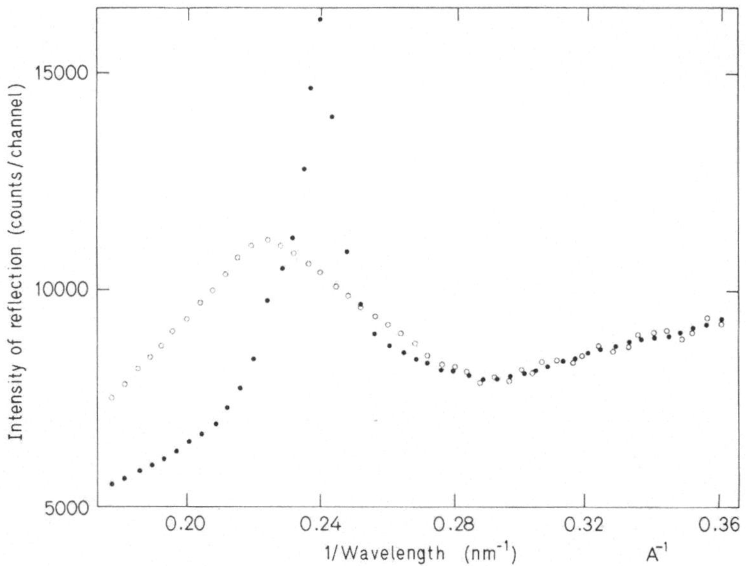

Fig. 2. High-angle X-ray diffraction of the total lipids extracted from cells grown in the presence of oleic acid (• = lipids at 2°C; o = lipids at 45°C). The lipid-water samples are prepared by mixing 40% dried lipids with 60% water. The mixture is allowed to equilibrate overnight. The same procedure was used as that described for Figure 1. The same types of spectra are observed for the other total lipid extracts

legend to Figure 3), are shown in Figures 3 through 5. Relevant parameters are given in Table 3. The following points are of interest:

- For the cytoplasmic membranes, the disorder-order transitions as a function of decreasing temperature display a gradual increase in the amount of ordered hydrocarbon chains. The broadness of the transition, ΔT, increases in the order elaidic, oleic, linoleic, and linolenic membranes.
- For the cytoplasmic membranes, the order-disorder transitions as a function of increasing temperature are similar to the disorder-order transitions except in the case of linolenic membranes where hysteresis is observed.
- The shape of the reflection characterizing the ordered hydrocarbon chains is similar to that of the standard (pure L_β structure; see legend to Figure 3), in the case of oleic and elaidic cytoplasmic membranes. In the case of linolenic (and to some extent linoleic) cytoplasmic membranes, the width of the reflection increases with increasing amount of ordered chains during the disorder-order transition (data not shown). During the first order-disorder

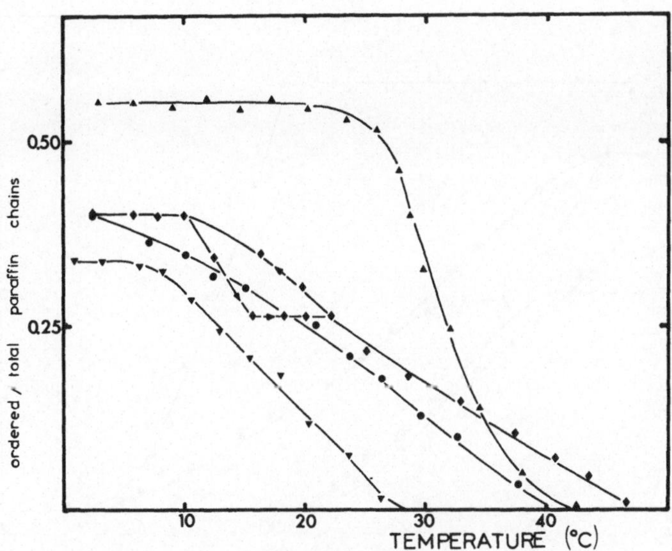

Fig. 3. Disorder-order and order-disorder transitions of the hydrocarbon chains of the lipids for the various cytoplasmic membranes as a function of temperature (▼,●,◆,▲ = membranes isolated from cells grown in the presence of oleic, linoleic, linolenic, and elaidic acids, respectively). The ratio of ordered-to-total hydrocarbon chains is determined by comparison of the integrated intensity of the 4.2 Å reflection to that of a standard consisting of the total lipid extract of elaidic membranes mixed with 5% water. Simultaneous low- and high-angle X-ray diffraction of this standard indicates that at 5°C, only one phase is present in which 100% of the hydrocarbon chains are in the β-type ordered conformation (L_β phase)

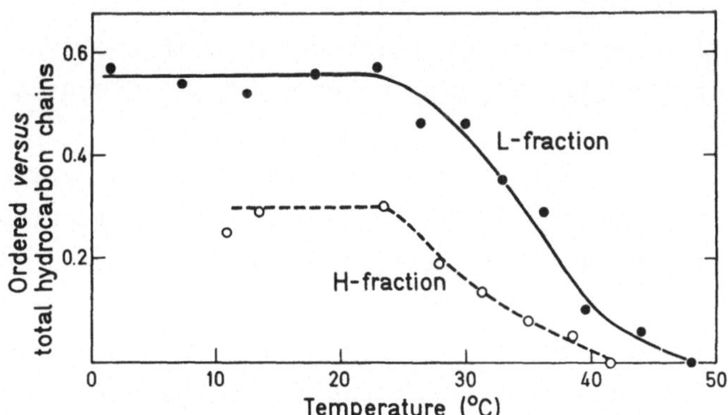

Fig. 4. Order-disorder transitions of the hydrocarbon chains of the lipids for inner and outer elaidic membranes. Both membranes are prepared according to the Osborn procedure, and the notation H and L (heavy and light fractions) of Osborn is used (OSBORN et al., 1972; OVERATH et al., in press). The transition for the inner membrane is identical to that of the cytoplasmic membrane prepared according to KABACK (1971). (See Figure 3 legend.)

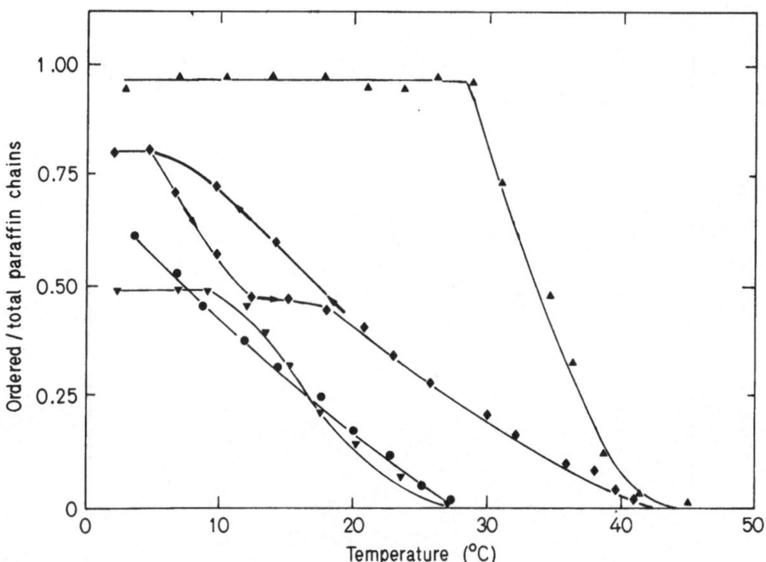

Fig. 5. Disorder-order and order-disorder transitions of the hydrocarbon chains of the total lipids extracted from cells grown in the presence of oleic (▼), linoleic (●), linolenic (◆), and elaidic (▲) acids. The transitions are similar for a given total lipid extract and the corresponding membrane. (See Figure 3 legend.)

transition as a function of increasing temperature, the reflection sharpens and remains subsequently similar to that of the standard.

- Qualitatively, the disorder-order transitions or order-disorder transitions are similar for a given membrane (cytoplasmic and outer) and the corresponding total lipid extract (parameters: T_t, mid-transition temperature; ΔT, broadness of the transition; sharpness of the reflection; presence or absence of an hysteresis).
- The amount of paraffin chains taking part in the various transitions increases in the following order: outer membranes, corresponding cytoplasmic membranes, corresponding total lipid extracts.

The similarity of each of the parameters (T_t, mid-transition temperature; ΔT, broadness of the transition; sharpness of the 4.2 Å reflection; presence or absence of an hysteresis) for a given membrane and the corresponding total lipid extract suggests that these transition characteristics are a consequence of the properties of the lipids themselves independently of the presence of the proteins. They can be explained and discussed in terms of a cocrystallization or a segregation of the different lipid species present: totally saturated lipids

Table 3. Characteristics of order-disorder transitions[a]

	T_t [b] (°C)	ΔT [c] (°C)	Ordered hydrocarbon chains at $T \ll T_t$ (%)
Elaidic			
Total lipid	34	12	95
Cytoplasmic	31	12	55
Outer	33	14	25
Oleic			
Total lipid	17	20	50
Cytoplasmic	18	20	35
Outer	nd[d]	nd	16
Linoleic			
Total lipid	nd	nd	>60
Cytoplasmic	24	24	40
Linolenic (disorder → order)			
Total lipid	22	32	80
Cytoplasmic	28	37	40
Linolenic (order → disorder)			
First transition:			
Total lipid	9	8	30
Cytoplasmic	13	6	15
Second transition:			
Total lipid	28	22	50
Cytoplasmic	34	22	25

[a]Data calculated from Figures 3 through 5 and taken from SHECHTER et al. (in press) and OVERATH et al. (in press).
[b]T_t is the temperature in a case in which half of the chains are ordered relative to the percentage of ordered chains at $T \ll T_t$.
[c]ΔT is defined as the interception of the tangent at T_t with the extrapolated horizontal lines above and below the transition.
[d]nd = not determined.

(lipids with two saturated chains); totally unsaturated lipids (lipids with two unsaturated chains); and mixed lipids (lipids with one saturated and one unsaturated chain).

The broadening of the disorder-order transition from elaidic, to oleic, linoleic, and linolenic membranes reflects the increasing difficulty in the cocrystallization of the mixed lipids with the totally saturated lipids and parallels the increasing structural difference between saturated and unsaturated fatty acids. In the case of linolenic (and to some extent linoleic) membranes and total lipid extracts, for which these differences are the greatest, this cocrystallization will actually perturb or even modify the structure of the ordered domains. This may be responsible for the broadening of the 4.2 Å reflection with decreasing temperature. The hysteresis in the case of linolenic membrane and total lipid extract is also a consequence of the great structural difference between a saturated and a linolenic fatty acid. Below a certain temperature, the most stable thermodynamic state would be one in which the different lipid species segregate in different types of ordered domains: those containing mainly the totally saturated lipids, and those containing mainly the mixed lipids. The further cocrystallization corresponds to metastable states and the segregation occurs only at the lower end of the disorder-order transition. The first order-disorder transition observed as a function of increasing temperature corresponds then to the disordering of the domains containing mainly the mixed lipids; the second transition corresponds to the disordering of the domains containing mainly the totally saturated lipids.

The amount of hydrocarbon chains taking part in the various transitions depends not only on the fatty acid composition and distribution, but also on the amount of membrane proteins. It decreases in the following order: total lipid extract, corresponding cytoplasmic membrane, corresponding outer membrane; and it parallels the increasing protein concentration. This decrease results from the interaction of membrane-penetrating proteins and the hydrocarbon chains of the lipids which will hinder the ordering of hydrocarbon chains that potentially are able to crystallize in the absence of proteins, i.e., in the total lipid extracts.

Freeze-Etch Electron Microscopy

The fracture surfaces corresponding to cytoplasmic membrane vesicles frozen from a temperature above the order-disorder transition are shown in parts A and B of Figures 6 through 9. They display a homogeneous distribution of densely packed particles on the convex face.

The fracture surfaces corresponding to cytoplasmic membrane vesicles frozen from a temperature below the order-disorder transition are shown in parts C and D of Figures 6 through 9. They all display more or less extended smooth surfaces separated by regions containing more or less aggregated particles. It is assumed that the smooth regions correspond to membrane regions containing ordered hydrocarbon chains from which the membrane-penetrating proteins have been excluded.

A qualitative correlation exists for these cytoplasmic membranes between the amount of ordered hydrocarbon chains as determined by X-ray

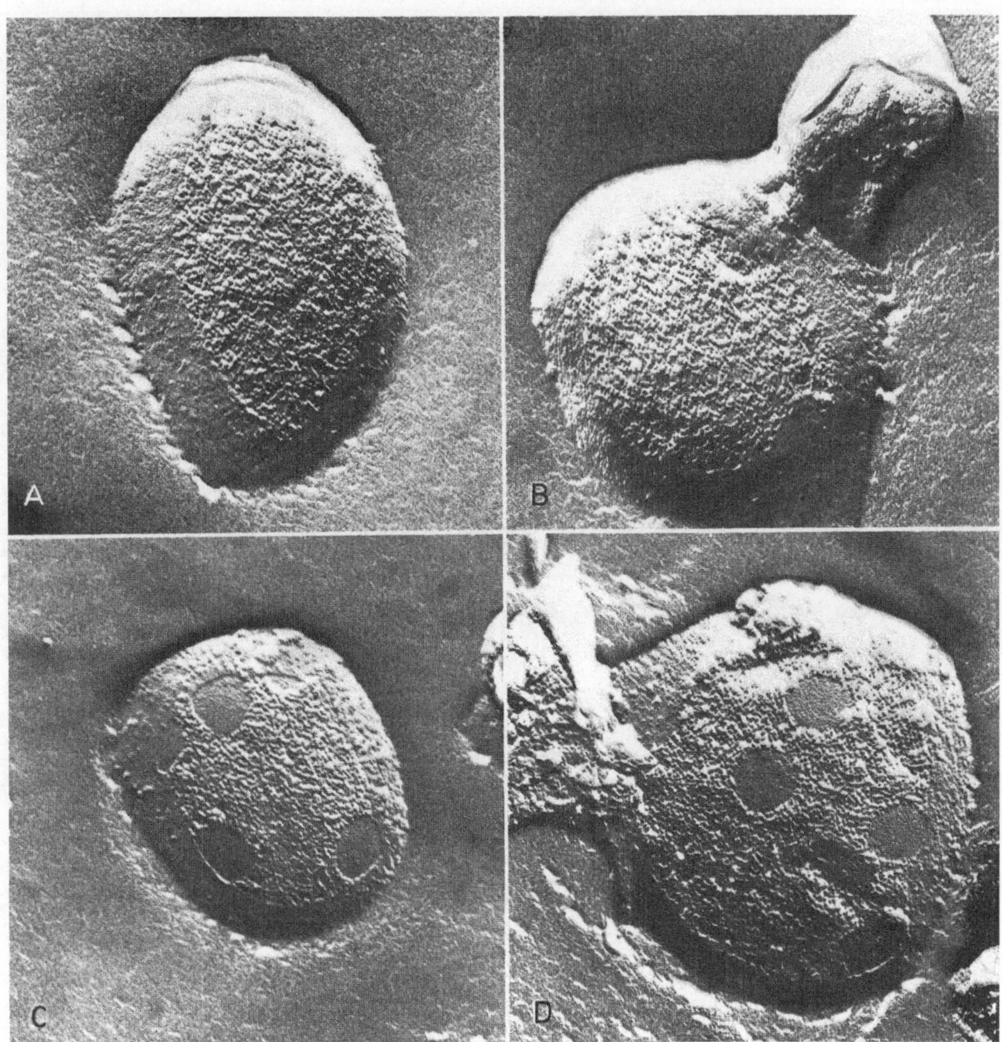

Fig. 6. Cytoplasmic membrane vesicles isolated from cells grown in the presence of oleic acid. Platinum-carbon replicas are obtained using a Balzer BAF 300 freeze-etch unit. The replicas are examined using a Siemens Elmiskop 102 electron microscope. The samples are frozen in Freon 22 in the absence of cryo-protective. The samples are freeze-fractured at -100°C and etched for 20 sec at 10^{-6} Torr. Shadowing is from bottom to top. Magnification, 90,000 x. Photomicrographs A and B correspond to samples frozen from 45°C; C and D correspond to samples frozen from 0°C. At high temperature, the fracture surfaces are covered with densely packed particles. At low temperature, the fracture surfaces display a mixture of smooth domains devoid of particles (more than 1,000 Å in diameter) separated by very densely packed particulated regions

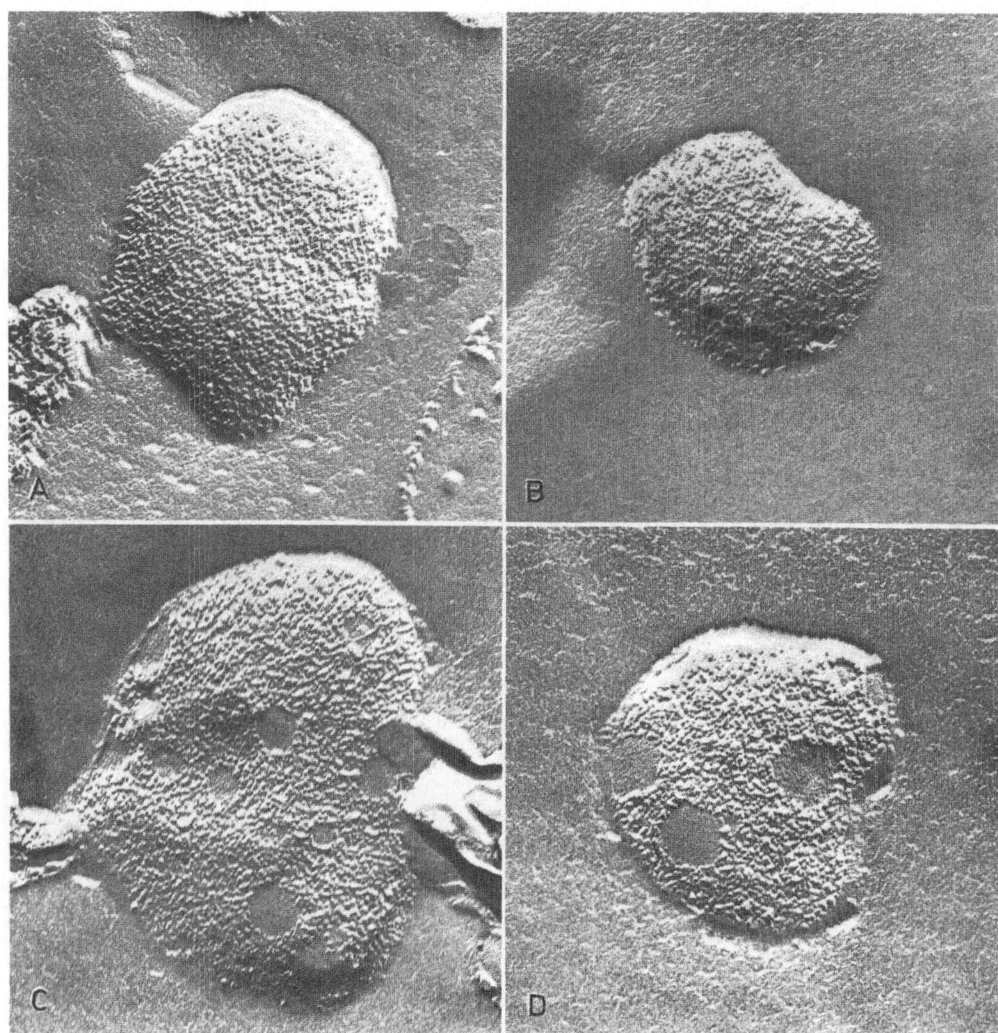

Fig. 7. Cytoplasmic membrane vesicles isolated from cells grown in the presence of linoleic acid. The experimental conditions are identical to those described for Figure 6, except for the absence of an etching. The fracture surfaces at high temperature (A and B) and at low temperature (C and D) are very similar to those observed for the membranes isolated from cells grown in the presence of oleic acid

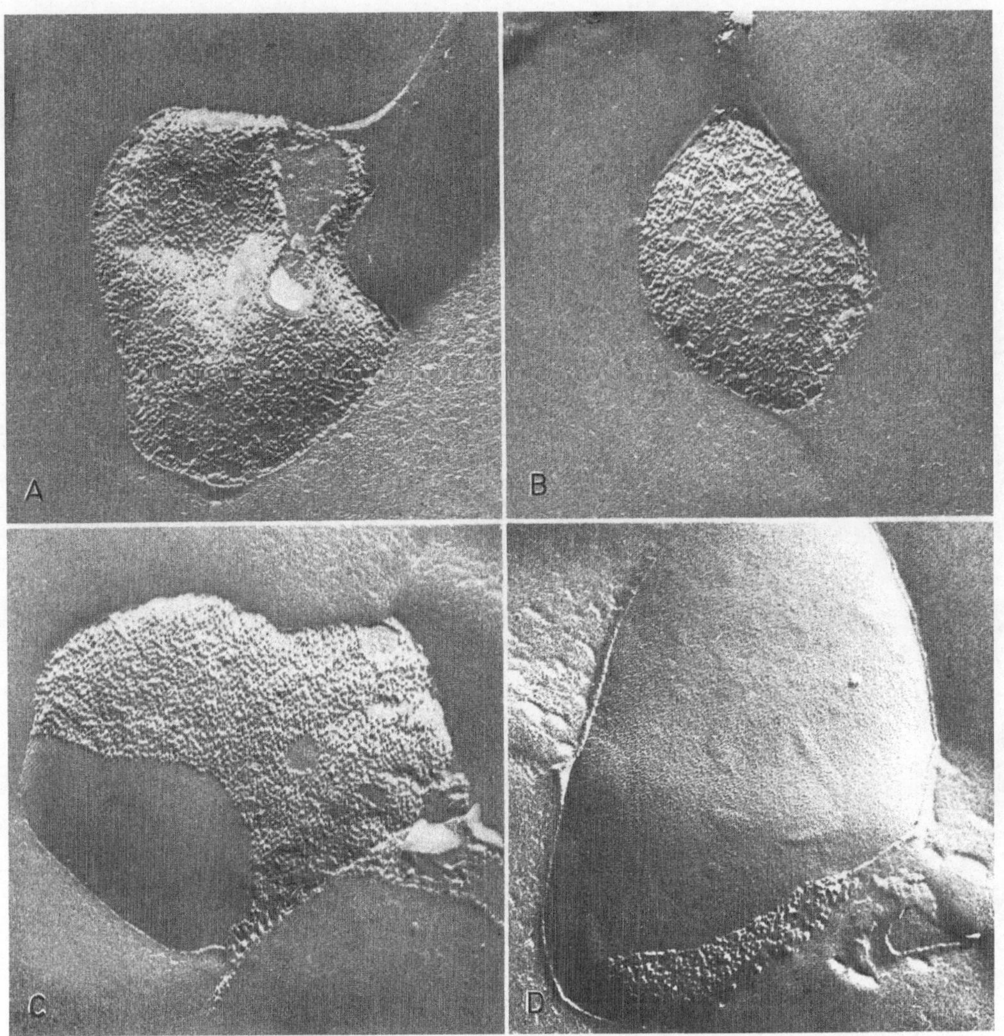

Fig. 8. Cytoplasmic membrane vesicles isolated from cells grown in the presence of linolenic acid. The experimental conditions are identical to those described for Figure 6, except for the absence of an etching. The fracture surfaces at high temperature (A and B) display a homogeneous but not quite statistical distribution of particles. The nonstatistical distribution is probably an experimental artifact, 45°C being the upper limit of the order-disorder transition. Glutaraldehyde prefixation eliminates the network-like distribution (KLEEMANN, McCONNEL, 1974). At low temperature (C and D), the fracture surfaces display very large, smooth domains, often extending over the entire fracture face (see D). In cases where the fracture occurs also along the particulated regions, small smooth domains (some 500 Å in diameter) are observed (see C)

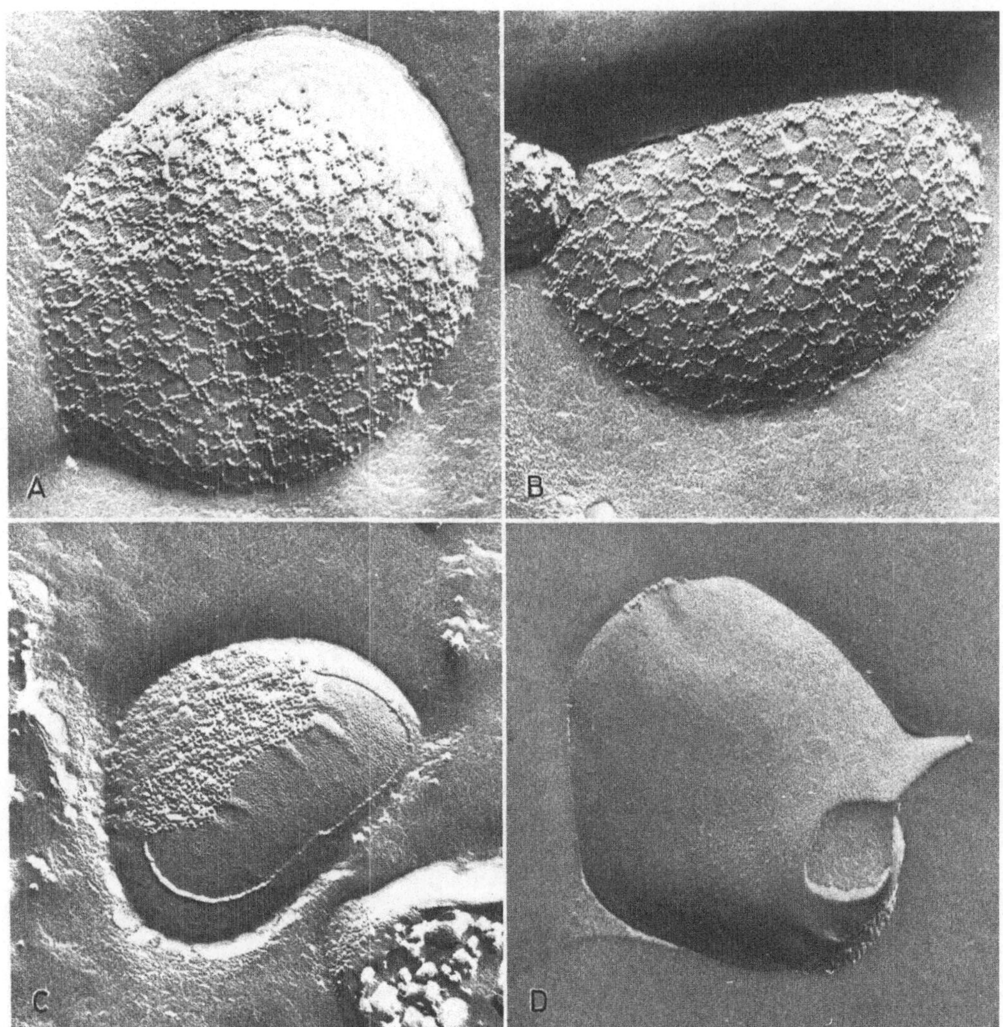

Fig. 9. Cytoplasmic membrane vesicles isolated from cells grown in the presence of elaidic acid. The experimental conditions are identical to those described for Figure 6, except for the absence of an etching in cases A, B, and D. At high temperature (A and B), the particle distribution is homogeneous but clearly not statistical (see legend for Figure 8). Moreover, the number of particles is smaller by a factor of 2 as compared to that for the other membranes. At low temperature, most of the fracture surfaces are completely smooth, with the particles being seen only at the edges (see D). In some rare cases, the fracture takes place also along particulated regions. In these domains the particles are strongly aggregated (see C). The aggregation is such that some of the particles seem to be excluded from the hydrophobic core toward the external membrane surface (note in C the particulated aspect of the etched membrane surface contiguous to the particulated fracture surface)

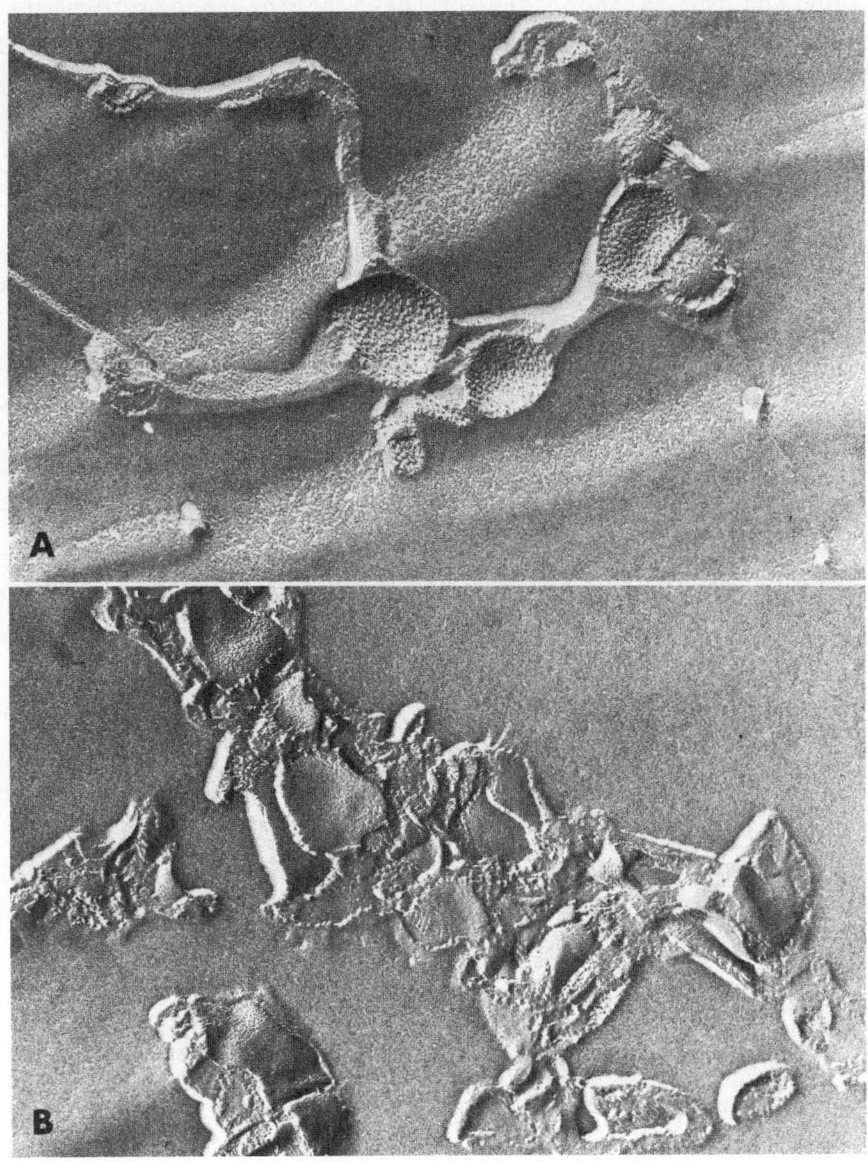

Fig. 10. Outer membrane vesicles isolated from cells grown in the presence of elaidic acid. The experimental conditions are identical to those described in Figure 6 except for the absence of an etching (A is a sample frozen from 45°C; B is a sample frozen from 0°C)

diffraction and the ratio of the surfaces of the smooth-to-particulated membrane regions. Precise quantitative values are hard to establish in view of the preferential fracturing along the large, smooth surfaces when they exist.

There seems also to exist a correlation between the size of the individual smooth domains and their fatty acid composition: the more structurally related the different fatty acids in a given domain are, the larger the smooth domains. Elaidic cytoplasmic membranes frozen from 0°C display only large, smooth domains: structurally the different fatty acids in these domains are similar (elaidic acid resembles more a saturated than an unsaturated fatty acid). Linolenic cytoplasmic membranes frozen from 0°C display a mixture of large and relatively small, smooth domains (see Figure 8); when frozen from 25°C, they display only the large, smooth domains (data not shown).

In view of the X-ray data presented above, the large, smooth domains should correspond to the ordered domains containing mainly saturated hydrocarbon chains (totally saturated lipids). The small, smooth domains that disappear at the end of the first order-disorder transition may correspond to ordered domains containing a mixture of saturated and linolenic fatty acids (mixed lipids). Oleic and linoleic cytoplasmic membranes frozen from 0°C display smooth surfaces whose sizes are to some extent intermediate between those observed for elaidic membranes and those observed on the particulated regions of linolenic membranes frozen from 0°C. This is to be expected since the structural difference between an oleic and a saturated or a linoleic and a saturated fatty acid is intermediate between that of an elaidic and a saturated hydrocarbon chain on the one hand, and a linolenic and saturated hydrocarbon chain on the other hand.

Fracture surfaces for elaidic outer membranes frozen from temperatures above and below the order-disorder transition are shown in Figure 10. At high temperature, they display densely particulated fracture surfaces. Contrary to the cytoplasmic membranes, however, the particles seem associated mainly with the concave face. At low temperature, they display mainly smooth surfaces.

Transport

The transport of substrates across *E. coli* cytoplasmic membranes has been well documented (KABACK, 1972). Two general types of transport systems have been described in detail: D-lactate dehydrogenase-coupled sugar and amino-acid transport, and phosphoenolpyruvate-dependent vectorial phosphorylation of certain monosaccharides. The data for D-lactate-dependent proline uptake and phosphoenolpyruvate-dependent glucose uptake are shown in an Arrhenius representation in Figure 11, together with relevant parameters (temperature of discontinuity and energy of activation).

We believe that a good correlation exists between transport of proline and order-disorder transition of lipid hydrocarbon chains as detected by X-ray diffraction. The discrepancies (in the case of oleic and elaidic cytoplasmic membranes, an apparent break in the Arrhenius plots at a temperature above that of the mid-transition temperature; absence of a break for linoleic cytoplasmic membranes; and presence of only one break in the case of linolenic cytoplasmic membranes for which two order-disorder transitions exist) are a consequence of the experimental conditions and are dealt with in the Appendix. On the other hand, vectorial phosphorylation of glucose clearly does not respond to the order-disorder transitions.

The transport of proline across the bacterial membrane is mediated by a protein carrier molecule which is presumed to move or to rotate within the membrane. As had been pointed out, the ordering of the hydrocarbon chains of the lipids will exclude penetrating proteins from the ordered domains. It follows, therefore, that proline transport at low temperature cannot take place through these domains, which correspond to the smooth areas observed on the freeze-fracture surfaces. Transport should take place through the particulated regions where the hydrocarbon chains of the lipids are disordered. Thus, the high energy of activation determined at low temperature is not the result of the inability of substrates to cross lipid bilayers with ordered hydrocarbon chains. It could be the consequence of a decreased mobility of the carriers upon aggregation of the membrane-penetrating proteins. Other possibilities cannot be ruled out (effect on the electron transport chain, effect on energy coupling, etc.).

Vectorial phosphorylation of glucose is probably also mediated by penetrating proteins (enzyme II) (KUNDIG, GHOSH, ROSEMAN, 1964). Therefore, glucose transport should also take place through the particulated regions. However, contrary to D-lactate-dependent active transport, vectorial phosphorylation is not a classic active transport and may not require mobile carriers. Our data indicate that the enzymatic activities required for this type of transport are not repressed by the aggregation of the membrane-penetrating proteins.

A last point to be mentioned concerns the absolute values of transport of glucose and proline at high temperature. At 40°C, a tem-

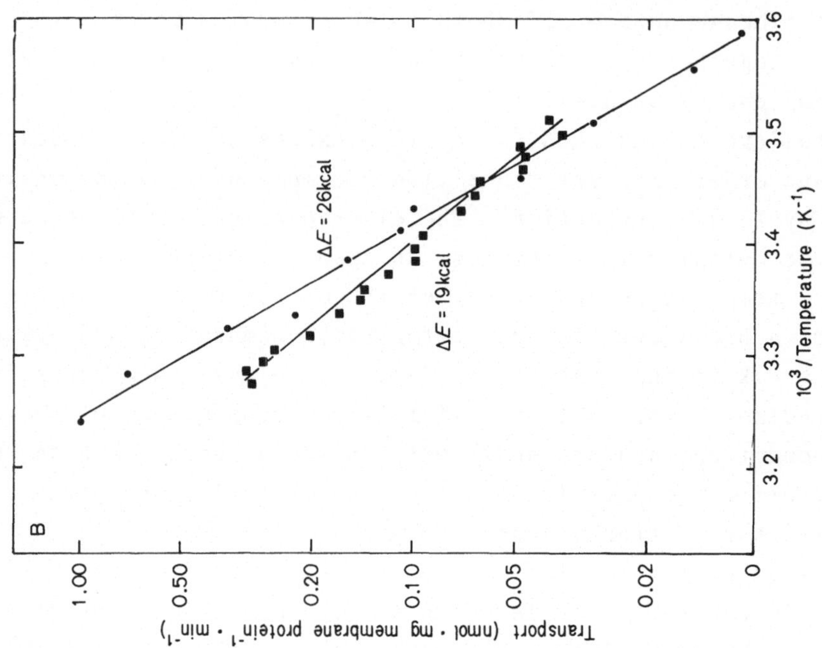
B
ΔE = 26kcal
ΔE = 19kcal
Transport (nmol · mg membrane protein⁻¹ · min⁻¹)
1.00
0.50
0.20
0.10
0.05
0.02
0
3.2
3.3
3.4
3.5
3.6
10³/Temperature (K⁻¹)

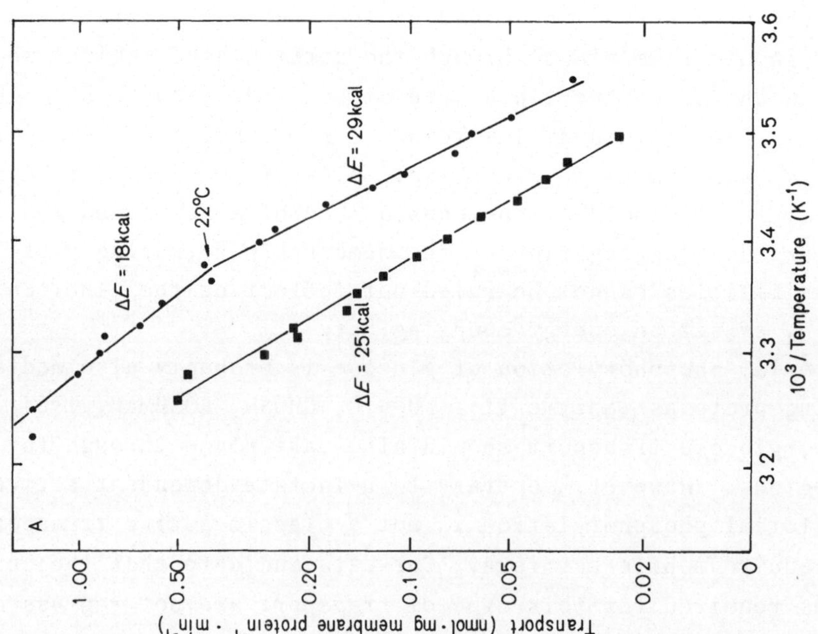
A
ΔE = 18kcal
22°C
ΔE = 29kcal
ΔE = 25kcal
Transport (nmol · mg membrane protein⁻¹ · min⁻¹)
1.00
0.50
0.20
0.10
0.05
0.02
0
3.2
3.3
3.4
3.5
3.6
10³/Temperature (K⁻¹)

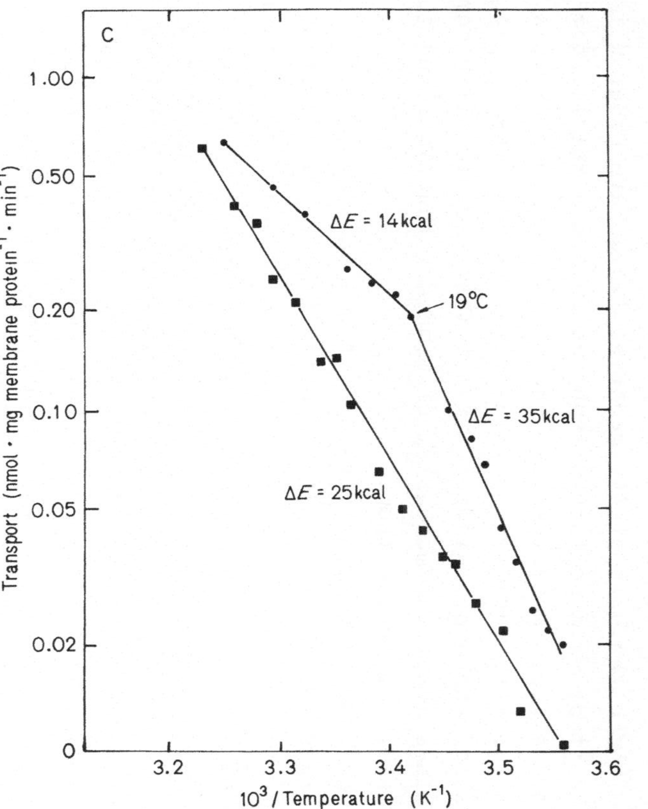

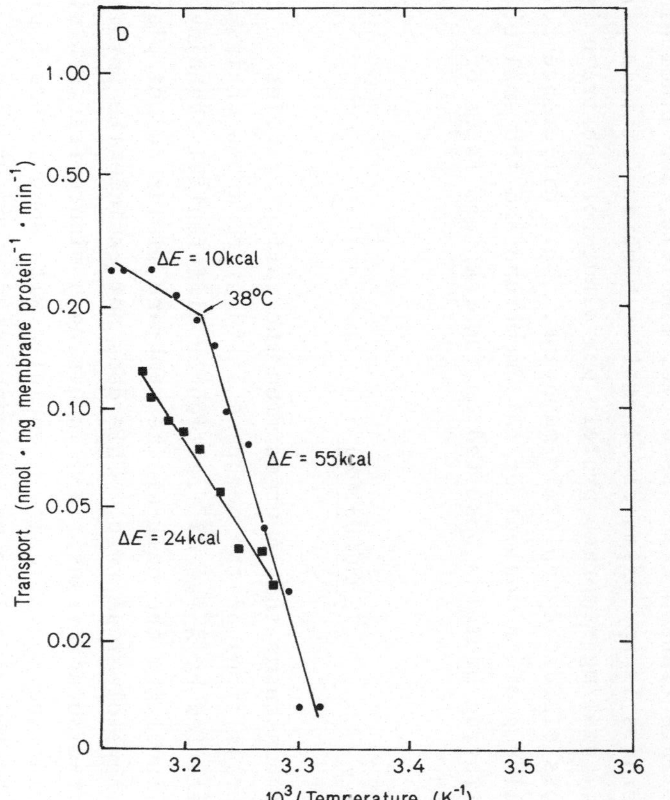

Fig. 11. Arrhenius plots for proline (●) and glucose (■) transport across cytoplasmic membrane vesicles isolated from cells grown in the presence of elaidic acid (A), oleic acid (B), linoleic acid (C), and linolenic acid (D). Transport of ^{14}C-proline (specific activity 290 mCi/mmole) and ^{14}C-glucose (specific activity 250 mCi/mmole) are performed according to KABACK (1971, in press). Initial rates of uptake at a given temperature are measured over the linear portion of the uptake curves

perature at which most of the lipid hydrocarbon chains are disordered, the rate of substrate transport across oleic, linoleic, and linolenic cytoplasmic membranes is approximately 1 nmol/mg membrane protein per minute. Only 0.2 nmol/mg membrane protein per minute of substrate is transported across elaidic cytoplasmic membrane. This difference could reflect the smaller protein content of this membrane and the smaller density of penetrating proteins observed on the fracture faces of elaidic membranes.

CONCLUSIONS

The general consensus that emerges from the various studies on the order-disorder transitions of the membrane lipid hydrocarbon chains is that the ones detected by X-ray diffraction or differential scanning calorimetry and involving the bulk of the membrane lipids do not play any significant physiological role. This does not exclude the possibility that order-disorder transitions involving a minute fraction of the membrane lipids and specifically associated with some membrane proteins regulate physiologically some membrane-associated functions. However, currently available techniques do not allow the determination of such relations.

It should be understood, therefore, that the study of a membrane under conditions where the bulk of the membrane lipids are ordered is equivalent to the study of a perturbed membrane in a reversible, partially denaturated state. Nevertheless, the studies of such systems, and the study of the structural and physiological changes taking place during the course of the order-disorder transitions, should allow some insight to be gained about the mechanism of membrane functions and the structural parameters governing them.

Some membrane-associated functions do not respond to the order-disorder transitions and their energy of activation remains constant throughout the transition. We have reported here the case of vectorial phosphorylation of glucose across the cytoplasmic membranes of *E. coli*. Other examples have been reported previously in mycoplasma membranes: NADH oxydase, p-nitrophenyl phosphorylase (DE KRUYFF et al., 1973) and phosphoenolpyruvate-dependent membrane functions (ROTTEM et al., 1973). The limited information that can be gained from such results is that the membrane functions do not respond to the ordering of the hydrocarbon chains (independent of the viscosity of the medium) and/or do not respond to the aggregation of the membrane-penetrating proteins (independent of the immobilization).

Some membrane-associated functions are well correlated with the order-disorder transition and display different energies of activation at temperatures above and below the transition , respectively. We have reported here the case of active transport of proline across the cytoplasmic membrane vesicles of *E. coli*. Other examples have been reported for mycoplasma membranes: ATPase-dependent functions (DE KRUYFF et al., 1973; ROTTEM et al., 1973). In some cases, this correlation could be a direct consequence of the ordering of the hydrocarbon chains (that is, a consequence of the formation of the ordered domains per se). Since membrane-penetrating proteins are generally excluded from these ordered domains, they probably do not play a role in this category of functions. In some other cases, the correlation could be an indirect consequence of the ordering and could result either from the aggregation and immobilization that is (as in our case) concomitant with the disorder-order transition, or from a physical separation of membrane-penetrating proteins from some proteins less embedded in the hydrophobic core of the membranes remaining in the ordered domains. It is probably a search for the exact reasons for these correlations that could lead to information regarding the mechanism of membrane-associated functions and the structural parameters governing them.

APPENDIX: ARRHENIUS PLOTS AND ORDER-DISORDER CONFORMATIONAL TRANSITIONS OF LIPID HYDROCARBON CHAINS

Calculated Arrhenius curves are constructed from the experimental X-ray data, assuming that for a given membrane the energy of activation E(T) at a temperature T is given by:

$$a/A = (E[T]-[T_h]) \; / \; (E[T_l]-E[T_h])$$

where *a* and *A* are the amounts of ordered hydrocarbon chains at a temperature T and the maximum amount of ordered hydrocarbon chains at temperatures below the transition temperature, respectively and $E(T_h)$ and $E(T_l)$ are the energies of activation at temperatures above and below the transition, respectively. This relation implies that the energy of activation is directly or indirectly related to the amount of ordered chains in the membrane, and thus that there exists a perfect correlation between the order-disorder transition and the membrane function under study. The values of *a* and *A* are taken from the X-ray data; the values of $E(T_h)$ and $E(T_l)$ are taken from the slopes of the observed Arrhenius plots for proline transport at the extreme temperatures. The one exception is linoleic membrane, for which the order-disorder transition extends over practically the whole temperature range studied, making it impossible to determine $E(T_h)$ and $E(T_l)$. For linoleic membrane, we have taken the values determined for oleic membrane. The rationale behind this originates from the freeze-etch electron microscopy data: whatever the temperature, very similar particle aggregation and distribution are observed on the fracture faces of the two membranes (see Figures 6 and 7*).

The Arrhenius plots constructed in this way are shown in Figure 11A. As expected from the underlying assumption, they display straight line portions in the temperature ranges of constant amount of ordered hydrocarbon chains (solid lines in Fig. 11A) and curvatures in the temperature ranges of the order-disorder transitions. The extrapolated straight lines intercept at temperatures corresponding to those of the mid-order-disorder transitions (18°C for oleic membrane, 31°C for elaidic membrane, 13°C and 34°C for linolenic membrane, and at a temperature not determinable for linoleic membrane).

The calculated Arrhenius plots appear different from the observed ones. However, the differences are mainly a result of the experimental conditions. Indeed, it is often impossible to follow a membrane function over a great temperature range. In the case of proline transport across cytoplasmic membranes, the determination of the initial rate of uptake is limited in the low temperature range by efflux (KABACK, BARNES, 1971) and in the high temperature range by irreversible denaturation of the membrane. If one limits the calculated Arrhenius plots to the temperature ranges accessible for the transport studies, they can be fitted with only straight lines (Figure 12, curve b) and they then resemble closely the observed ones. It seems unrealistic, in view of the experimental precision, to search for more complex behavior. Thus, according to Arrhenius plots, one should conclude that a good correlation exists between active transport of proline and the order-disorder transition of the membrane lipid hydrocarbon chains.

The following remarks should be made:

(1) In the case of a membrane displaying a single and well-defined order-disorder transition (oleic and elaidic membranes), the apparent break in the Arrhenius plot takes place at a temperature higher than that corresponding to the mid-order-disorder transition (37°C instead of 31°C for elaidic membrane, 22°C instead of 18°C for

*With the exception of Figure 12, all figures referenced in this Appendix can be found in the body of the paper.

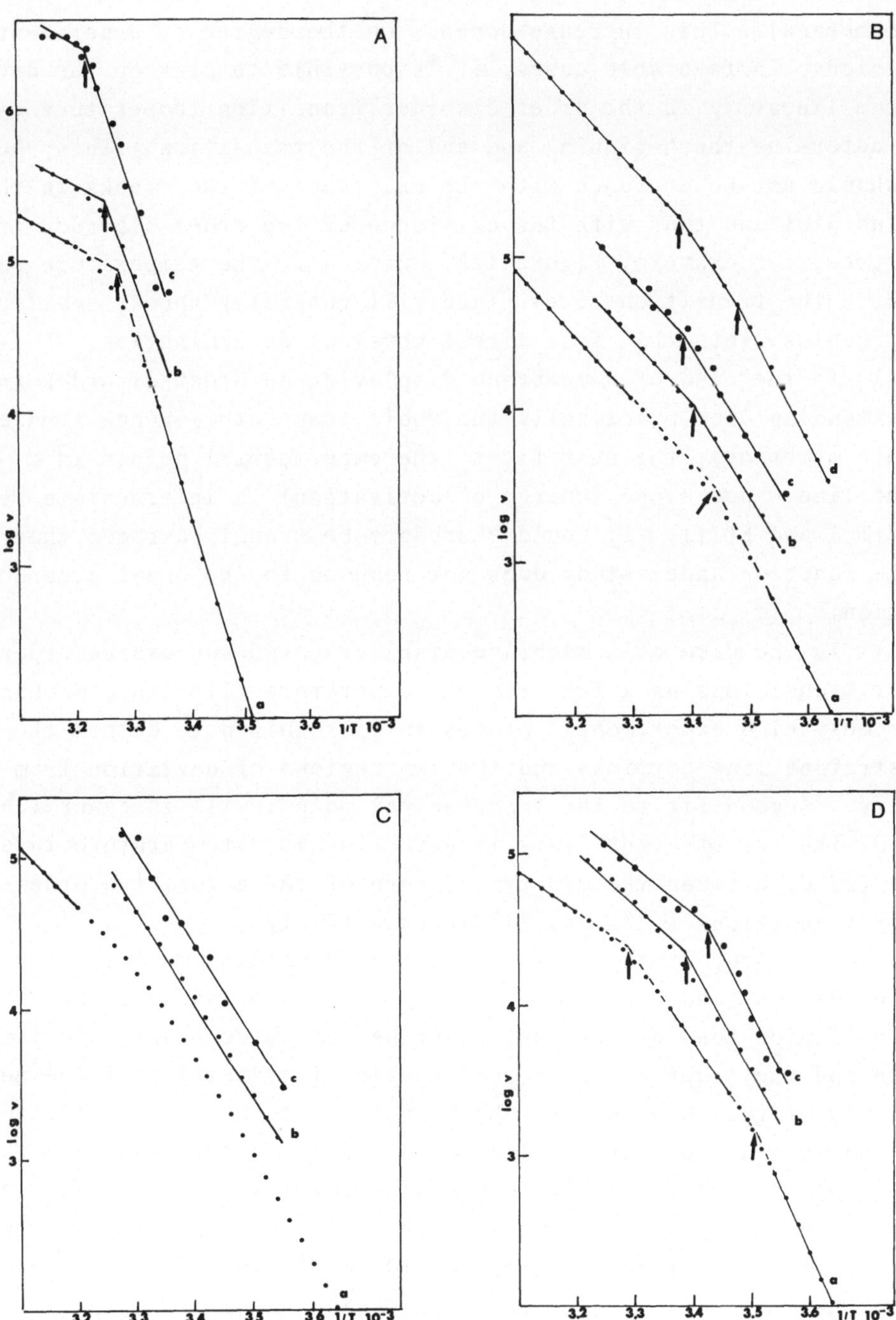

Fig. 12. Arrhenius plots of the initial rate of proline uptake (arbitrary scale) across elaidic membrane (A), oleic membrane (B), linoleic membrane (C), and linolenic membrane (D). Curve a = calculated from the X-ray data (see text). Curve b = calculated curve limited to the temperature range over which the transport studies are performed. Curve c = experimentally determined curve (the data are taken from Figure 11). Curve d (Figure 12B) = generation of two apparent breaks for a single and well-defined transition. The data are taken from SHECHTER (in press)

oleic membrane). This increase depends on the degree of experimental imprecision. In favorable cases, it is possible to pick up the deviation from linearity in the order-disorder transition temperature range and to determine the beginning and end of the transition. This, however, should not be confused with the existence of two breaks in the Arrhenius plot and thus with the existence of two order-disorder transitions (see, for example, Figure 12B, curve d). The temperature range over which the transition takes place will generally appear smaller from Arrhenius plots than from direct physical determination.

(2) In the case of a membrane displaying an order-disorder transition extending over practically the whole temperature range studied (linoleic membrane), the best fit to the experimental points is a straight line whose slope (energy of activation) is intermediate between $E(T_h)$ and $E(T_l)$. It could therefore be wrongly assumed that the membrane function under study does not respond to the order-disorder transition.

(3) In the case of a membrane displaying two successive order-disorder transitions as a function of temperature (linolenic membrane), an extremely high experimental precision is required to detect the three straight line portions and the two regions of deviation from linearity. A good fit to the experimental points will in general be obtained with two straight lines intersecting at a temperature intermediate (22°C) between the mid-temperature of the actual two order-disorder transitions (13°C and 34°C, respectively).

In conclusion, it should be clear that depending on the experimental conditions and precision, it is possible to assume either that a good correlation does or does not exist between a membrane-associated function and the order-disorder conformational transition of the membrane lipid hydrocarbon chains. Even in the presence of correlation, a single and well-defined transition could result in two breaks in the Arrhenius plots; two successive order-disorder transitions could result in only one break in the Arrhenius plots; a transition extending over a temperature range that is too great could result in no apparent break in the Arrhenius plots. Thus, the interpretation of Arrhenius plots in the membrane field should be handled with great care.

REFERENCES

DE KRUYFF, B., VAN DIJCK, P. W. M., GOLDBACH. R. W., DEMEL, R. A., VAN DEENEN, L. L. M.: Biochim Biophys. Acta 330, 269-282 (1973).

DUPONT, Y., GABRIEL, A., CHABRE, M., GULIK-KRZYWICKI, T., SHECHTER, E.: Nature 238, 331-333 (1972).

KABACK, H. R.: Methods Enzymol. 22, 99-120 (1971).
KABACK, H. R.: Biochim. Biophys. Acta 265, 367-416 (1972).
KABACK, H. R.: Methods Enzymol. (in press).
KABACK, H. R., BARNES, E. M., Jr.: J. Biol. Chem. 246, 5523-5531 (1971).
KLEEMANN, W., McCONNEL, H. M.: Biochim. Biophys. Acta 345, 220-230 (1974).
KUNDIG, W., GHOSH, S., ROSEMAN, S.: Proc. Nat. Acad. Sci. U.S.A. 52, 1067-1074 (1964).
LEE, A. G., BIRDSALL, N. J. M., METCALFE, J. C., TOON, P. A., WARREN, G. B.: Biochemistry 13, 3699-3705 (1974).
LUZZATI, V., TARDIEU, A., GULIK-KRZYWICKI, T., MATEU, L., SHECHTER, E., CHABRE, M., CARON, F.: Proceedings of the 8th Meeting of the Federation of European Biochemical Societies, Vol. XXVIII, pp. 173-183. Amsterdam: Elsevier 1972.
OSBORN, M. J., GANDER, J. E., PARISI, E., CARSON, J.: J. Biol. Chem. 247, 3962-3972 (1972).
OVERATH, P., BRENNER, M., GULIK-KRZYWICKI, T., SHECHTER, E., LETELLIER, L.: Biochim. Biophys. Acta (in press).
RANCK, J. L., MATEU, L., SADLER, D. M., TARDIEU, A., GULIK-KRZYWICKI, T., LUZZATI, V.: J. Mol. Biol. 85, 249-277 (1974).
ROTTEM, S., YASHOUV, J., NÉEMAN, Z., RAZIN, S.: Biochim. Biophys. Acta 323, 495-508 (1973).
SHECHTER, E.: FEBS Letters (in press).
SHECHTER, E., LETELLIER, L., GULIK-KRZYWICKI, T.: Eur. J. Biochem. 49, 61-76 (1974).
STEIM, J. M.: Proceedings of the 8th Meeting of the Federation of European Biochemical Societies, Vol. XXVIII, pp. 185-196. Amsterdam: Elsevier 1972.

Changes in Receptor Immunoglobin Turnover during B-Lymphocyte Differentiation

Fritz Melchers
Basel Institute for Immunology, 487 Grenzacherstraße, CH–4058 Basel, Switzerland

INTRODUCTION

Lymphocytes recognize antigens with immunoglobulin (Ig) molecules that are located in their surface membrane. Binding of antigens to variable regions of Ig receptor molecules (HILSCHMANN, CRAIG, 1965) specific for these antigens is believed to be the first step in a chain of reactions that may lead to either an immune response or to paralysis. One lymphocyte expresses Ig molecules of only one antigen specificity, i.e., of one given structure (MÄKELÄ, PASANEN, SORVAS, 1971). The many foreign antigenic determinants that can evoke an immune response, and many of the body's own determinants to which tolerance is induced are recognized by many different lymphocytes possessing receptor Ig molecules of differing structures. Recognition of an antigenic determinant by any of these lymphocytes leading to an immune response involves proliferation of antigen-specific clones of lymphocytes and differentiation of cells in these clones to Ig-secreting plasma cells. Thymus-derived (T-) and bone-marrow-derived (B-) lymphocytes cooperate in this immune response (MILLER, MITCHELL, 1969). B-cells produce and, in the end cell, secrete Ig molecules, while T-cells cooperate in the induction of B-cells to proliferation and differentiation. Recognition of the body's own determinants and the reactions that lead to paralysis of lymphocytes are much less understood.

Since both immune induction and tolerance induction are believed to be initiated by the recognition of an antigenic determinant by receptor Ig molecules on the surface of lymphocytes, knowledge of the structure of receptor Ig, its synthesis, turnover, and representation in the lipid bilayer of the surface membrane will help us to understand the different biological reactivities of B-lymphocytes.

The dual recognition of determinants leading to either tolerance or to an immune response may occur in the same cell through the binding of the same surface-bound receptor Ig. A detailed model has been proposed (BRETSCHER, COHN, 1972), in which different conformations of the receptor complex in the surface membrane determine reactions leading either to tolerance or to immunity. It is, however, also conceivable that induction of tolerance and induction of immunity occur at different stages of lymphocyte differentiation within a clone of cells producing Ig of a given structure and antigen specificity. In that case, the receptor complex and the Ig molecule in it could change its conformation during differentiation of lymphoctyes and thereby change its functional role from tolerance induction to immune induction.

This report summarizes our attempts to characterize the receptor complex on small, resting, Ig-bearing B-lymphocytes involved in reactions leading to proliferation and differentiation of Ig-secreting plasma cells. Synthesis, turnover, and surface representation of the Ig molecules on B-lymphocyte subpopulations are used to characterize different possible stages of B-lymphocyte differentiation. Our studies indicate that the Ig-receptor complex in the surface membrane of B-lymphocytes may change its representation by changing the turnover of the antigen-recognizing Ig molecules. Experimental data substantiating the results described in this paper have been published (ANDERSSON, LAFLEUR, MELCHERS, 1974; MELCHERS, ANDERSSON, 1973, 1974, in press [a], [b]; ANDERSSON, MELCHERS, 1974; ANDERSSON, BULLOCK, MELCHERS, in press; MELCHERS, CONE, in press; ANDERSSON et al., 1974) or are being prepared for publication (MELCHERS et al., in preparation).

METHODS FOR THE DETECTION OF SYNTHESIS, TURNOVER, STRUCTURE, AND SURFACE REPRESENTATION OF Ig MOLECULES IN B-LYMPHOCYTES

Synthesis of Ig molecules in B-lymphocytes is measured through biosynthetic incorporation of radioactive leucine into proteins including Ig. The radioactive Ig is detected by precipitation with Ig-specific antisera. Separation of the labeled cells from the supernatant medium lets us distinguish cell-associated Ig from released Ig. Size and polypeptide composition of the radioactive Ig molecules are determined by electrophoresis in polyacrylamide gels (MELCHERS, ANDERSSON, 1973).

For turnover measurements of Ig molecules, radio-labeled B-lymphocytes are cultured for various periods of time in a nonradioactive ("chase") medium. Ig molecules, unlike other mammalian membrane proteins (SIEKEVITZ, 1972), are not degraded but are released into the extracellular fluids. Turnover of Ig therefore shows itself as a state of dynamic equilibrium in which there is a release of Ig molecules from the cell surface into the extracellular fluids and replacement by newly synthesized molecules.

Two methods are used to radioactively label Ig molecules. The first method is described above and constitutes biosynthetic incorporation of radioactive leucine into newly synthesized Ig. In the second method, surface-bound protein and Ig molecules are radioiodinated by the lactoperoxidase-catalyzed radioiodination reaction. The first method detects *only* Ig molecules that are synthesized by the lymphocytes but not those that may only be cytophilically attached to, but not synthesized by, the lymphocytes. The second method detects *only* Ig molecules that are accessible on the outside of lymphocytes to radioiodination by the lactoperoxidase-catalyzed reaction. The biosynthetic (first) method, in contrast to the radioiodination (second) method, *also* detects Ig molecules that are synthesized and contained inside the cells, may either be found only shortly on the surface, or may not be exposed to the outside at all and therefore are not approachable for external labeling.

SOURCES OF B-LYMPHOCYTES

Different lymphoid organs at different stages of the development of the mouse contain lymphocyte populations at different stages of their functional maturation, probably at different stages of differentiation. Fetal liver at between 13 and 18 days of gestation and bone marrow contain most of the precursors of B-lymphocytes found at that time of development in the mouse. After birth, the spleen develops into a lymphoid organ containing a mixture of B-lymphocytes in precursor and in more differentiated states. Lymphocytes in the lymph nodes and ductus thoracicus of adult mice predominantly contain the recirculating small, resting lymphocytes thought to represent antigen-experienced, antigen-sensitive differentiated B-cells. Stimulation of these small lymphocytes by antigen and T-cells or by mitogens (MÖLLER, 1972) leads to proliferation of clones of cells and to differentiation into Ig-secreting plasma cells, thought to be end cells in the differentiation of B-cells.

The contribution of T-cells in measurements on synthesis, turnover, and surface representation of Ig on B-lymphocytes can be minimized when lymphocytes from genetically athymic, so-called "nude" mice (PANTELOURIS, 1968) are used. These mice lack the differentiated forms of T-cells and can therefore be regarded as sources for lymphocytes containing only B-lymphocytes in their differentiated forms.

Heterogeneities in B-lymphocyte populations make measurements on synthesis, turnover, and surface representation of Ig difficult to interpret. Studies with B-lymphocytes from different lymphoid organs (see above) show that at least cells from lymph node and ductus thoracicus are enriched for small, resting cells. Cells from bone marrow or spleen must, however, be separated according to their size and surface charge by velocity sedimentation (MILLER, PHILLIPS, 1969) and free-flow electrophoresis (SCHLEGEL, VON BOEHMER, SHORTMAN, in press) to yield lymphocyte populations enriched for certain biological functions such as reactivities with antigens and mitogens. Lymphocytes from different lymphoid organs, separated by size and surface charge, were used in the studies described in this chapter on synthesis, turnover, and surface representation of Ig.

Ig molecules found in B-lymphocytes from "nude" mice and in most B-lymphocytes from unprimed normal mice are of IgM class. It should be emphasized that our serological methods of detection of Ig do not distinguish between IgM and other Ig classes cross-reacting with the μ-specific antisera (subclasses of IgM) as long as the H-chains of any cross-reacting Ig molecules are of the size of carbohydrate-free or carbohydrate-containing Hμ-chains of IgM.

SMALL, RESTING G_0-B-LYMPHOCYTES

Small, resting B-lymphocytes occur in the spleen, lymph nodes, and thoracic duct and in small amounts in the bone marrow of adult mice. Fetal liver, at up to 17 days of gestation, has no detectable small, resting lymphocytes, which are characterized as follows: Small B-lymphocytes contain approximately 10^5 surface-bound Ig molecules (RABELLINO et al., 1971) as 7-8S IgM subunits, with core but not branch sugars attached to them. One small, resting lymphocyte synthesizes between 250 and 500 IgM molecules per hour and turns them over by shedding the 7-8S IgM subunits into the extracellular fluids. The half-disappearance time of this IgM is 10 to 30 hr; synthesis of the IgM is Actinomycin D-sensitive--even more so than the sum of the syntheses of all cellular proteins. Lactoperoxidase-catalyzed radioiodination detects this slowly released IgM on small B-cells (MELCHERS,

ANDERSSON, 1973; ANDERSSON, LAFLEUR, MELCHERS, 1974; MELCHERS, CONE, in press). Small B-cells do not synthesize DNA (i.e., do not incorporate radioactive thymidine into acid-precipitable material and do not divide). Small, resting B-lymphocytes are the main antigen- and mitogen-sensitive cells which develop into IgM-secreting, Jerne plaque-forming cells after stimulation. They can be enriched from contaminating large cells by velocity sedimentation (MILLER, PHILLIPS, 1969; ANDERSSON, LAFLEUR, MELCHERS, 1974).

Exposure of small B-cells to anti-Ig antibodies with specificities for Hμ- and L-κ- chains at high concentration prior to the addition of mitogen will inhibit the clonal proliferation and differentiation into Ig-secreting plasma cells (ANDERSSON, BULLOCK, MELCHERS, in press). Inhibition of DNA synthesis by hydroxyurea or cytosinearabinoside in small B-cells and subsequent stimulation by mitogens inhibits proliferation but leads to the development of plasmablastoid cells containing surface-bound Ig molecules and cytoplasm filled with polyribosomes but without developed rough or smooth endoplasmic reticulum. Synthesis of IgM in hydroxyurea-inhibited, mitogen-stimulated small lymphocytes is increased 25-fold one day after stimulation. Secretion of 19S IgM molecules can be detected; Jerne plaque-forming cells develop (ANDERSSON, MELCHERS, 1974).

CHANGES IN SMALL, RESTING B-LYMPHOCYTES AFTER MITOGENIC STIMULATION

Within 20 minutes of the addition of LPS 7-8S, IgM molecules on the surface membrane aggregate to form detergent-insoluble complexes much larger than the 19S IgM pentamer. These complexes contain other cellular material, and it is not known whether they also contain LPS. Binding of anti-Ig antibodies to mitogen-treated small cells is reduced to 10% within 30 minutes of that originally found with untreated small lymphocytes (MELCHERS, ANDERSSON, 1973, 1974). Biosynthetically labeled, surface-bound IgM molecules can no longer be precipitated in the surface membrane of intact small B-cells by anti-Ig antibodies. Consequently, inhibition of mitogenic stimulation by anti-Ig antibodies can no longer be achieved (MELCHERS, ANDERSSON, 1974; ANDERSSON, BULLOCK, MELCHERS, in press).

The aggregated-surface IgM is progressively degraded by proteases. Activation of surface-located proteases by mitogens may be a very early event in B-lymphocyte stimulation.

Concentrations of cyclic AMP inside B-cells do not change significantly within 4 hr of stimulation nor up to 72 hr after stimulation.

LPS stimulates the rate of IgM synthesis in small, resting B-lymphocytes twofold to threefold within the first hour. This stimulation is mitogen-dose-dependent and can also be observed with PPD or FCS as mitogens. IgM molecules synthesized at this increased rate are actively secreted from the cells with a medium disappearance time of 2 to 4 hr, mainly as 19S IgM pentamers (MELCHERS, ANDERSSON, in press [b]).

The initial change in the rate of IgM synthesis after mitogenic stimulation can also be observed in the presence of 5 μg of Actinomycin D per milliliter. IgM synthesis, which is more sensitive to inhibition by Actinomycin D than the sum of all cellular protein synthesis in small B-cells (see above), is rendered more resistant to this inhibitor immediately after mitogenic stimulation. Stimulation of small B-cells, in the presence or absence of Actinomycin D, leads to a redistribution of ribosomes from monoribosomes to polyribosomes within the first hour of stimulation. Stimulation of B-cells by mitogen therefore may lead to stabilization of RNA-synthesis-dependent components of IgM synthesis from degradation, through the formation of polyribosomes, and may reprogram IgM synthesis from a synthesis of membrane-bound, receptor-type IgM to a synthesis of actively secreted IgM (MELCHERS, ANDERSSON, in press [b]).

The ratios of the rates of synthesis or of secretion of IgM, over those of all proteins made in the cell, increase within the first hours of stimulation and continuously over a 3-day period of stimulation. Proteins other than IgM are made and secreted by B-cells after activation, but IgM synthesis and secretion increases selectively over synthesis and secretion of other proteins in the cell. Given values for these ratios characterize the time period of mitogenic stimulation which a small B-cell has experienced (MELCHERS, ANDERSSON, 1973, 1974). The increased IgM synthesis is directed entirely toward a type of IgM that is actively secreted from the cells as 19S pentamer at an increased rate with a half-disappearance time of IgM from the cells of 2 to 4 hr. The secreted 19S IgM molecules contain not only the core but also the branch carbohydrate moieties attached to their Hμ-chains. Lactoperoxidase-catalyzed radioiodination does not detect these actively secreted 19S IgM molecules (MELCHERS, CONE, in press). Secretion of IgM appears to occur in such a way that the IgM molecules to be secreted as 19S IgM pentamers are never exposed to the outside of the cell surface. This indicates that secretion of IgM by plasma cells and release of IgM by small lymphocytes at a slow rate follow two different pathways of exteriorization.

B-Cells stimulated for 3 days by mitogens now synthesize and secrete as much IgM as does an IgM-producing plasma cell tumor. Synthesis of IgM is now resistant to Actinomycin D for up to 5 hr. The expansion of the phenotypic expression of IgM after mitogenic stimulation through clonal proliferation and differentiation into secreting cells is approximately 10^4-fold.

Many of the early changes observed in small B-lymphocytes after stimulation can occur in the absence of either RNA or DNA synthesis. Biochemical parameters connected with IgM synthesis in particular show this. Many of these parameters may in fact be viewed as signs for the differentiation of B-cells into secreting plasma cells rather than for the induction of B-cells to proliferation. While it seems desirable to distinguish between reactions leading to and regulating the proliferative response of B-cells and those leading to and regulating the differentiation of B-cells, we cannot attribute any of the early changes to reactions leading to proliferation. Our data suggest that, since many early changes characteristic of the differentiation of B-cells occur in the absence of proliferation, proliferation and differentiation of B-lymphocytes may be independently induced and regulated. Mitogens thus could play a dual role in the induction of B-lymphocytes, and it will be interesting to see whether this dual role can be attributed to different structural components of the mitogens.

In support of the idea of an independent regulation of proliferation and of differentiation in B-cells are the morphological and biochemical characterizations of five IgM-producing tumor cell lines (ANDERSSON et al., 1974). The five tumor cell lines can be compared by morphological and biochemical criteria with mitogen-stimulated B-cell cultures. They resemble B-cells that have been stimulated for periods of 10 to 25 hr, 20 to 35 hr, 45 to 65 hr, 80 to 110 hr, and 100 to 130 hr. The cell lines that may be regarded as deregulated in their proliferation are, however, stable upon repeated transplantation for a characteristic state in the differentiation from a small B-lymphocyte to a plasma cell.

EARLIER STAGES OF B-CELL DIFFERENTIATION BEFORE THE STAGE OF THE SMALL, RESTING B-LYMPHOCYTE

The preceding studies indicate that we have begun to define biochemically the status of a small, resting B-cell and that we are able to monitor changes that occur after stimulation. Our knowledge for earlier stages of B-cell differentiation is, however, limited. Studies of LAFLEUR, MILLER, and PHILLIPS (1973), OSMOND and NOSSAL (1974), SCHLEGEL, VON BOEHMER, and SHORTMAN (in press), and RYSER and VASSALLI (1974) all suggest that precursor B-cells exist in bone marrow, spleen, and fetal liver, which are antigen-reactive but physically and functionally different from the antigen-sensitive small, resting B-cells. These studies also suggest that the precursor B-cells develop into resting, small B-cells.

Precursor B-cells have been defined by their physicochemical parameters as larger and more negatively charged compared with resting, small B-cells. These physicochemical properties allow precursor B-cells to be separated by velocity sedimentation and by free-flow electrophoresis from small, resting B-lymphocytes. Such physicochemical separations of spleen cells yield precursor B-cell fractions that unfortunately are contaminated with "background" plaque-forming plasma cells. Since these plaque-forming cells possess a very high capacity to synthesize and secrete Ig, they obscure measurements on Ig synthesis and release in precursor B-cells from the spleen. Bone marrow and fetal liver cells, however, are devoid of "background" plaque-forming cells, and can thus be used as convenient sources for precursor B-cells to be purified by velocity sedimentation and free-flow electrophoresis.

Results so far obtained in our laboratory (MELCHERS, CONE, in press; MELCHERS et al., in preparation) suggest that these large, negatively charged B-cells can be converted into small, resting B-cells through in vivo transfer into irradiated hosts or through in vitro incubation in tissue culture, both in a time period of 24 to 48 hr and in the absence of antigen or mitogens. It is not known whether the large, negatively charged B-cells are antigen-sensitive. They synthesize Ig, apparently of IgM class, which is detectable by lactoperoxidase-catalyzed radioiodination on their surface. Release of IgM is rapid, with half-disappearance times between 1 and 4 hr. The IgM appears released as the 7-8S IgM subunit form. The rapid release of surface-bound IgM may render these cells Ig-negative in staining reactions with fluorescein-coupled anti-Ig antibodies or with radioactive antigen.

It is tempting to speculate that diversification of antigen-combining site structures in Ig molecules and tolerance induction (JERNE, 1969) may occur in these large precursor B-cells.

REFERENCES

ANDERSSON, J., LAFLEUR, L., MELCHERS, F.: Eur. J. Immunol. 4, 170-180 (1974).

ANDERSSON, J., MELCHERS, F.: Eur. J. Immunol. 4, 533-540 (1974).

ANDERSSON, J., BULLOCK, W., MELCHERS, F.: Eur. J. Immunol. (in press).

ANDERSSON, J., BUXBAUM, J., CITRONBAUM, R., DOUGLAS, S., FORNI, L., MELCHERS, F., PERNIS, B., SCOTT, D.: J. Exp. Med. 140, 742-763 (1974).

BRETSCHER, P., COHN, M.: Transplant. Rev. 11, 1 (1972).

HILSCHMANN, N., CRAIG, L. C.: Proc. Nat. Acad. Sci. U.S.A. 53, 1403-1409 (1965).

JERNE, N. K.: Eur. J. Immunol. 1, 1 (1969).

LAFLEUR, L., MILLER, R. G., PHILLIPS, R. A.: J. Exp. Med. 137, 954 (1973).

MÄKELÄ, O., PASANEN, V., SORVAS, V.: In: Cell Interactions and Receptor Antibodies in Immune Responses (ed. O. MÄKELÄ, A. CROSS, T. V. KOSUNEN), pp. 243-247. New York: Academic Press 1971.

MELCHERS, F., ANDERSSON, J.: Transplant. Rev. 14, 76-130 (1973).

MELCHERS, F., ANDERSSON, J.: Eur. J. Immunol. 4, 181-188 (1974).

MELCHERS, F., ANDERSSON, J.: Eur. J. Immunol. (in press [a]).

MELCHERS, F., ANDERSSON, J.: Biochemistry (in press [b]).

MELCHERS, F., CONE, R. E.: Eur. J. Immunol. (in press).

MELCHERS, F., VON BOEHMER, H., CONE, R. E., SPRENT, J.: (in preparation).

MILLER, J. F. A. P., MITCHELL, G. F.: Transplant. Rev. 1, 3-42 (1969).

MILLER, R. G., PHILLIPS, R. A.: J. Cell. Physiol. 73, 191-198 (1969).

MÖLLER, G. (ed.) Transplant. Rev. 11 (1972).

OSMOND, D. G., NOSSAL, G. J. V.: Cellular Immunol. 13, 117-131 and 132-145 (1974).

PANTELOURIS, E. M.: Nature 217, 370 (1968).

RABELLINO, E., COLON, S., GREY, H. M., UNANUE, E. R.: J. Exp. Med. 133, 156-167 (1971).

RYSER, J-E., VASSALLI, P.: J. Immunol. 113, 719-728 (1974).

SCHLEGEL, R. A., VON BOEHMER, H., SHORTMAN, K.: Cellular Immunol. (in press).

SIEKEVITZ, P.: Physiol. Rev. 34, 117 (1972).

Time Domains in Membrane Kinetics

Britton Chance
Johnson Research Foundation, University of Pennsylvania, Philadelphia, Pennsylvania 19174

My interest in the dynamic properties of membranes concerns the small effects on the rates of enzymatic reactions that are observed in "membrane transitions" in the Arrhenius profile in the temperature range from 10°C to 40°C. Alternatively, there may be shifts of the rate-limiting step from one process to another of a different energy of activation--a real possibility in the more complicated enzymatic sequences; hence, we find support from the general idea that nature avoids a single rate-limiting step or "master reaction" and employs instead multi-enzyme sequences with a distribution of the rate-limiting steps over the reaction pathway.

I have taken the mitochondrial membrane as a model for dynamic processes and have studied one-turnover electron transfer reactions where the elementary step is under direct observation and the possibility of a shift to another step is unlikely. There are eight levels at which motional properties of the membrane or its components may be involved in the temperature range from 10°C to -269°C, near absolute zero.

We will consider those rate processes that can be attributed to a single function, a single site, or a single reaction. As the temperature is lowered, the first clear response of mitochondrial function is the loss of ADP phosphorylation due to inactivation of the translocase for adenine nucleotides from one side of the membrane to the other.

The second temperature-sensitive process is calcium translocation. These two carrier-related processes lose their effective function in the temperature range from 10°C to 4°C, presumably due to inadequate mobility of a mobile carrier.

In the remaining cases, the electron transport reactions afford a further discrete process which can be explored at much lower temperatures.

Third, NADH oxidation in the matrix space of the mitochondria depends upon diffusional properties in a medium containing high protein concentrations. NADH oxidation is no longer observable below -20°C; apparently the matrix space becomes viscous or, indeed, "freezes out" so that this phenomenon can no longer be observed.

Fourth, in the hydrocarbon phase of the lipids, motional processes are still feasible at -20°C. Ubiquinone, a membrane component of the hydrocarbon phase, collides actively and effectively with 12-(9-anthroyl)-stearic acid or 2-(9-anthroyl)-palmitic acid (fluorescent probes mentioned in this volume by George Radda), and quenches the fluorescence down to about -30°C. This temperature coincides with the low temperature end of Steim's thermogram of extracted mitochondrial lipids which shows the heat of fusion to become negligible at -30°C. Below that, fast diffusion processes probably do not occur.

Fifth, an especially important reaction to membrane biochemists is lateral diffusion in the plane of the membrane. For example, cytochrome *c* diffuses laterally between its reductase and its oxidase in a process to which Dutton will refer in more detail in this book. This reaction tails off at about -40°C; apparently lateral diffusion of cytochrome *c* between reductase and oxidase depends upon a water phase, probably a eutectic. Its existence is identified by proton nmr, which shows that the water protons are no longer detectable at -40°C. This is a highly solvent-dependent temperature and, depending upon the salt composition of the eutectics, may vary from -30°C to -50°C.

Sixth, the rotational relaxation of the electron carriers as a factor in electron transport between cytochromes *c* and a_3 via cytochrome *a* becomes ineffective at -40°C to -50°C, a phenomenon attributed to inadequate motion. This is an interesting observation because cytochrome *a* is considered to be a membrane protein but seems to be subject to the same motional limitations as cytochrome *c*, which would be considered to be a surface protein in which lateral diffusion is required.

The seventh category involves small-scale motions. We have found an example of this in the ability of copper and iron components of cytochrome oxidase to donate electrons to reduce oxygen to water. This process is undetectable below -80°C. Apparently, the oxidase relaxes to a configuration that is important in electron transfer, at a significant rate at about -80°C. A second small-scale motional property, for which the postulated role of the valine-13 in the hemoglobin-oxygen reaction affords a possible model, is the reaction of cytochrome oxidase with oxygen or carbon monoxide which behaves as if the ligand were caged below -120°C.

Eighth is the electron transfer process itself. Electron transfer reactions between cytochrome *c* and chlorophyll are observed in bacteria down to -269°C, where electron tunneling from donor to acceptor is probably involved.

So, there is a whole spectrum of processes that involve motional properties either on the scale of the membrane, or of the molecule, or of parts of the molecule, and finally, down to that of the electron itself. I believe we must be watchful to see that these processes, which can be separated so nicely on a temperature scale in mitochondria, are adequately taken into account in our considerations of membrane-linked functions at higher temperatures.

A Genetic Approach to the Study of Oxidative Phosphorylation in Bacteria*

David L. Gutnick
The George Wise Center for the Study of Life Sciences, Department of Microbiology, Tel Aviv University, Ramat Aviv, Israel

INTRODUCTION

A great deal of progress has been made in recent years in the elucidation of the process by which electron transport is coupled to the conservation of energy in subcellular systems. Much of this information stems from the analysis of energy transducing processes in mitochondria and chloroplasts, but only recently has research with bacterial systems yielded new and important information in this area.

Despite the relatively slow start, the study of bacteria offers a number of advantages to the student of bioenergetics: (1) because of the rapid doubling, large populations of relatively homogeneous populations of cells can be obtained quickly; (2) with the development of techniques for the isolation of bacterial membrane vesicles, coupled processes such as the active transport of solutes can be studied in much greater detail than could be studied in whole cell systems; and finally (3) bacteria lend themselves to convenient genetic analysis in which mutants with specific lesions in a biochemical process can be selected and subsequently characterized.

The third approach has recently been applied to a study of oxidative phosphorylation in *E. coli*. It is the purpose of this review to describe briefly the various experimental approaches leading to the isolation and biochemical characterization of mutants of *E. coli* with specific defects in the ATP synthetase complex. The concluding section will attempt to outline some of the directions that will be taken in this research in the future.

*This research was supported in part by a grant from the United States Israel Binational Science Foundation (BSF), Jerusalem, Israel.

MUTANT ISOLATION AND CHARACTERIZATION

Selection Techniques

The selection procedures for the isolation of mutants defective in energy transduction rely heavily on assumptions concerning the probable physiological behavior of such mutants. Selection is made for mutant cells exhibiting the postulated physiological characteristics. Mutants of *E. coli* defective in oxidative phosphorylation would thus be expected to utilize fermentable carbon compounds as sole sources of carbon and energy, but should not be able to grow at the expense of carbon sources such as Krebs-cycle intermediates. BUTLIN, COX, and GIBSON (1971) isolated a large number of mutants from *E. coli* K12, which could grow on glucose but not on succinate, malate, etc., by exposing a mutagenized culture to penicillin together with succinate as a sole source of carbon and energy. The growing cells were killed by the penicillin and the mutants survived. This isolation and selection procedure led to a number of different mutants, some of which were shown to be defective in oxidative phosphorylation. These strains were termed *unc* (for uncoupled).

A second general approach to the isolation of mutants is based on reports that mutants resistant to the antibiotic neomycin that were blocked in the biosynthesis of hemin could be isolated (SASARMAN et al., 1968, 1970). This was surprising in view of the action of neomycin as an inhibitor of protein synthesis. One possible explanation for the fact that many neomycin-resistant mutants are hemin-deficient may be that the toxic antibiotic is not taken up in such cells; perhaps because there is no respiratory chain in hemin-deficient strains. If neomycin uptake into cells were an active process requiring energy, then strains lacking a functional respiratory chain might not be able to generate sufficient energy to concentrate the neomycin. One prediction of this explanation is that other classes of mutants with other defects in the energy transduction process might also be expected to be resistant to neomycin. This hypothesis was tested by isolating a series of neomycin-resistant mutants, many of which were able to grow on glucose but not on Krebs-cycle intermediates (KANNER, GUTNICK, 1972).

Both the penicillin enrichment, as well as the selection for antibiotic resistance, have been utilized in the isolation of mutants of *E. coli* defective in oxidative phosphorylation (BUTLIN, COX, GIBSON, 1971; KANNER, GUTNICK, 1972a; KANNER, GUTNICK, 1972b; TURNOCK et al., 1972;

SCHAIRER, HADDOCK, 1972; SIMONI, SHALLENBERGER, 1972; HONG, KABACK, 1972; GUTNICK, KANNER, POSTMA, 1972; BUTLIN, COX, GIBSON, 1973; SCHAIRER, GRUBER, 1973; YAMAMOTO, MÉVEL-NINIO, VALENTINE, 1973; ROSEN, 1973; COX, GIBSON, 1974; COX, GIBSON, McCANN, 1974). Table 1 lists the various mutants isolated to date and illustrates many of the biochemical and physiological properties that have been examined. It is particularly interesting that all of the mutants examined map in the same area of the chromosome around 73.5 min, near the genes for isoleucine-valine. Throughout this chapter, reference will be made to Table 1 in an attempt to compare the results obtained with various mutants at different laboratories.

Mutants lacking ATPase activity have been termed *uncA* (BUTLIN, COX, GIBSON, 1971), while those retaining the ATPase activity have been termed *uncB* (COX, GIBSON, 1974; BUTLIN, COX, GIBSON, 1973). As will be discussed, mutants that appear to fall into these two general classes exhibit many different properties depending on the kinds of biochemical analyses to which they have been subjected. For this reason, the mutants listed in Table 1 have not been subdivided.

Physiological and Biochemical Characterization

Growth

In general, mutants defective in oxidative phosphorylation exhibit a much lower growth yield on limiting concentrations of glucose than do the parental cells (BUTLIN, COX, GIBSON, 1971; KANNER, GUTNICK, 1972; SIMONI, SHALLENBERGER, 1972; BUTLIN, COX, GIBSON, 1973). In fact, the aerobic growth yields of these mutants generally resemble the anaerobic growth yields of the parental cells under similar conditions. The mutant *uncB* (Table 1) can be differentiated from the others in that it grows quite well anaerobically, while N_{I44}, B_{V4}, K_{I1}, and A_{I44}, as well as *uncA*, grow anaerobically only in the presence of an electron acceptor such as nitrate or fumarate (COX, GIBSON, 1974). The inability of most mutants to grow anaerobically in the absence of an electron acceptor probably reflects the inability of such mutants to generate a high-energy membrane state in the absence of a coupled electron transport system; such a state may be required for certain energy-driven cellular processes such as active transport (SCHAIRER, HADDOCK, 1972; SIMONI, SHALLENBERGER, 1972; PARNES, BOOS, 1973; YAMAMOTO, MÉVEL-NINIO, VALENTINE, 1973; OR, KANNER, GUTNICK, 1973; BERGER, 1973; COX, GIBSON, 1974).

ATPase

Most of the mutants isolated thus far contain defects associated in some way with the membrane ATPase complex. In some cases the mutants lack the ATPase activity itself (see Table 1), while in other cases the mutants retain ATPase catalytic activity, but the complex is altered in such a way as to cause a defect in oxidative phosphorylation. These results demonstrate the requirement for a functional membrane ATPase for oxidative phosphorylation. As shown in Table 1, a number of mutants defective in oxidative phosphorylation actually retain catalytic activity. In some of those mutants, the ATPase activity was found to be resistant to the inhibitor N,N'-dicyclohexylcarbodiimide (DCCD) (KANNER, GUTNICK, 1972b). The ability of this compound to inhibit ATPase is generally dependent on the association of the enzyme with the membrane (EVANS, 1969a, b) and thus resembles the action of oligomycin or DCCD on F_1 in mitochondria (LARDY, JOHNSON, McMURRAY, 1958). It is of interest that even though the ATPase in B_{V4}, K_{I1}, and A_{I44} is membrane bound, the activity is resistant to DCCD. It appears likely, therefore, that these mutations are affected somehow in the binding of the ATPase complex to the membrane, and that the altered binding properties are reflected in the sensitivity to DCCD (KANNER, GUTNICK, 1972; GUTNICK, KANNER, POSTMA, 1972).

Since it is likely that mutants with altered binding of the ATPase might release the ATPase prematurely into the supernatant fraction following preparation of the membrane particles, it is necessary to establish that the presumed ATPase-negative mutant is in fact lacking the activity in all of the subcellular fractions.

Energy-Dependent Reactions

In addition to the observations of alterations in the catalytic activity of ATPase leading to the defect in oxidative phosphorylation, a number of energy-dependent reactions have also been measured (Table 1). Thus, it has been shown that mutants defective in oxidative phosphorylation lack the ability to couple ATP hydrolysis to the reduction of NADP by NADH (ATP-driven transhydrogenase), but they retain the respiration-linked activity (COX et al., 1971; KANNER, GUTNICK, 1972; GUTNICK, KANNER, POSTMA, 1972; BRAGG, HOU, 1973; COX, GIBSON, McCANN, 1973, 1974). Most of the mutants, although unable to couple respiration to the synthesis of ATP because of lesions associated with the ATP synthesizing complex, are in fact able to couple respiration to the generation of a high energy state or intermediate as evidenced by their

Table 1. Characterization of mutants of *E.coli* defective in energy conservation

Strains	ATPase	Transhydrogenase Resp.-	Transhydrogenase ATP-	Ox-Phos	Quenching of fluorescence Resp.-	Quenching of fluorescence ATP-	Defective membranes	Defective C.F.
UncA[a]	-	+	-	-				+
N_{I44}[b]	-	+	-	-	+	-	+	+
B_{V4}[b]	+[c]	+	-	-	low	-		+
UncB[d]	+	+	-	-			+	
Unc 405[e]	-	+	-	-				+
DL54[f]	-							+
NR70[g]	-							+
Etc[h]	+[c]							+
K_{I1}[b]	+	low	-	-	low	-	+	
A_{I44}[b]	+	low	-	-	low	-	+	
UncA 103C[i]	-		-	-				
Unc 253[j]	+[c]		-	-				

[a]BUTLIN, COX, and GIBSON (1971); COX et al. (1971); COX, GIBSON and McCANN (1973); PREZIOSO et al. (1973); COX and GIBSON (1974).

[b]KANNER and GUTNICK (1972 a, b); GUTNICK, KANNER, and POSTMA (1972); NIEUWENHUIS et al. (1973); OR, KANNER, and GUTNICK (1973); VAN THIENEN and POSTMA (1973).

[c]Resistant to DCCD.

[d]COX, GIBSON, and McCANN (1973); BUTLIN, COX, and GIBSON (1973); COX and GIBSON (1974).

[e]COX, GIBSON, and McCANN (1974).

ability to carry out uncoupler-sensitive, respiratory-driven transhydrogenase.

A second in vitro assay for the characterization of mutants involves the ability of the membranes to catalyze the quenching of fluorescence of a fluorescent dye (KRAAYENHOF, 1970; EILERMANN, 1970; LEE, 1971; KRAAYENHOF, IZAWA, CHANCE, 1972; SCHULDINER, ROTTENBERG, AVRON, 1972; NIEUWENHUIS et al., 1973). The dye, 9-amino-6-chloro-2-methoxy acridine (ACMA), a derivative of 9-amino acridine, becomes fluorescent upon addition to a suspension of membrane particles, and the fluorescence is rapidly quenched upon addition of a respirable substrate such as succinate. Anaerobiosis or the addition of KCN abolishes the quenching, which can be restored by the addition of ATP. Addition of

Transport Aerobic	Transport Anaerobic	Proton permeability	Effect of low N,N'-dicyclohexylcarbodiimide (DCCD) conc.
Normal	None		
Normal	None		Stimulates fluorescence quenching
Normal	None		Stimulates fluorescence quenching
Normal	None		
Normal	None		
Abnormal	None	Enhanced	Decreases proton permeability Stimulates transport
Abnormal	None	Enhanced	Decreases proton permeability Stimulates transport
Abnormal	None		Stimulates transport and fluorescence quenching
Abnormal	None		Stimulates transport and fluorescence quenching
Abnormal	None		
Normal	None		
Normal	None		

[f] SIMONI and SHALLENBERGER (1972); ALTENDORF, HAROLD, and SIMONI (1974).

[g] ROSEN (1973).

[h] HONG and KABACK (1972); BRAGG, DAVIES, and HOU (1973); KANNER (unpublished observations).

[i] SCHAIRER and HADDOCK (1972).

[j] SCHAIRER and GRUBER (1973).

the ATPase inhibitor DCCD restores the fluorescence which is inhibited under all conditions by the addition of uncoupler. The mutants N_{I44}, B_{V4}, K_{I1}, and A_{I44} retain the ability to couple succinate oxidation to fluorescence quenching, but the ATP-driven activity is missing.

Another aspect of the fluorescence quenching is the effect of low concentrations of DCCD on the respiration-driven activity. In the case of the mutants, the addition of low concentrations of DCCD, much below that required for the inhibition of membrane ATPase, causes a significant enhancement of the quenching (NIEUWENHUIS et al., 1973). In this regard, the effect of DCCD resembles the effect of oligomycin in submitochondrial particles (LEE, ERNSTER, 1965; RACKER, HORSTMAN, 1967; McCARTY, RACKER, 1967). This stimulation can also be achieved in membrane particles prepared from the parent strain, following the removal of the coupling factor (KRAAYENHOF, 1970; EILERMANN, 1970; LEE, 1971; KRAAYENHOF, IZAWA, CHANCE, 1972; SCHULDINER, ROTTENBERG, AVRON, 1972; NIEUWENHUIS et al., 1973). The ability of DCCD to "recouple" respiratory activities in membrane preparations and in whole cells has been found for the mutant DL-54, originally isolated by SIMONI and SHALLENBERGER (1972), and subsequently investigated for its transport properties in isolated membrane vesicles (ALTENDORF, HAROLD, SIMONI, 1974). In addition, ROSEN (1973) described properties of a mutant NR70, which was simultaneously ATPase negative and defective in active transport. The mutant was found to yield membranes with an increased permeability for protons, and this could be reduced by the addition of DCCD which simultaneously cured the mutant of its defect in transport. The implications of these studies in active transport will be discussed in a subsequent section.

Reconstitution

The membrane fraction from *E. coli* obtained by cell rupture and high-speed centrifugation catalyzes both respiratory- and ATP-driven transhydrogenase activities (FISHER, SANADI, 1971; BRAGG, HOU, 1972). Simply washing these particles in a buffer of low ionic strength in the absence of magnesium yields a particulate preparation almost totally devoid of ATPase activity, which no longer catalyzes either the respiratory- or the ATP-driven transhydrogenase activities (BRAGG, HOU, 1972). Both energy-dependent activities can be reconstituted by the mixing of the supernatant fraction with the "stripped" particles (BRAGG, HOU, 1972).

In a subsequent study, BRAGG and HOU (1973) found that, whereas membranes of mutant DL54 could be reconstituted by the wild-type coupling factor, similar preparations from mutant N_{I44} do not respond to the wild-type factor. In addition, it was found that the membranes from N_{I44} retained the respiratory-driven transhydrogenase even after extensive washing. These results were essentially confirmed using the fluorescence quenching assay described above (NIEUWENHUIS et al., 1973). Of particular interest was the finding that, whereas mutant N_{I44} could not be reconstituted by soluble coupling factor, the quenching of ACMA fluorescence was enhanced by the addition of low concentrations of DCCD. The mutation in N_{I44} apparently results in an altered complex which somehow remains attached to the membrane and cannot be solubilized by the usual procedures. This tightly associated complex presumably blocks the site of attachment of the wild-type coupling factor.

Reconstitution with *unc* Mutants

Gibson and Cox, and co-workers, have carried out a series of studies on reconstitution, using various mutants defective in oxidative phosphorylation (COX, GIBSON, McCANN, 1973; COX, GIBSON, 1974). Reconstitution of membrane preparations from different *unc* mutants was examined for ATP-driven transhydrogenase activity, and in some cases for oxidative phosphorylation (COX, GIBSON, 1974). Reconstitution was examined among three different *unc* mutants: *uncA*, *unc* 405, and *uncB*. In *uncA*, no catalytic activity for ATPase is found, but this mutant can be reconstituted by wild-type coupling factor provided that the membranes are washed in low ionic strength buffer before the addition of the coupling factor. *Unc* 405, a second type of ATPase-less mutant, yields membrane preparations that can be reconstituted for ATP-driven transhydrogenase, even without prior removal of the inactive complex. The wash fluid from the *uncA* membranes did not exhibit any ATPase activity, but a protein peak could be eluted from a Bio-Gel column corresponding in mobility with the ATPase activity from the wild-type. In contrast, however, the wash fluid from *unc* 405 membranes appeared to be specifically lacking this protein fraction. The authors concluded that mutant *unc* 405 was lacking the ATPase aggregate, and thus could be reconstituted without prior washing of the membranes. The membranes from *uncB* mutant could not be reconstituted by wild-type ATPase, but did yield an active ATPase fraction which could bring about reconstitution of ATP-driven transhydrogenase in appropriate membrane preparations from *uncA* and *unc* 405.

Mutants N_{I44}, B_{V4}, K_{I1}, A_{I44}

In a previous section it was mentioned that energized quenching of fluorescence of ACMA using the mutants K_{I1}, B_{V4}, A_{I44}, and N_{I44} could be recoupled by low concentrations of the energy transfer inhibitor DCCD. This respiratory-driven quenching of fluorescence, as well as ATP-driven activity, can also be enhanced in mutant B_{V4} by the addition of crude coupling factor from the parent. In contrast, however, none of the other mutants responds to the parental coupling factor. Initial attempts at mixing of soluble fractions and membrane residues from the various mutants also failed to detect any reconstitution. However, when whole cells from mutants B_{V4} and cells of either K_{I1} or A_{I44} are mixed prior to the preparation of the membrane particles, a "hybrid" membrane preparation is obtained that is able to carry out ATP-driven quenching of ACMA fluorescence to the same extent as the wild-type vesicles. It is of interest that the hybrid particles exhibit ATP-driven quenching that is inhibited by normal concentrations of DCCD (KANNER, NELSON, GUTNICK, in preparation). By examining all of the fractions of K_{I1} following cell rupture and throughout the subsequent stages of differential centrifugation, it was determined that the factor from K_{I1} catalyzing the reconstitution of ATP-driven activity in membranes of B_{V4} is in the original supernatant fluid obtained following high-speed centrifugation of the cell-free extract. Whereas only about 10% to 20% of the parental ATPase activity is located in this supernatant fraction, about 90% of the ATPase activity in mutants K_{I1} and B_{V4} remains in the supernatant following high-speed centrifugation of the cell-free extract. Nevertheless, this fraction from K_{I1} could reconstitute the ATP-driven quenching of fluorescence in intact membranes of B_{V4} as well as in depleted parental particles, whereas a similar fraction from B_{V4} is inactive (KANNER, NELSON, GUTNICK, in preparation). Although the factor from K_{I1} has ATPase activity, its ability to reconstitute ATP-driven quenching is very unstable as compared with ATP hydrolase activity. One possible explanation for this may be that the ATPase complex is altered in such a way as to be defective in binding to the membrane, which brings about a dissociation of the complex such that subunit(s) required for binding are no longer associated with the catalytic activity. This would predict that not all of the subunits of the ATPase complex are necessary for hydrolytic function (see subsequent section, and NELSON, KANNER, GUTNICK, 1974).

The results on reconstitution indicate that the defects in mutants K_{I1} and B_{V4} are in fact due to lesions in two different cistrons. In this regard, the properties of the *uncB* mutant resemble those of K_{I1} in that they both have residual ATPase activity, they each contain a soluble factor that can reconstitute ATP-dependent activities, and yet neither the membranes from K_{I1} nor those from *uncB* can themselves be reconstituted. One difference between K_{I1} and *uncB* is that the coupling factor from K_{I1} is very loosely bound, whereas the factor from the *uncB* mutant is normally associated with the membrane. Although 90% of the ATPase activity of K_{I1} and B_{V4} is found to be in the soluble fraction, there remains considerable ATPase activity associated with the membranes. It is of interest that this activity in both mutants is resistant to DCCD, unlike the parental membrane-bound activity. Since the lesions in K_{I1} and B_{V4} are in different cistrons, it would appear that the sensitivity to this energy transfer inhibitor in *E. coli* is due to the interaction of at least two polypeptides, probably in a fashion similar to ATPase from mitochondria and *S. faecalis* (CATTELL et al., 1971; TZAGALOFF, 1971; BARON, ABRAMS, 1971; ABRAMS, SMITH, BARON, 1972; HORSTMAN, RACKER, 1970; MacLENNAN, TZAGALOFF, 1968).

ATPase: Purification and Properties

A number of investigators have studied the properties of membrane-bound and purified ATPase preparations from *E. coli* (KOBAYASHI, ANRAKU, 1972; EVANS, 1969a, b; ROISIN, KEPES, 1973; GIORDANO, RIVIERE, AZOULAY, 1973; CARRIERA et al., 1973; HANSON, KENNEDY, 1973; BRAGG, HOU, 1972; NELSON, KANNER, GUTNICK, 1974; FUTAI, STERNWEIS, HEPPEL, 1974). The enzyme has been purified to homogeneity by a variety of techniques and varies in specific activity as well as subunit composition (KOBAYASHI, ANRAKU, 1972; HANSON, KENNEDY, 1973; NELSON, KANNER, GUTNICK, 1974; FUTAI, STERNWEIS, HEPPEL, 1974). Bragg and co-workers have described the properties of a homogeneous preparation of ATPase consisting of five subunits (α, β, γ, ρ, ε) which was able to reconstitute depleted membrane particles for both respiratory and ATP-driven transhydrogenase activities. The five subunits of the enzyme have molecular weights of 58,000 (α), 52,000 (β), 31,000 (γ), 20,000 (ρ), and 12,000 (ε), respectively (BRAGG, HOU, 1972). The preparations of HANSON and KENNEDY (1973) and NELSON, KANNER, and GUTNICK (1974) lack the ρ subunit and are inactive in either binding to the membrane or in reconstitution of energy-dependent reactions. A modification of the purification procedure by NELSON, KANNER, and GUTNICK (1974) has been described which yielded a highly active ATPase consisting of five subunits and which retained the ability to reconstitute ATP-dependent transhydrogenase activity (FUTAI, STERNWEIS, HEPPEL, 1974). One important feature emerging from all of this work is the great similarity between the various ATPase preparations from bacterial sources and those from mitochondria and chloroplasts. In addition to the subunit composition and molecular weights, other common features have recently been described (NELSON, KANNER, GUTNICK, 1974). Thus, trypsin treatment of the purified enzyme yields a fraction consisting exclusively of the α and β subunits which retain catalytic activity. Of interest also is the fact that the ATPase inhibitor 7-chloro-4-nitrobenzo-2-oxa-1,3-diazole (NBD-Cl) (FERGUSON, LLOYD, RADDA, 1974) also inhibits the bacterial ATPase, and that the inhibitor appears to bind preferentially to the β subunit of the enzyme (DETERS et al., in preparation; NELSON, KANNER, GUTNICK, 1974; FERGUSON, LLOYD, RADDA, 1974). Antibodies prepared against the trypsin-treated *E. coli* enzyme [anti-(α+β)] inhibit not only the ATP hydrolytic activity but the various ATP-driven reactions of the membrane particles themselves. It is of interest, in this regard, that the trypsin-treated enzyme does not bind to the particles. Anti-(α+β) inhibits all of the ATP-driven reactions in membrane particles as well as the various crude and purified preparations of ATPase themselves (NELSON, KANNER, GUTNICK, 1974). Recently, antibodies have been prepared against the purified denatured subunits α, β, and γ, and the effects of these antibodies on a variety of ATP-driven reactions have been examined (KANNER, NELSON, GUTNICK, in preparation). One of these systems is the ATP-P_i exchange reaction which is inhibited by the anti-(α+β) as well as by the anti-α alone. A small but reproducible inhibition is observed with the anti-β and the anti-γ preparations as well. The various anti-sera also inhibit ATPase activity under the same ATP-P_i exchange conditions (pH 7), as well as the ATPase activity at optimum pH.

The membrane particles from K_{I1}, B_{V4}, and A_{I44} all show cross-reacting material against the anti-sera of the three denatured subunits as well as material that cross-reacts with anti-(α+β). In sharp contrast, however, no cross-reacting material against any of the antibody preparations is found with mutant N_{I44}, either in the membranes or in the supernatant fluid (KANNER, NELSON, GUTNICK, in preparation).

It has become increasingly clear that the role of the delta subunit is somehow involved in the binding as well as in the ability of the ATPase to reconstitute energy-dependent activities with washed membranes of *E. coli*, although the dependence on the delta subunit appears to be more critical for the ATPase from K12 than for the enzyme from ML (FUTAI, STERNWEIS, HEPPEL, 1974; BRAGG, DAVIES, HOU, 1973). In addition, BRAGG, DAVIES, and HOU (1973) have described some of the properties of the ATPase from a mutant termed *etc* (HONG, KABACK, 1972), in which membrane vesicles are defective for the active uptake of solutes coupled to D-lactate oxidation. The ATPase from the *etc* mutant was not as effective in reconstituting energy-dependent transhydrogenase to washed particles, and appeared to contain an altered gamma subunit (BRAGG, DAVIES, HOU, 1973).

Active Transport

The use of mutants with specific defects in the ATPase complex offers a tool for studying the respiration-coupled transport of solutes into whole cells or vesicular preparations in the absence of coupled ATP synthesis or hydrolysis. A number of mutants defective in oxidative phosphorylation have been isolated in which active transport is normal under aerobic conditions, indicating that a functional ATP hydrolase is not essential for aerobic transport of most amino acids, galactosides, and galactose (SCHAIRER, HADDOCK, 1972; PARNES, BOOS, 1973; PREZIOSO et al., 1973; OR, KANNER, GUTNICK, 1973; SCHAIRER, GRUBER, 1973). Transport in these mutants is generally inhibited completely by KCN, in contrast to a partial inhibition observed with parental preparations. Nevertheless, certain classes of ATPase mutants (DL54, NR70, K_{I1}, A_{I44}, and *etc*) (SIMONI, SHALLENBERGER, 1972; HONG, KABACK, 1972; VAN THIENEN, POSTMA, 1973; ROSEN, 1973; ALTENDORF, HAROLD, SIMONI, 1974) are inhibited for transport even under aerobic conditions. One explanation for the transport behavior of these mutants emerges from the studies of VAN THIENEN and POSTMA (1973), ROSEN (1973), and ALTENDORF, HAROLD, and SIMONI (1974), who found that DCCD in low concentrations was able to stimulate transport in these mutants. These investigators have shown that the effect of the presence of DCCD is to lower the proton permeability of the mutant membranes in such a way as to allow for the efficient coupling of electron transport to the generation of a membrane potential which drives the aerobic transport. Under anaerobic conditions this potential is presumably generated by the hydrolysis of ATP. It is interesting to note that the membrane potential can be induced in parental membrane vesicles by treatment with valinomycin, which causes the potassium to leak out, generating a gradient capable of driving the transport (HIRATA, ALTENDORF, HAROLD, 1974; KASHKET, WILSON, 1972, 1973; LOMBARDI et al., 1974). This property is missing in DL54, but can be induced by treatment with DCCD (ALTENDORF, HAROLD, SIMONI, 1974). The defect in those mutations specifically affecting the proton permeability has been postulated to be due to an exposed proton channel in the membrane (ROSEN, 1973; ALTENDORF, HAROLD, SIMONI, 1974), which can be sealed by DCCD. In the absence of DCCD, the channel is exposed, presumably because of a defect in the binding of the ATPase to the membrane. The observations with mutants K_{I1} and A_{I44}, which in fact possess ATPase hydrolytic activity but in which transport is defective, and which appear to have lesions in the binding of the ATPase (see above), are consistent with this hypothesis. On the other hand, it is interesting that mutant B_{V4}, which also has an ATPase exhibiting defective binding, has normal transport under aerobic conditions in intact cells (OR, KANNER, GUTNICK, 1973).

Finally, it is of interest that a study with DL54 has elucidated a new energy coupling process for the uptake of glutamine (BERGER, 1973). Thus, whereas proline uptake in whole cells of DL54 was completely sensitive to KCN and uncoupling agents, with glutamine the respiration-driven transport could only be driven by glucose and was sensitive to arsenate but not to uncoupling agents. It was concluded that the uptake of glutamine is apparently driven directly by phosphate-bond energy at the substrate level and not via a membrane potential or ATP hydrolysis. This mode of energy coupling has since been observed for a number of different solutes, all of which interact with shockable binding proteins normally located in the periplasmic space (BERGER, 1973; BERGER, HEPPEL, 1974; COWELL, 1974; CURTIS, 1974).

CONCLUSIONS

The study of mutants blocked in energy conservation, while incomplete, and in many cases inconclusive, has led to certain experimental directions. Mutants have been described that are defective in oxidative phosphorylation by virtue of a defect in the membrane, while others appear to be defective in the ATPase itself. These kinds of mutations sometimes retain ATPase activity, while in other cases the enzyme is inactive. In some *uncB*-type mutants, the binding of ATPase to the membrane is altered and the enzyme is easily solubilized even in the presence of magnesium ion. In addition, an *uncA* mutant has been described (N_{I44}) which cannot be reconstituted, nor can parental ATPase bind to its membranes. In all instances where they have been examined, the mutants are defective in oxidative phosphorylation. It appears, therefore, that single-point mutations may lead to a number of different biochemical phenotypes, each of which leads to a defect in ATP synthesis. The large variety is presumably due to the fact that ATPase is a complex of at least five subunits that must interact with specific sites in the membrane in order for oxidative phosphorylation to take place. A mutation in any one of these components might thus be expected to affect not only oxidative phosphorylation, but might have some consequences relating to the binding or the catalytic activity of the protein as well. Indeed, it is perhaps the central challenge of work with the mutants to identify more precisely the nature of these lesions in an attempt to define the kinds of interactions that take place.

The ATPase of *E. coli* consists of five subunits that are present in unequal amounts, and this poses interesting questions concerning the regulation of synthesis and assembly of this enzyme complex. The availability of specific antibodies against the subunits might prove useful in studying these questions.

It is not certain from the reports that have emerged thus far whether all of the cistrons coding for polypeptides that play a direct role in oxidative phosphorylation map in the 73- to 74-minute region on the *E. coli* chromosome. It will be important, therefore, to try to isolate other classes of mutants that map in other regions but that exhibit phenotypes expected from mutants defective in oxidative phosphorylation.

Finally, it appears that the information emerging from the work with mutants is largely confirmatory, owing perhaps to the nature of the assumptions leading to their isolation as well as to the kinds of analyses to which they have been subjected. Nevertheless, new insights should be forthcoming as a result of a more detailed investigation of the genetic alterations in the specific proteins associated with the coupling process and oxidative phosphorylation.

ACKNOWLEDGMENTS

I would like to thank Drs. S. Schuldiner and W. Konings for their critical reading of the manuscript, and Ms. Katherine Tighe for her assistance in its preparation.

REFERENCES

ABRAMS, A., SMITH, J. B., BARON, C. J.: J. Biol. Chem. 247, 1484-1488 (1972).

ALTENDORF, K., HAROLD, F. M., SIMONI, R. D.: J. Biol. Chem. 249, 4587-4593 (1974).

BARON, C., ABRAMS, A.: J. Biol. Chem. 247, 1542-1544 (1971).

BERGER, E. A.: Proc. Nat. Acad. Sci. U.S.A. 70, 1514-1518 (1973).

BRAGG, P. D., DAVIES, P. L., HOU, C.: Arch. Biochem. Biophys. 159, 664-670 (1973).

BRAGG, P. D., HOU, C.: FEBS Letters 28, 309-312 (1972).

BRAGG, P. D., HOU, C.: Biochem. Biophys. Res. Commun. 50, 729-736 (1973).

BUTLIN, J. D., COX, G. B., GIBSON, F.: Biochem. J. 124, 75-81 (1971).

BUTLIN, J. D., COX, G. B., GIBSON, F.: Biochim. Biophys. Acta 292, 366-375 (1973).

CARREIRA, J., LEAL, J. A., ROJAS, M., MUÑOZ, E.: Biochim. Biophys. Acta 307, 541-556 (1973).

CATTELL, K., LINDOP, C., KNIGHT, I., BEECHEY, R.: Biochem. J. 125, 169-177 (1971).

COWELL, J. L.: J. Bacteriol. 120, 139-145 (1974).

COX, G. B., GIBSON, F.: Biochim. Biophys. Acta 346, 1-25 (1974).

COX, G. B., GIBSON, F., McCANN, L.: Biochem. J. 134, 1015-1021 (1973).

COX, G. B., GIBSON, F., McCANN, L.: Biochem. J. 138, 211-215 (1974).

COX, G. B., NEWTON, N. A., BUTLIN, J. D., GIBSON, F.: Biochem. J. 125, 489-493 (1971).

CURTIS, S. J.: J. Bacteriol. 120, 295-303 (1974).

EILERMANN, L. J. M.: Biochim. Biophys. Acta 216, 231-233 (1970).

EVANS, D. J.: J. Bacteriol. 100, 914 (1969a).

EVANS, D. J.: J. Bacteriol. 104, 1203 (1969b).
FERGUSON, S. J., LLOYD, W. J., RADDA, G. K.: FEBS Letters 38, 234-236 (1974).
FISHER, R. J., SANADI, D. R.: Biochim. Biophys. Acta 245, 34-41 (1971).
FUTAI, M., STERNWEIS, P. C., HEPPEL, L. A.: Proc. Nat. Acad. Sci. U.S.A. 71, 2725-2729 (1974).
GIORDANO, G., RIVIERE, C., AZOULAY, E.: Biochim. Biophys. Acta 307, 513-524 (1973).
GUTNICK, D. L., KANNER, B. I., POSTMA, P. W.: Biochim. Biophys. Acta 283, 217-222 (1972).
HANSON, R. L., KENNEDY, E. P.: J. Bacteriol. 114, 772-781 (1973).
HIRATA, H., ALTENDORF, K., HAROLD, F. M.: J. Biol. Chem. 249, 2939-2945 (1974).
HONG, J.-s., KABACK, H. R.: Proc. Nat. Acad. Sci. U.S.A. 69, 3336-3340 (1972).
HORSTMAN, L. L., RACKER, E.: J. Biol. Chem. 245, 1336-1344 (1970).
KANNER, B. I., GUTNICK, D. L.: J. Bacteriol. 111, 287-289 (1972a).
KANNER, B. I., GUTNICK, D. L.: FEBS Letters 22, 197-199 (1972b).
KASHKET, E. R., WILSON, T. H.: Biochem. Biophys. Res. Commun. 49, 615-620 (1972).
KASHKET, E. R., WILSON, T. H.: Proc. Nat. Acad. Sci. U.S.A. 70, 2866-2869 (1973).
KOBAYASHI, H., ANRAKU, Y.: J. Biochem. 71, 387-399 (1972).
KRAAYENHOF, R.: FEBS Letters 6, 161-165 (1970).
KRAAYENHOF, R., IZAWA, S., CHANCE, B.: Plant Physiol. 50, 713-718 (1972).
LARDY, H. A., JOHNSON, D., McMURRAY, W. C.: Arch. Biochem. Biophys. 78, 582 (1958).
LEE, C. P.: Biochemistry 10, 4375-4381 (1971).
LEE, C. P., ERNSTER, L.: Biochem. Biophys. Res. Commun. 18, 523-529 (1965).
LOMBARDI, F. J., REEVES, J. P., SHORT, S. A., KABACK, H. R.: Ann. N.Y. Acad. Sci. 227, 312-322 (1974).
MacLENNAN, D. H., TZAGALOFF, A.: Biochemistry 7, 1603-1610 (1968).
McCARTY, R. E., RACKER, E.: J. Biol. Chem. 242, 3435-3439 (1967).
NELSON, N., DETERS, P. W., NELSON, H., RACKER, E.: J. Biol. Chem. 248, 2049-2055 (1973).
NELSON, N., KANNER, B. I., GUTNICK, D. L.: Proc. Nat. Acad. Sci. U.S.A. 71, 2720-2724 (1974).
NIEUWENHUIS, F. J. R. M., KANNER, B. I., GUTNICK, D. L., POSTMA, P. W., VAN DAM, K.: Biochim. Biophys. Acta 325, 62-71 (1973).
OR, A., KANNER, B. I., GUTNICK, D. L.: FEBS Letters 35, 217-219 (1973).
PARNES, J. R., BOOS, W.: J. Biol. Chem. 248, 4429-4435 (1973).
PREZIOSO, G., HONG, J.-s., KERWAR, G. K., KABACK, H. R.: Arch. Biochem. Biophys. 154, 575-582 (1973).
RACKER, E., HORSTMAN, L. L.: J. Biol. Chem. 242, 2547-2551 (1967).
ROISIN, M. P., KEPES, A.: Biochim. Biophys. Acta 305, 249-259 (1973).
ROSEN, B. P.: J. Bacteriol. 116, 1124-1129 (1973).
SASARMAN, A., SANDERSON, K., SURDEANU, M., SONEA, S.: J. Bacteriol. 102, 531-536 (1970).
SASARMAN, A., SURDEANU, M., SZEGLI, G., HORODNICEANU, T., GRECEANU, V., DUMITRESCU, A.: J. Bacteriol. 96, 570-572 (1968).
SCHAIRER, H. U., GRUBER, D.: Eur. J. Biochem. 37, 282 (1973).
SCHAIRER, H. U., HADDOCK, B. A.: Biochem. Biophys. Res. Commun. 48, 544-551 (1972).
SCHULDINER, S., ROTTENBERG, H., AVRON, M.: Eur. J. Biochem. 25, 64-70 (1972).
SIMONI, R. D., SHALLENBERGER, M. K.: Proc. Nat. Acad. Sci. U.S.A. 69, 2663-2667 (1972).

TURNOCK, G., ERICKSON, S. K., ACKRELL, B. A. L., BIRCH, B.: J. Gen. Microbiol. 70, 507-515 (1972).
TZAGALOFF, A.: In: Current Topics in Membranes and Transport (ed. F. BRONNER, A. KLEINZELLER) Vol. II, pp. 157-205. New York: Academic Press 1971.
VAN THIENEN, G., POSTMA, P. W.: Biochim. Biophys. Acta 323, 429-440 (1973).
YAMAMOTO, T. H., MÉVEL-NINIO, M., VALENTINE, R. C.: Biochim. Biophys. Acta 314, 267-275 (1973).

Adenosine Triphosphate Synthesis by Sodium, Potassium Adenosine Triphosphatase

Robert L. Post, Kazuya Taniguchi, and Gotaro Toda
Department of Physiology, Vanderbilt University Medical School, Nashville, Tennessee 37232

INTRODUCTION

Purpose, Scope, and Procedures

To characterize the mechanism of the sodium and potassium ion pump of plasma membranes of animal cells, we have investigated the sodium and potassium ion-transport adenosine triphosphatase (Na, K-ATPase). This enzyme expresses the activity of the pump in preparations of broken or leaky membranes. We have investigated specifically the sequence of addition and release of ligands and have become interested in the interaction-free energies of their binding (WEBER, 1974). In particular, binding of Na^+ appears to increase the free energy of hydrolysis of a phosphate group at an active site on the enzyme. To distinguish between effects due to binding and those due to translocation of ions, we have preferred to work with leaky membranes, across which concentration gradients do not persist. It has also been an advantage to conduct the hydrolytic reaction of this enzyme in the backward direction. In a first step, the enzyme was phosphorylated from inorganic phosphate (P_i). In a second step, addition of ADP with a high concentration of Na^+ produced ATP.

The Pump

The pump is composed of phosphatidyl serine and at least one protein (molecular weight about 100,000) and possibly also a glycoprotein (molecular weight about 50,000) which are intrinsic to the plasma membrane. The pump has access to the solutions in contact with the intracellular and extracellular faces of the membrane. Per cycle it transports approximately 3 Na^+ outward and 2 K^+ inward, thus generating an electric current (KERKUT, YORK, 1971). For net outward transport, Na^+ is a unique substrate, but for net inward transport, K^+

has as congeners Li^+, NH_4^+, Rb^+, Cs^+, and Tl^+. Per cycle, the pump hydrolyzes the terminal bond of one molecule of intracellular ATP with intracellular release of products. Intracellular Mg^{2+} is required. The pump is thus an endo-ATPase with intracellular Na^+ and extracellular K^+ as additional substrates and with extracellular Na^+ and intracellular K^+ as additional products (Figure 1).

In the presence of abnormally large concentration gradients of Na^+ and K^+, opposing transport in the forward direction, the pump runs backward and synthesizes ATP. In preparations of broken or leaky membranes, where gradients cannot persist, the activity of the pump is an ATPase (EC 3.6.1.3) that requires Mg^{2+}, Na^+, and K^+ simultaneously and that is inhibited by cardioactive steroids. This is the Na, K-ATPase. For a review, see HOKIN and DAHL (1972) or SCHWARTZ, LINDENMEYER, and ALLEN (1972).

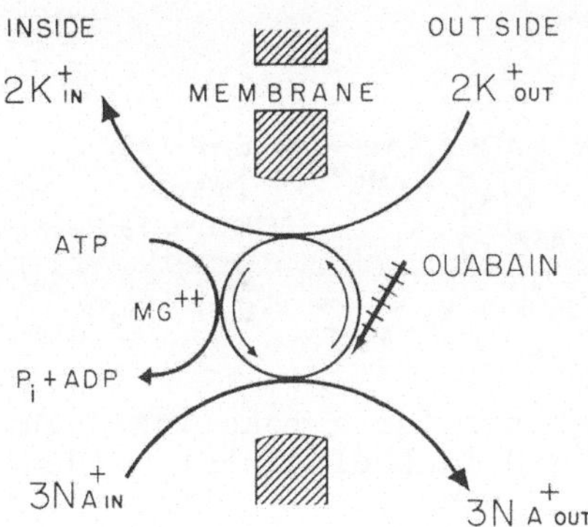

Fig. 1. Stoichiometry and sidedness of the sodium and potassium ion pump of plasma membranes. Ouabain is a specific cardioactive steroid inhibitor. (From POST, 1968.)

The Phosphoenzyme

The larger protein in the pump is phosphorylated from ATP in the presence of Na^+ uniquely. In the presence of K^+ or its congeners, it is phosphorylated from P_i; and Na^+ is uniquely inhibitory (Table 1). It is also phosphorylated from P_i when complexed with ouabain. In all cases, Mg^{2+} is required for phosphorylation. In each case the active site of phosphorylation is the β-carboxyl group of a specific aspartyl residue (POST, KUME, 1973; NISHIGAKI, CHEN, HOKIN, 1974; POST, TODA, ROGERS, 1975).

A working hypothesis of the reaction sequence for phosphorylation and dephosphorylation is shown in Figure 2. It treats phosphorylation

Table 1. Effect of monovalent cations on phosphorylation of Na, K-ATPase from ATP or from P_i in the presence of Mg^{2+}*

Monovalent cation	Phosphoenzyme (%) (from ATP)	(from P_i)
Li^+	7	18
Na^+	100	0.4
NH_4^+	0	14
K^+	2	10
Rb^+	5	13
Cs^+	0	12
Tl^+	-	8
None	14	44†

*From POST, SEN, ROSENTHAL (1965); POST, KUME, ROGERS (1973).

†This reactive state of the phosphoenzyme is insensitive to K^+ or ADP.

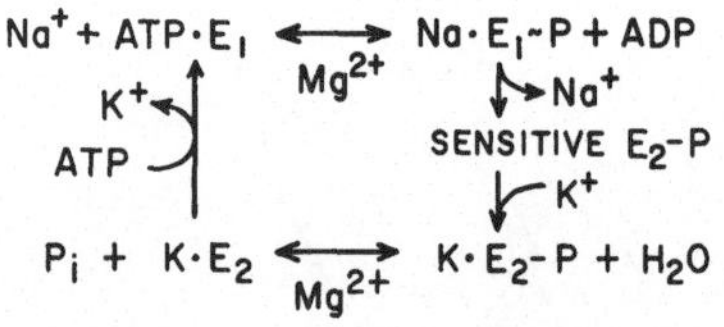

$Na^+ + ATP \cdot E_1 \xleftrightarrow{Mg^{2+}} Na \cdot E_1{\sim}P + ADP$

K^+, ATP; Na^+; SENSITIVE E_2-P; K^+

$P_i + K \cdot E_2 \xleftrightarrow{Mg^{2+}} K \cdot E_2{-}P + H_2O$

Fig. 2. A working hypothesis for a reaction sequence of the phosphoenzyme. For simplicity and lack of direct evidence, the stoichiometry of Na^+ and K^+ has not been specified. Similarly, the actions of Mg^{2+} are incomplete

from ATP or from P_i as functions of alternative reactive forms of the dephosphoenzyme. These two forms of dephosphoenzyme are designated E_1 and E_2, respectively. The corresponding phosphorylated products are designated $E_1 \sim P$ and $E_2 - P$. At physiological concentrations of Na^+ (up to 160 mM), $E_1 \sim P$ appears only transiently, but a related reactive state is stabilized by partial inhibition of the enzyme with N-ethylmaleimide or oligomycin. Thus, phosphorylation from ATP ordinarily yields sensitive $E_2 - P$ which equilibrates with P_i (POST, TODA, ROGERS, 1975); it is extremely sensitive to attack by K^+, as will be shown, but ordinarily it is insensitive to ADP. K-complexed phosphoenzyme equilibrates with P_i much more rapidly (but less completely) than do the other reactive states of $E_2 - P$ (POST, TODA, ROGERS, 1975).

This hypothesis is based on evidence from isolated reaction steps. For example, the attack of ATP or MgATP on $K \cdot E_2$ is based partly on experiments of POST, HEGYVARY, and KUME (1972). It is not established that this hypothetical pathway is necessarily predominant under physiological conditions.

RESULTS

K^+-Sensitive Phosphoenzyme from ATP

The enzyme was in a crude membrane suspension obtained from a homogenate of guinea pig kidney by differential centrifugation. It was phosphorylated from [^{32}P]ATP in the presence of Mg^{2+} and Na^+ at 0°C and neutral pH. To expose the rate of dephosphorylation, an excess of EDTA was added at zero time to chelate the free Mg^{2+}. The reaction was stopped with acid, and the denatured membranes were washed and counted. Background labeling, appearing when K^+ replaced Na^+, was subtracted from the labeling in all of the samples. Dephosphorylation was exponential, with a time constant of about 16 sec. When K^+ was added at 4 sec, dephosphorylation was immediate (Figure 3). The sensitivity to K^+ is shown by the ratio of Na^+ to K^+ (namely, 160 to 1).

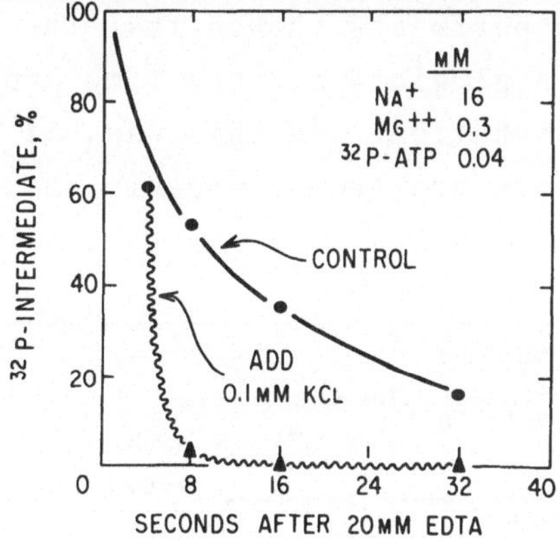

Fig. 3. Dephosphorylation of K^+-sensitive phosphoenzyme from ATP. (From POST, 1968.)

K^+-Sensitive Phosphoenzyme from Inorganic Phosphate

The strategy for synthesis of ATP was first to prepare sensitive phosphoenzyme from P_i and second to treat this phosphoenzyme with ADP and Na^+. Because Na^+ inhibits phosphorylation from P_i in a steady state (Table 1), the enzyme cannot be recycled and the recovery of ATP does not exceed the initial amount of phosphoenzyme.

The procedure for synthesis of sensitive E_2 - P from P_i was discovered by Dr. Toda in this laboratory. The membranes are first washed in 1 mM $MgCl_2$ by centrifugation and resuspension at 4°C and neutral pH and then the Mg^{2+} is removed. The effect of washing persists for at least 18 hr at 4°C. It is not difficult to phosphorylate the native enzyme from P_i in the absence of K^+ (Table 1). The difficulty is that Mg^{2+} rapidly attacks sensitive phosphoenzyme in the absence of monovalent cations. This attack makes the enzyme insensitive to K^+ (POST, TODA, ROGERS, 1975). Fortunately, a low concentration of Na^+ protects sensitive phosphoenzyme from Mg^{2+} for about 30 sec without itself immediately inhibiting phosphorylation from P_i.

To form sensitive phosphoenzyme, P_i and Na^+ were added to washed membranes followed by Mg^{2+} 5 sec later to initiate phosphorylation at zero time. Under these conditions, 30% of the enzyme was phosphorylated and the level was stable for at least 21 sec (Figure 4). If the enzyme had not been washed with $MgCl_2$ previously, no significant amount of phosphoenzyme would have appeared. Addition of CDTA (cyclohexylenedinitrilotetraacetic acid) to the phosphoenzyme chelated the free Mg^{2+} and exposed a slow dephosphorylation. Addition of K^+ with the CDTA showed that half of the phosphoenzyme was sensitive and that half was not (Figure 4). Other experiments showed that the amount of the insensitive component was negligible at zero time and increased progressively during the incubation. In this way, it was possible to prepare sensitive phosphoenzyme from inorganic phosphate.

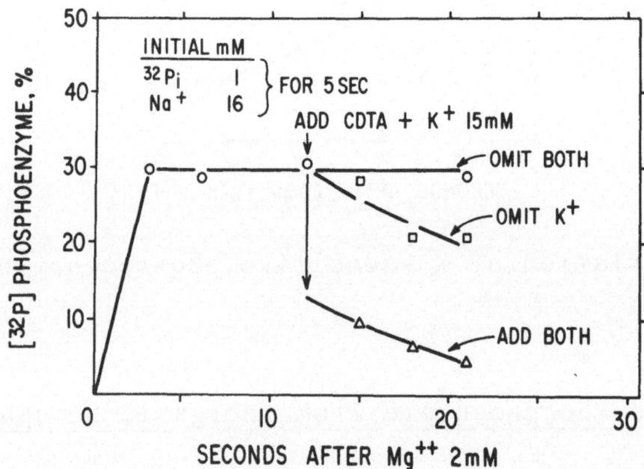

Fig. 4. Formation of K^+-sensitive phosphoenzyme from inorganic phosphate. (From POST, TANIGUCHI, TODA, 1975, by permission of Elsevier and the Publishing House of the Hungarian Academy of Sciences.)

Synthesis of ATP from K^+-Sensitive Phosphoenzyme

The membranes were first treated with concentrated NaI to remove or inactivate a nonspecific ATP synthetase. Then they were washed--first with 1 mM $MgCl_2$ and second with Mg^{2+}-free solution. The membranes were incubated at neutral pH and at 0°C with 1 mM $^{32}P_i$ and 0.1 mM CDTA (to prevent premature phosphorylation). At zero time, 0.5 mM $MgCl_2$ with 16 mM NaCl was added. At 4 sec, 1 mM ADP with 20 mM $(Tris)_3$CDTA and various concentrations of NaCl were added. At 6 sec or later times, the reaction was stopped with acid. Phosphoenzyme was estimated in the precipitate and [^{32}P]ATP was estimated in the supernatant (POST, TANIGUCHI, TODA, 1974). The 20 mM CDTA was a convenience; it prevented destruction of the synthesized ATP and reduced the rate of synthesis by only 25%. In the absence of extra Na^+ the phosphoenzyme disappeared exponentially with a time constant of about 40 sec. (It was stabilized partly by the ADP and partly probably by a greater freedom from traces of K^+ in the reaction mixture.) Very little ATP appeared. As the concentration of NaCl increased, so did the rate of dephosphorylation. The rates of appearance of ATP also increased. The increase in ATP corresponded fairly closely to the loss in phosphoenzyme. The concentration of NaCl for a half-maximal effect was remarkably high, about 0.6 M (Figure 5). Further experiments are described by TANIGUCHI and POST (1975).

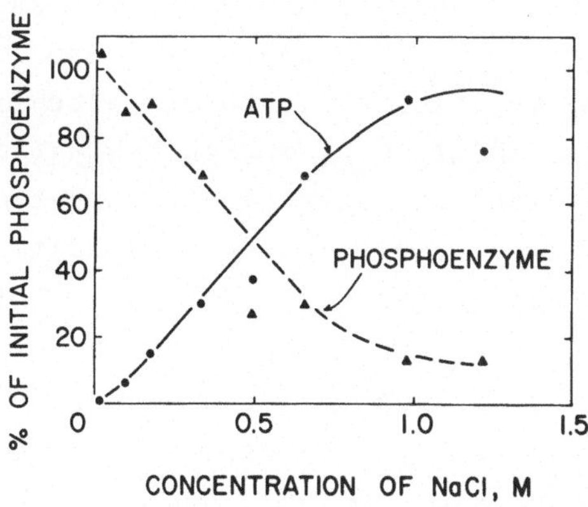

Fig. 5. Dephosphorylation of sensitive phosphoenzyme and synthesis of ATP in the presence of increasing concentrations of Na^+. The reaction was stopped 2 sec after the addition of ADP and NaCl to the phosphoenzyme at 0°C. (From POST, TANIGUCHI, TODA, 1975, by permission of Elsevier and the Publishing House of the Hungarian Academy of Sciences.)

Control Experiments

The significance of the synthesis is shown by experiments in which ATP did not appear (TANIGUCHI, POST, 1975). When the initial amount of NaCl (16 mM) was incubated with the enzyme for 10 minutes before addition of P_i, neither phosphoenzyme nor ATP appeared. When the initial amount of NaCl was replaced with KCl, K^+-complexed phosphoenzyme appeared (Table 1) but ATP did not. When 0.25 mM ouabain was present in the initial reaction mixture and CDTA was omitted, a large amount of stable phosphoenzyme appeared but no ATP. It is well known that the complex of the enzyme with the inhibitor ouabain is easily phosphorylated from P_i (HOKIN, DAHL, 1972; SCHWARTZ, LINDENMEYER, ALLEN, 1972). When 16 mM K^+ was added with 0.7 M Na^+ and CDTA to sensitive phosphoenzyme, the phosphoenzyme was immediately dephosphorylated and no ATP appeared. When ADP was added to the initial mixture, phosphorylation from P_i was inhibited and ATP did not appear.

DISCUSSION

Is Binding of Na^+ Sufficient or is Translocation of Na^+ Necessary?

Electron microscopic examination of the membrane preparation showed membrane fragments with free edges and also apparently unbroken membrane vesicles. The vesicles could be resealed fragments of homogenized plasma membrane. Addition of a high concentration of NaCl could possibly produce a transient concentration gradient of Na^+ across the membranes of these vesicles. This could translocate Na^+ and translocation could provide energy for the synthesis. This mechanism requires not only binding of Na^+ to the pump on the exterior face of the membrane but also release of bound Na^+ from the pump on the interior face of the membrane. To test the necessity for a gradient, we added agents intended to dissipate a gradient, namely, the ionophore gramicidin or the detergent Lubrol WX. In neither case was the recovery of ATP from the phosphoenzyme significantly affected, although the rate of synthesis was slowed in the presence of Lubrol WX. We also estimated the rate at which Na^+ reached the site at which it catalyzes phosphorylation from ATP. There was no significant delay within a period of 3 sec. This is presumably the site from which it would have to be released if translocation were necessary for synthesis. We concluded that translocation was probably not necessary and that binding was sufficient (TANIGUCHI, POST, 1975).

Energy Coupling

Since binding of Na^+ seems to be sufficient for synthesis of ATP, it is appropriate to estimate a free energy of interaction (WEBER, 1974). In principle, the change in the free energy of hydrolysis of the phosphate group produced by the binding of Na^+ should equal the change in the free energy of binding of Na^+ produced by phosphorylation. The free energy of hydrolysis of $E_1 \sim P$ can be estimated in a form stabilized by inhibition with N-ethylmaleimide. The free energy of hydrolysis of $E_2 - P$ can be estimated from the half-maximal concentration of P_i required for phosphorylation of insensitive $E_2 - P$. The free energy of binding of Na^+ to the dephosphoenzyme can be estimated from the stoichiometry of transport and the half-maximal concentration for inhibition of phosphorylation from P_i or for catalysis of phosphorylation from ATP; these values are about the same. Similarly, the free energy of binding of Na^+ to the phosphoenzyme can be estimated from the concentration for a half-maximal effect in Figure 5 and similar experiments. A diagram of the relationships shows that the two estimates of the interaction free energy are similar (Figure 6). The similarity permits consideration of the idea that binding of Na^+ provides energy for the synthesis of ATP.

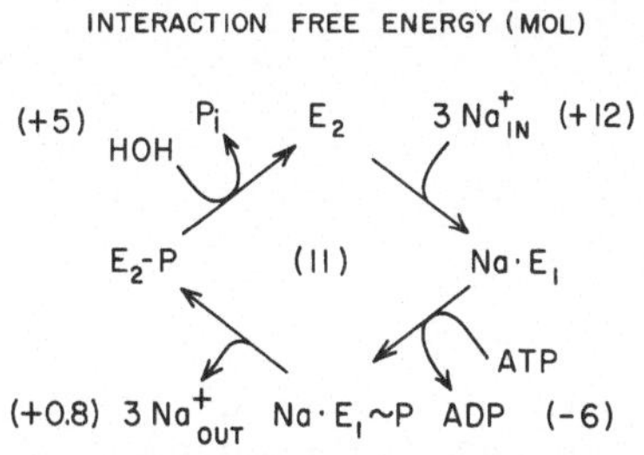

INTERACTION FREE ENERGY (MOL)

(+5) HOH P_i E_2 $3\,Na^+_{IN}$ (+12)

E_2-P (11) Na·E_1

ATP

(+0.8) $3\,Na^+_{OUT}$ Na·$E_1 \sim$P ADP (-6)

Fig. 6. Estimation of interaction free energy between binding of Na^+ and hydrolysis of the phosphoenzyme. The symbols are those of Figure 2. The numbers in parentheses are estimates of free energy of hydrolysis of the phosphate group or of dissociation of Na^+ from the enzyme in kcal/mole at 0°C. The number in the center is the interaction free energy. The stoichiometry is that of transport. Details are given by TANIGUCHI and POST (1975). To consider the free energy change around the cycle, the signs of the numbers on the right-hand side should be reversed

The Mechanism of the Pump

Transition States

The limited data available allow a variety of interpretations. As a point of insight, one may consider that the transition state for transphosphorylation and the transition state for translocation may be the same. An advantage of this point of view is that the similarity is as appropriate for translocation of Na^+ with transphosphorylation from ATP as for translocation of K^+ with transphosphorylation from P_i (Figure 7). Some of the partial reactions that favor this dichotomy of functional states are listed in Table 2. It is a clear implication of this hypothesis that the switching of the covalent bond of the phosphorus atom on the enzyme from the oxygen of the β-aspartyl carboxyl group to the oxygen of ADP or that of HOH is tightly linked to the switching of the access pathway of the bound 3 Na^+ or 2 K^+, respectively, from the solution in contact with the extracellular face of the membrane to the solution in contact with the intracellular face of the membrane.

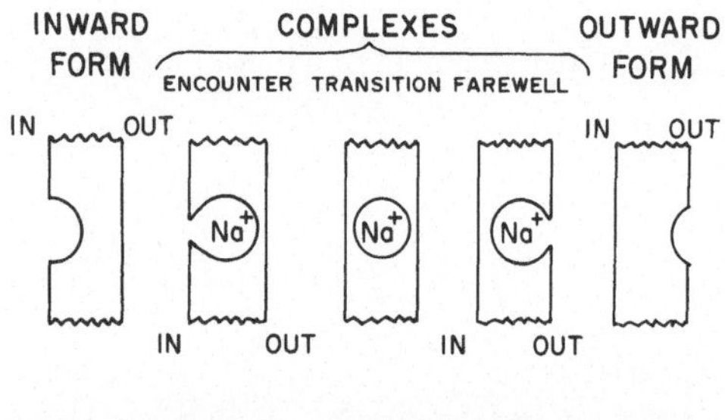

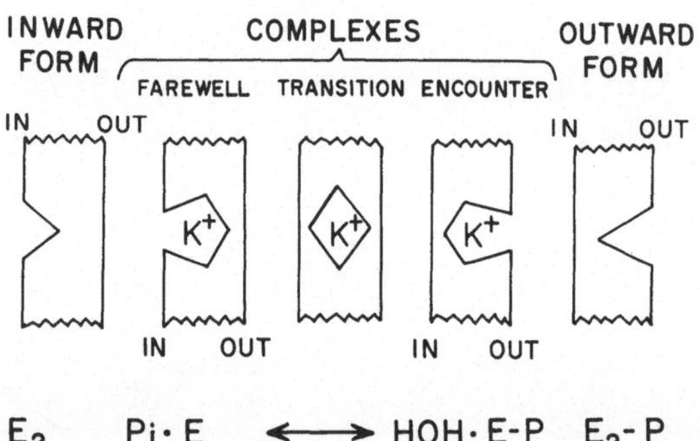

Fig. 7. Diagram of a hypothesis that the transition states for ion translocation and those for the corresponding ion activated transphosphorylation are the same. Reactions with respect to Na^+ are shown in the *upper panel* and those with respect to K^+ are shown in the *lower panel*. In each panel, the upper part portrays steps in ion translocation according to a model of molecular peristalsis (PATLAK, 1957; JARDETZKY, 1966). The central occluded conformation is the transition state. In each panel, the lower part shows corresponding complexes for transphosphorylation except that the transition state itself is represented only by a *double-headed arrow*. For simplicity, the stoichiometry of monovalent cation binding is not specified. Also, Mg^{2+} is omitted

Table 2. Contrasting actions of Na^+ and K^+*

Cation discrimination	Na^{+}†	K^{+}†
Net transport	out	in
In $\rightleftharpoons$ out exchange needs	ADP	P_i
ATP $\rightleftharpoons$ ADP exchange	+	-
HOH $\rightleftharpoons P_i$ exchange	-	+
Phosphate donor	ATP	P_i
ATP binding	strong	weak
Be^{2+} poisoning	-	+

*HOKIN, DAHL (1972).

†+ indicates an activator and - indicates an inhibitor. The first two reactions pertain to the pump. The others pertain to Na, K-ATPase.

Overlapping Transport

As represented in Figure 7, this hypothesis is equivalent to sequential translocation of Na^+ and K^+. This is because the ion-free enzyme is shown for the transition states between complexes with Na^+ and those with K^+. This representation is most likely incorrect since the kinetics of Na,K-transport in erythrocytes is incompatible with sequential transport (GARAY, GARRAHAN, 1973). Another state of transition from enzyme complexed with 3 Na^+ to enzyme complexed with 2 K^+ is one binding both Na^+ and K^+. This transition state implies binding of at least one K^+ before the last Na^+ comes off with respect to the extracellular solution or binding of at least one Na^+ before the last K^+ comes off with respect to the intracellular solution. In such cases, translocation of Na^+ and K^+ overlap and the reaction sequence of Na,K-translocation is neither sequential nor simultaneous (SKOU, 1971) but overlapping. In the future, overlapping transport may be found to resolve apparent conflicts between sequential and simultaneous models.

SUMMARY

ATP was synthesized from ADP and inorganic phosphate by partial reversal of the forward reaction of sodium and potassium ion-transport ATPase. In a transient reaction, the recovery of ATP was less than or equal to the quantity of enzyme. In a first step, potassium-sensitive phosphoenzyme was synthesized from inorganic phosphate. In a second step, addition of ADP and up to 1 M Na^+ produced ATP. This enzyme is intrinsic to plasma membranes. However, a concentration gradient of sodium ions across the membrane did not appear to be necessary for synthesis. It was suggested that free energy of interaction between phosphorylation and binding of sodium ions supplied the energy for the synthesis.

ACKNOWLEDGMENTS

This paper is a revised version of a paper published in the *Proceedings of the 9th Meeting of the Federation of European Biochemical Societies* (POST, TANIGUCHI, TODA, 1975). It is republished here with the permission of North Holland Publishing Company and the Publishing House of the Hungarian Academy of Sciences.

This research was supported by grants from the National Institutes of Health, U.S. Public Health Service. The grants were No. 5R01-HL01974 from the National Heart and Lung Institute and No. 5P01-AM07462 from the National Institute of Arthritis and Metabolic Diseases.

The authors thank Dr. Sidney Fleischer for providing electron micrographs of the enzyme preparation.

REFERENCES

GARAY, R. P., GARRAHAN, P. J.: The interaction of sodium and potassium with the sodium pump in red cells. J. Physiol. 231, 297-325.

HOKIN, L. E., DAHL, J .L.: In: Metabolic Pathways, 3rd Ed. (ed. L. E. HOKIN) Vol. VI, p. 269. New York and London: Academic Press 1972.

JARDETZKY, O.: Simple allosteric model for membrane pumps. Nature (London) 211, 969 (1966).

KERKUT, G. A., YORK, B.: The Electrogenic Sodium Pump. Bristol: Scientechnica Ltd. 1971.

NISHIGAKI, I., CHEN, F. T., HOKIN, L. E.: Studies on the characterization of the sodium-potassium transport adenosine triphosphatase. XV. Direct chemical characterization of the acyl phosphate in the enzyme as an aspartyl β-phosphate residue. J. Biol. Chem. 249, 4911-4916 (1974).

PATLAK, C. S.: Contributions to the theory of active transport. II. The gate-type non-carrier mechanism and generalizations concerning tracer flow, efficiency and measurement of energy expenditure. Bull. Math. Biophys. 19, 209-235 (1957).

POST, R. L.: In: Regulatory Functions of Biological Membranes (ed. J. JÄRNEFELT), p. 163. Amsterdam, London, and New York: Elsevier Publishing Co. 1968.

POST, R. L., KUME, S.: Evidence for an aspartyl phosphate residue at the active site of sodium and potassium ion transport adenosine triphosphatase. J. Biol. Chem. 248, 6993-7000 (1973).

POST, R. L., HEGYVARY, C., KUME, S.: Activation by adenosine triphosphate in the phosphorylation kinetics of sodium and potassium ion transport adenosine triphosphatase. J. Biol. Chem. 247, 6530-6540 (1972).

POST, R. L., KUME, S., ROGERS, F. N.: In: Mechanisms in Bioenergetics (ed. G. F. AZZONE, L. ERNSTER, S. PAPA, E. QUAGLIARIELLO, N. SILIPRANDI), p. 203. New York and London: Academic Press 1973.

POST, R. L., SEN, A. K., ROSENTHAL, A. S.: A phosphorylated intermediate in adenosine triphosphate-dependent sodium and potassium transport across kidney membranes. J. Biol. Chem. 240, 1437-1445 (1965).

POST, R. L., TANIGUCHI, K., TODA, G.: Synthesis of adenosine triphosphate by sodium, potassium and adenosine triphosphatase. In: Proceedings of the 9th Meeting of the Federation of European Biochemical Societies. Budapest: Akadémiai Kiadó; Amsterdam and London: North Holland Publishing Co. (in press).

POST, R. L., TODA, G., ROGERS, F. N.: Phosphorylation by inorganic phosphate of sodium plus potassium ion-transport adenosine triphosphatase: Four reactive states. J. Biol. Chem. (in press).

SCHWARTZ, A., LINDENMEYER, G. E., ALLEN, J. C.: In: Current Topics in Membranes and Transport (ed. F. BRONNER, A. KLEINZELLER) Vol. III, p. 1. New York and London: Academic Press 1972.

SKOU, J. C.: In: Current Topics in Membranes and Transport (ed. D. R. SANADI) Vol. IV, p. 357. New York and London: Academic Press 1971.

TANIGUCHI, K., POST, R. L.: Synthesis of adenosine triphosphate and exchange between inorganic phosphate and adenosine triphosphate in sodium, potassium ion-transport adenosine triphosphatase. J. Biol. Chem. (in press).

WEBER, G.: Addition of chemical and osmotic energies by ligand-protein interactions. Ann. N.Y. Acad. Sci. 227, 486-496 (1974).

Bilayer Dynamics Studies Using Capacitance Relaxation*

David F. Sargent
Institute of Molecular Biology and Biophysics, Eidg. Technische Hochschule, 8049 Zürich, Switzerland

INTRODUCTION

This paper introduces a new technique for studying the dynamics of artificial lipid bilayers. The technique can be used to study planar bilayers, which, because both sides of the bilayer are freely accessible, have advantages over liposomes in many research cases. In addition, the bilayer structure is not strained like the highly curved sonicated liposomes (SHEETZ, CHAN, 1972). Thus, the planar bilayer is not only often more convenient to use, but its structure may be expected to be closer to that of most biological membranes.

Unfortunately, the minute amount of material in a bilayer film approximately 1 mm in diameter (about 10^{-11} moles) puts direct measurement of the molecular properties of individual bilayers beyond the capabilities of most methods (e.g., nuclear magnetic resonance [nmr], X-ray diffraction, infrared [IR] spectroscopy, differential thermal analysis, etc.). An exception is fluorescence spectroscopy, which can be used in simple bilayer vesicles (YGUERABIDE, STRYER, 1971).

The technique I will present provides information about the molecular dynamics of bilayers without knowledge of the exact origin of the dynamic parameters obtained. The method exploits the electrical capacitance of the bilayer. By studying the time course of changes in the capacitance following sudden changes in the voltage applied across the bilayer, one may obtain information about molecular mobility, cluster formation, phase changes, surface adsorption, and so on.

---

*This project was carried out in the laboratories of Professor R. Schwyzer. Thanks are due to the Institute of Molecular Biology and Biophysics, Swiss Federal Institute of Technology (Zürich), for space and supplies. The work was financially supported by research grants from the Swiss National Foundation and the Medical Research Council of Canada. The author gratefully acknowledges his MRCC Postdoctoral Fellowship.

TECHNIQUE

This section presents a synopsis of the technique; a detailed account will appear elsewhere (SARGENT, in preparation [a], [b]).

Theory: Voltage and Time Dependence of Membrane Capacitance

The capacitance (C_m) of a bilayer is given by the formula

$$C_m \propto \varepsilon_m \frac{A}{d} \tag{1}$$

where ε_m = dielectric constant of bilayer material, A = area of bilayer region, and d = thickness of bilayer. Since an electric field can affect all three parameters, we must consider each one in turn.

Dielectric Constant

The dielectric constant of the membrane reflects both the dielectric properties of the individual molecules (e.g., dipole moment, polarizability), and the orientation and packing of the molecules in the bilayer. All of these parameters could conceivably be affected by an applied field, but it has been suggested that voltage has ". . . no direct or large effect on the static dielectric coefficient . . ." (WHITE, 1970). The presence of time-dependent reorientations is to be expected, however, giving rise to characteristic dispersion effects (Figure 1).

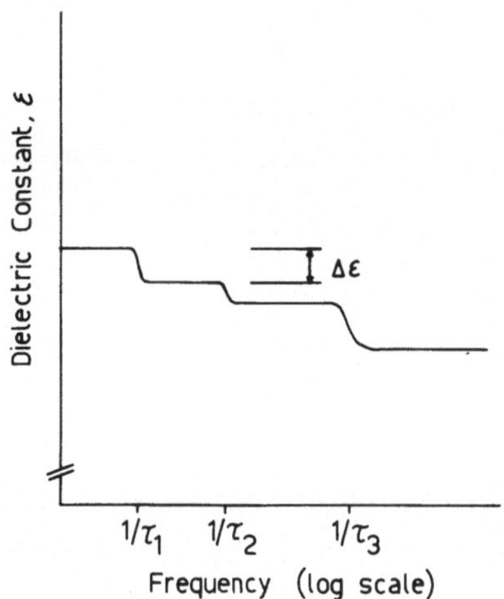

Fig. 1. A generalized representation of the dielectric constant of a material as a function of frequency. At certain characteristic frequencies (with corresponding time constants, τ_i), various dipole reorientation (or other) processes can no longer keep pace, resulting in a lowering of the dielectric constant by an amount known as the dielectric increment, $\Delta\varepsilon$

The frequency-dependent capacitance can be attributed to molecular rearrangement or reorientation, resulting in the presence of oriented electric dipoles in the medium. This in turn causes an apparent net surface charge, which is dependent on the number of dipoles aligned and hence on the applied voltage. If the intensity or direction of the external field is changed, the surface charge associated with the dipoles will also change, causing a current ("displacement current") to flow in the external circuit. The time course of the current will depend on the time taken for the dipolar reorientation.

The total charge transferred will be proportional to the Langevin function

$$L(p \cdot E/kT) = \cot(p \cdot E/kT) - (p \cdot E/kT)^{-1} \tag{2}$$

where p = dipole moment, E = field strength, k = Boltzmann constant, and T = absolute temperature. The Langevin function saturates at high field strengths: in liquids, saturation effects can be found at intensities of several hundred thousand volts per centimeter (BOTTCHER, 1952). Such field strengths are readily attained in bilayers (100 mV across a 50 Å layer gives 2×10^5 V/cm), so that saturation effects may be expected.

It should be noted that the Langevin function involves the scalar product of two vectors (p and E), so that the displacement current depends on the field direction as well as its magnitude. Such behavior will be called "asymmetric" with respect to voltage. For comparison with the next section, we note here that ΔQ_d, the total charge transferred due to dipole reorientation, will be the same for a voltage change of V volts, independent of the initial voltage. Specifically, this implies that, apart from saturation phenomena,

$$\Delta Q_d(-V/2 \rightarrow +V/2) = \Delta Q_d(0 \rightarrow +V) \tag{3}$$

and that $\Delta Q_d \propto \Delta V$.

Thickness and Area

A voltage applied between the plates of a parallel plate capacitor (i.e., across the planar bilayer in this case) exerts a compressive force proportional to the square of the applied voltage:

$$F_c = C_m{}^2 V^2 / 2\varepsilon_m \varepsilon_o$$

where C_m = membrane capacitance, V = applied potential, ε_m = dielectric coefficient of bilayer, and ε_o = dielectric constant of free space. For typical values, F_c will be a fraction of 1 atm (about 0.1 atm at 125 mV), so that geometric changes brought about by the compression should depend linearly on F_c. Thus $\Delta C_m = \Delta Q_c/V \propto V^2$, and ΔQ_c, the charge that flows as a result of the electrostrictive force, will be given by $\Delta Q_c \propto V^3$, in contrast to the case of dipole reorientation ($\Delta Q_d \propto \Delta V$ for low voltages). Note that F_c depends only on the magnitude of the applied voltage, not its sign, so that suddenly changing the membrane voltage from -V/2 to +V/2 should not result in a (further) ΔC_m, and no relaxation currents would be expected to result from geometric changes. Such a "symmetric" response to the applied voltage is also in contrast to the case of dipole reorientation (eq. 3). This electrostrictive force may cause the bilayer to become thinner and/or to extend further into the boundary region, and both effects have been reported (e.g., BABAKOV, ERMISHKIN, LIBERMAN, 1966; WHITE, 1970; WHITE, THOMPSON, 1973). Both dipole reorientation and membrane thinning affect the specific capacitance of the bilayer; i.e., the corresponding ΔQ's will be proportional to membrane area. Any increase in bilayer area at the bilayer/torus boundary will depend approximately linearly on the bilayer circumference.

Two fundamental types of time-dependent processes could contribute to the geometric effects. First, reorganization of the molecular packing could lead to a thinning of the bilayer (Figure 2). Excess bilayer area caused by an increased area per molecule would be pulled

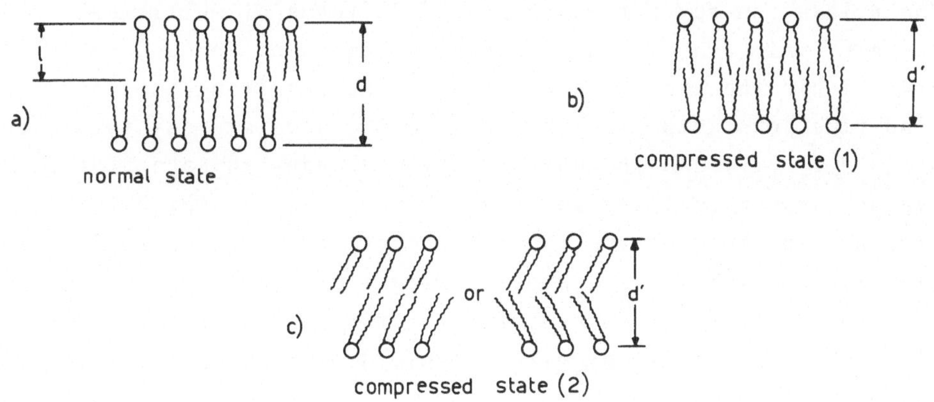

Fig. 2. Some possible configurations of phospholipids in bilayer lipid membrane. (a) Normal state (no applied voltage): little interpenetration of hydrophobic regions of the two opposed monolayers. $d \simeq 2 \cdot l$ where l = length of phospholipid molecules. (b) Compressed state (1): hydrophobic regions of the two monolayers intermingle, allowing a total thickness $d' < 2 \cdot l$. Hydrophilic heads are less densely packed than in case (a). (c) Compressed state (2): little or no interpenetration of the hydrophobic regions, but lipid molecules are aligned at an angle to the perpendicular. As in case (b), $d' < 2 \cdot l$, and the heads are less densely packed than in case (a)

into the torus within microseconds by surface tension (SARGENT, in preparation [b]).

A second class of effects would be the squeezing out of residual solvent molecules, which might pass into the aqueous phase or diffuse to the border. The time constant of the latter process is given by

$$\tau_d \simeq r^2/6D$$

where r = radius of bilayer region and D = diffusion constant (WHITE, 1970). Reasonable values for r and D give time constants on the order of minutes.

Apparatus: Current Relaxation Measurements

Time-dependent changes in capacitance ("capacitance relaxation") following a step change in the potential applied to the bilayer were determined by measuring the time course of the displacement current. A detailed description of the construction and operation of the apparatus will be given elsewhere (SARGENT, in preparation [a]). An example of the types of curves that were obtained is shown in Figure 3, where an RC test circuit having a known "relaxation" (I_o = 200 nA, τ = 22 μsec) was placed in series with the electrodes. The measured values (I_o = 190 ± 30 nA, τ = 21 ± 2 μsec) agree very well with expectations.

The relaxation current was analyzed in terms of multiple-exponentials, using the minimum number of components needed to describe the curve within the experimental error. The resulting amplitudes (I_i^o) and time constants (τ_i) can be used to calculate the total charge passed ($Q_i = I_i^o \cdot \tau_i$) and the corresponding changes in capacitance, ΔC_i. These values were normalized for the bilayer area, of which the simplest measure is the membrane capacitance at zero voltage (C_m'). Thus, the relaxation amplitude is expressed as a fraction or percentage of change in capacitance. The resolution of the apparatus allows the detection of changes of 1 per mil or less, depending on the time constant.

Materials and Methods

Bilayers were composed of either cholesterol, oxidized according to TIEN, CARBONE, and DAWIDOWICZ (1966), or dioleoyllecithin, in decane. The aqueous phase was generally 0.2 M NaCl, buffered with 8 mM phosphate (pH = 7).

Membranes were formed on a 0.15-cm-diameter hole in a Teflon cup. The temperature of the experimental chamber could be controlled between 12°C and 50°C. Electrical measurements were made with a single pair of Ag/AgCl electrodes. The run shown in Figure 3 shows that diffusion polarization is not significant. Indeed, no time-dependent current changes were observed down to a resolution of 10^{-10} Amp for time constants longer than 20 μsec.

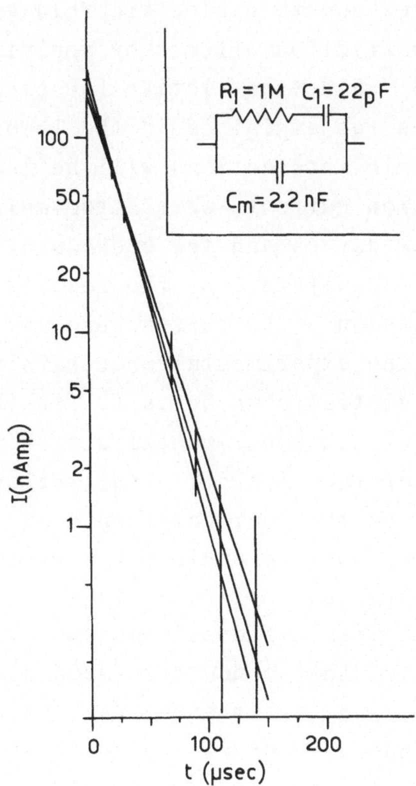

Fig. 3. Test of the relaxation apparatus and effects of diffusion polarization. The test circuit shown in the inset was connected in series with the measuring chamber (*no* membrane present). The electrodes were raised to give a minimum area of contact with the salt solution (0.2 M NaCl) and thus accentuate any diffusion polarization effects. With a solution/electrode resistance of about 1200 Ω and an applied voltage of 200 mV, the following values hold: charging time constant of C_m is R_sC_m = 2.65 μsec; charging time constant of C_1 (= "relaxation time constant") is R_1C_1 = 22 μsec; initial current to be compensated by the analog circuit is $\Delta V/R_s \simeq 1.6 \times 10^{-4}$ A; initial "relaxation" current is $\Delta V/R_1 = 2 \times 10^{-7}$ A. Measured values: $R_1C_1 = 21 \pm 2$ μsec, $\Delta V/R_1 = (1.9 \pm 0.3) \times 10^{-7}$ A. Averages and standard deviations of four trials are shown. (Inset: $M = 10^6$ Ω, $nF = 10^{-9}$ F, $pF = 10^{-12}$ F)

RESULTS

Oxidized Cholesterol Bilayers

A detailed description of the basic relaxation properties of oxidized cholesterol bilayers is currently in preparation (SARGENT, in preparation [b]); they are summarized here for comparison with the new results that are presented in the following sections.

Dielectric Relaxation

Capacitative changes due to dielectric relaxations can be distinguished from electrostriction effects by applying a voltage jump symmetrical around zero. Electrostrictive forces are then equal in the initial and final states (as explained in the Theory section above) and only changes in asymmetric interactions will be induced.

Asymmetric relaxation currents were detectable to beyond 10 msec. Examples of both the raw curves and the subsequent exponential analysis are given in Figure 4. As with all of the results, several curves were recorded in quick succession. The curves were averaged and the standard deviation taken as the experimental uncertainty. A formal description required four exponential components to fit the region between 10 μsec and 10 msec to within experimental accuracy. A study of the temperature dependence of the dielectric relaxations allowed a definite identification of three of the four components as separate processes. All four components were found to exhibit the expected linear dependence of displacement charge on applied voltage (Figure 5).

Control experiments with different aqueous solutions and different organic solvents indicated that conductive mechanisms were not involved in the observed effects, and that the relaxations were indeed characteristic of the membrane and not of the adjacent aqueous layers. Thus, these asymmetric relaxation phenomena can only be explained by molecular reorientation in the bilayer.

The time constants of the four exponential components at room temperature, and their Q_{10}'s, are shown in Table 1. The tentative identification of the components is as follows: component 4--a fast "rocking" motion of the individual molecules about an axis perpendicular to the long molecular axis; component 3--molecular rotation about the long axis; components 1 and 2--a slight, cooperative motion, similar to that of component 4, of statistically changing clusters of molecules. The time constants of components 3 and 4 may appear impossibly long for molecular motion, especially as figures found by mag-

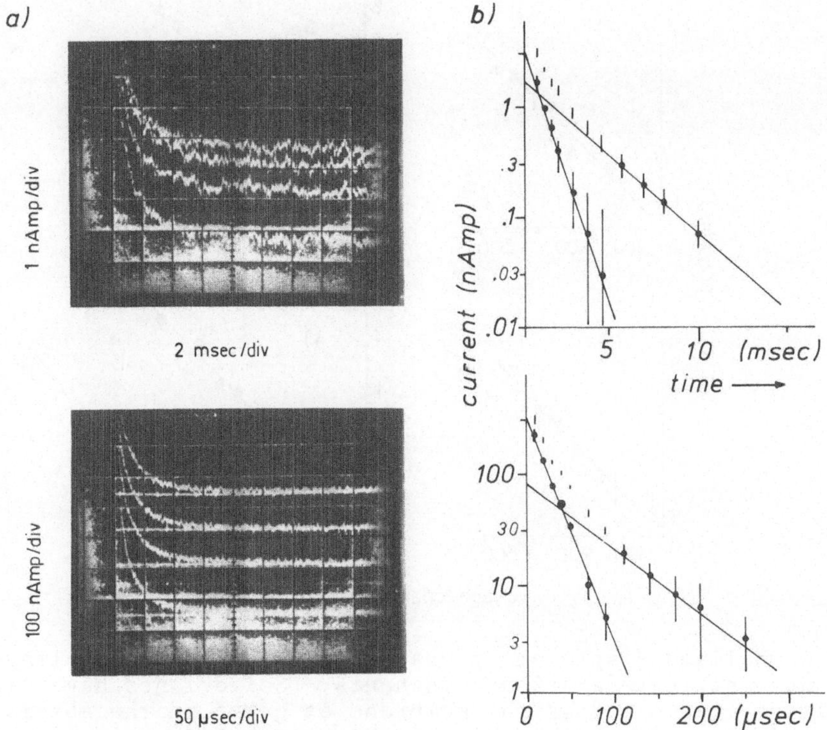

Fig. 4. (a) Raw current relaxation curves following a change of voltage from -150 mV to +150 mV (square wave with period of about 0.2 sec). Temperature = 33.7°C. (b) Exponential analysis of the relaxation curves, showing sequential subtraction of the identified components. Upper: longer times, yielding two components with $I^o_1 = 1.8 \pm 0.6$ nA, $\tau_1 = 3.3 \pm 0.6$ msec and $I^o_2 = 3.6 \pm 0.1$ nA, $\tau_2 = 0.9 \pm 0.2$ msec. Lower: shorter time scale, showing two further components with $I^o_3 = 85 \pm 15$ nA, $\tau_3 = 73 \pm 10$ µsec and $I^o_4 = 330 \pm 20$ nA, $\tau_4 = 21 \pm 4$ µsec. The means and standard deviations of the digitized curves are indicated: where no error bars are shown, the uncertainty limits correspond to the size of the symbols. Judging from the consistency of the points, uncertainties in the I's and τ's are considered generous (only the main curves are shown for the sake of clarity)

netic resonance methods are of the order of 10^{-7} to 10^{-10} sec. It is therefore instructive to look at the Debye formula relating the relaxation time constant of a (spherical) molecule with radius r to the viscosity, η, of the medium (e.g., DANIEL, 1967):

$$\tau = (4\pi/kT)\eta r^3 \qquad (4)$$

The main uncertainty in the evaluation of this expression is the value of the viscosity, η. A system with certain similarities to bilayer lipid membrane (BLM) is found in liquid crystals, where anisotropic molecules arrange themselves with a high degree of order. BAESSLER, BEARD, and LABES (1970) used a value of η = 100 poise in an analysis of

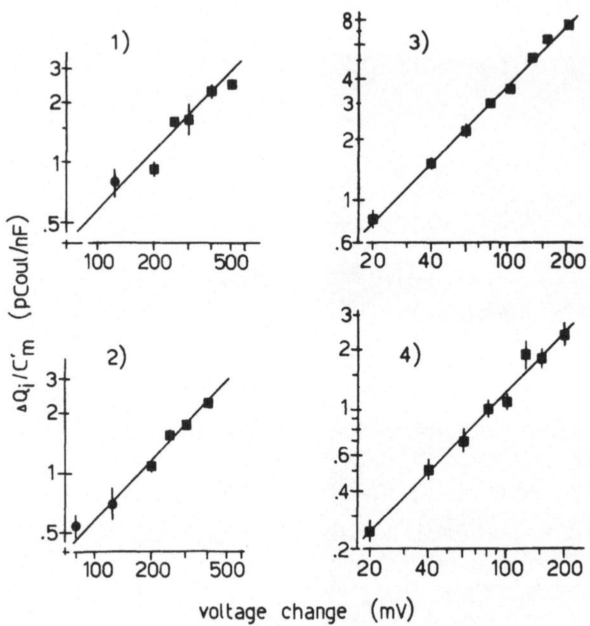

Fig. 5. Normalized displacement charge as a function of voltage for the four asymmetric relaxation components. Solid lines have slope = 1. Total voltage change across the membrane is given on the abscissa (e.g., for jump from -150 mV to +150 mV, V_{tot} = 300 mV). Temperature is 27.5°C. For components 1 and 2, results for two different membranes formed during the same set of experiments are shown

Table 1. Asymmetric capacitance relaxation components of oxidized cholesterol membranes at room temperature. Uncertainties represent standard deviations (at least six trials)

Component	ΔC/C (%)	Time constant, τ	$Q_{10}(\tau)$
1	1.2 ± 0.2	3.4 ± 0.8 msec	1 - 1.3
2	0.6 ± 0.1	0.7 ± 0.1 msec	∿ 1
3	2.8 ± 0.2	150 ± 30 μsec	∿ 1.3
4	1.0 ± 0.3	17 ± 3 μsec	∿ 1

dipole relaxation times in liquid crystals formed from a mixture of cholesteryl chloride and cholesteryl myristate. A viscosity of about 300 poise can be estimated from a combination of studies of the lateral diffusion times in bilayers and the effectiveness of valinomycin in promoting cation transport in bilayer lipid membrane (SZABO et al., 1972). An approximate value for r can be taken from the molecular volume of cholesterol, $(4/3)\pi r^3 = 10^{-21}\text{cm}^3$. Substituting these values into the Debye equation, which was derived for dilute solutions of noninteracting molecules, yields $\tau \simeq 20$ µsec. For tightly packed, interacting dipoles the expected relaxation times would be considerably longer. The foregoing assignments thus appear quite feasible. Also, the lack of saturation found for all components is understandable on the basis of restricted motion (components 1, 2, and 4) or of a reduced effective dipole moment in the direction of motion (component 3) (SARGENT, in preparation [b]).

YGUERABIDE and STRYER (1971) suggested that the rotational motion of fluorescent probes in cholesterol bilayers occurred on a time scale of nanoseconds. This is hard to understand in the light of the present study and the implications of the Debye equation presented above. The resolution of this problem must await further study.

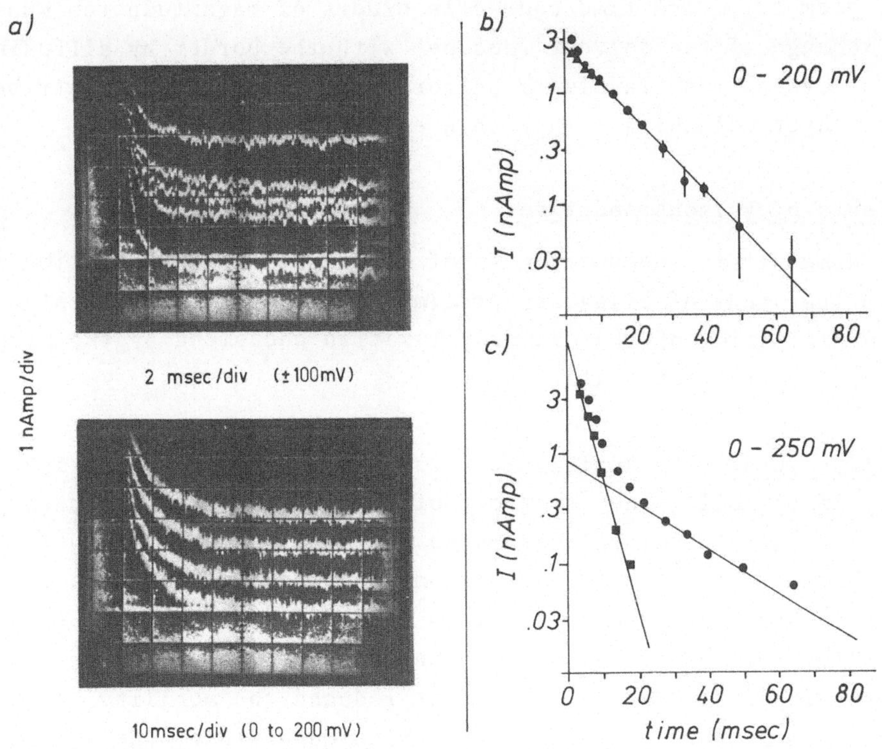

Fig. 6. Electrostrictive relaxation phenomena at T = 16°C. (a) Examples of raw data. (b) Relaxation analysis: (o) = points direct from 0 to +200 mV curve; (Δ) = effect of correction for $\Delta V = \pm 100$ mV (of significance only for $t < 8$ msec). The curve is well described by a single exponential $I^{o}_{1} = 2.7 \pm 0.2$ nA, $\tau_1 = 14 \pm 1$ msec). (c) Relaxation analysis for $\Delta V = 0$ to $+250$ mV, corrected for ±125 mV (only one trace could be made before bilayer broke): the curve requires a double exponential fit ($I^{o}_{1} = 0.9 \pm 0.2$ nA, $\tau_1 = 42 \pm 6$ msec; $I^{o}_{2} = 13 \pm 3$ nA, $\tau_2 = 7 \pm 2$ msec)

Geometrical Effects

Electrostrictive effects were also easily seen in oxidized cholesterol bilayers. Figure 6 compares the current relaxation curves for a symmetrical (±V/2) and a monophasic (0 to +V) voltage jump and shows the subsequent analysis. Curve (c) for 0 to 250 mV shows a more complicated time course than curve (b) (0 to 200 mV), but the total relaxation amplitude follows the expected V^3 dependence.

A complete analysis of the electrostrictive effects has not yet been possible because there are much slower relaxations, with time constants on the order of tenths of 1 second, which were not detected in the original study. For this reason, this aspect of the capacitance relaxation technique will not be discussed further except to say that the relatively quick relaxations shown here must again be interpreted on the basis of molecular reorientations associated with the changing bilayer geometry. The time course is orders of magnitude too short to allow exchange of the solvent (decane) with the border by diffusion, so that the effects must involve a uniform reorganization (redistribution) of the bilayer molecules rather than exclusion of solvent.

The Effects of Various Additives

To demonstrate the usefulness of the capacitance relaxation technique in the study of bilayers, preliminary results are presented below on the modification of the native relaxation phenomena by the presence of membrane-active materials.

Valinomycin. The antibiotic valinomycin is interesting for its activity in promoting the transport of alkali metal cations through both artificial and natural membranes. An nmr study (HSU, CHAN, 1973) has indicated that uncomplexed valinomycin binds to the surface of bilayers rather than penetrating into the hydrophobic core. The data also implied that bound valinomycin increased the mobility of some of the phospholipid head groups, while it reduced the mobility of others. Finally, from a study of the concentration dependence, the authors suggested that each valinomycin molecule affects the mobility of up to 1000 neighboring phospholipid molecules. This effect was presumed to be caused by cooperative effects among the bilayer constituents.

Figure 7 shows the capacitance relaxations found for oxidized cholesterol membranes, both with and without valinomycin, in the presence of various alkali cations. (The relaxation components attributable to the *conduction* by valinomycin occur at much longer time con-

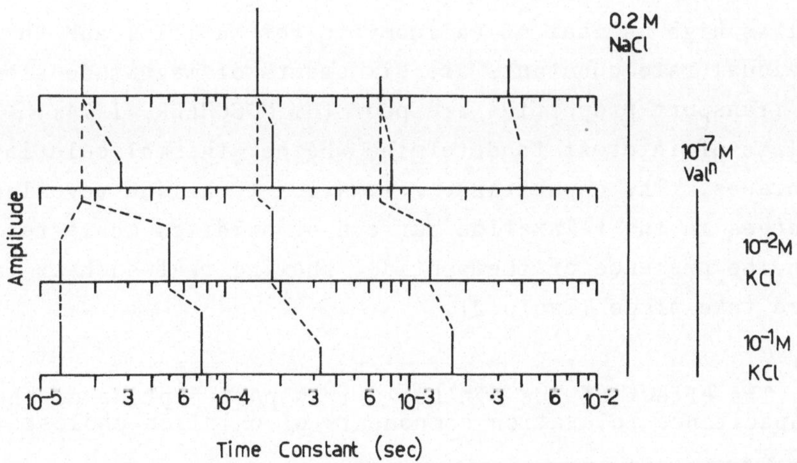

Fig. 7. Asymmetric capacitance relaxation patterns in oxidized cholesterol membranes. Concentrations of substances added to the aqueous phases are indicated beside the curves. The height of the bars indicates the amplitude of the relaxation component. Dashed lines show suggested relationships between the individual patterns

stants [τ > msec] and are not shown here.) The relaxation pattern in the presence of valinomycin depends strongly on the cation present. With Na^+, the whole spectrum shifts toward slower time constants, consistent with a general increase in effective viscosity. With K^+, on the other hand, a qualitatively different pattern is found. In this case, the results are consistent with the suggestions of the nmr study mentioned above, in which a double effect of valinomycin--liquifying some regions while rigidifying others--was postulated. The relaxation spectrum indicates both a speeding up of some components (e.g., τ = 13 μsec) and a slowing down of others (τ = 50 μsec, 180 μsec, and 1.3 msec at 10^{-2} M KCl). A possible relationship between the various components is suggested in the figure.

While there has not yet been time to conduct a more detailed analysis, it is undoubtedly relevant to the qualitatively different relaxation patterns found with Na^+ and K^+ to recall the work of GRELL, FUNCK, and EGGERS (1972) who, among others, have shown that the structure of the valinomycin/cation complex is significantly different for Na^+ than for K^+ and other avidly complexed ions.

Artificial Transport Peptide. The synthetic compound S,S'-bis (cycloglycyl-L-hemicystylsarcosylsarcosyl-L-prolyl) (see, e.g., SCHWYZER et al., 1970) was conceived in Professor R. Schwyzer's laboratory as a possible model for both passive and active cation transport through membranes. Its cation-complexation equilibrium constant is almost as high as that of valinomycin for Na^+, K^+, and Rb^+; but, as the individual rate constants are six orders of magnitude slower, the membrane transport properties are poor (H. MÖSCHLER, 1974). It was nevertheless of interest to determine whether the molecule interacted with membranes. The capacitance relaxation technique revealed significant changes in the relaxation pattern of oxidized cholesterol bilayers in the presence of the peptide, showing that such an interaction did indeed take place (Table 2).

Table 2. The effect of the synthetic transport peptide on the asymmetric capacitance relaxation components of oxidized cholesterol membranes[a]

Native membrane[b]		With 2.10^{-4} M peptide	
τ	ΔC/C (%)	τ	ΔC/C (%)
3.4 msec	1.2	1.2 msec	1.2
0.7 msec	0.6		
150 μsec	2.8	210 μsec	1.1
17 μsec	1	36 μsec	1.3

[a]Aqueous phase 0.2 M NaCl; temperature = 21°C.
[b]See Table 1.

Phospholipid Bilayers: Dioleoyllecithin

The preceding sections have shown that capacitance relaxation studies can provide information for both fundamental and applied studies using oxidized cholesterol bilayers. Of even greater interest is the extension of the technique to phospholipid bilayers, which are both more widely used and more relevant biologically. Previous reports in the literature have suggested that the capacitance of phospholipid bilayers is generally frequency independent up to 10^7 Hz (e.g., HANAI, HAYDON, TAYLOR, 1965), but improved accuracy might reveal either dielectric or geometric effects of the kind discussed here. Indeed, BAMBERG and LAUGER (1973) reported a capacitance change with a time constant of about 3 msec in dioleoyllecithin membranes, and BABAKOV, ERMISHKIN, and LIBERMAN (1966) had a similar result with phospholipid bilayers.

Initial capacitance relaxation results on dioleoyllecithin bilayers are shown in Table 3. Only two dielectric components were detected: the slower component (τ = 1.4 msec) has a Q_{10} of about 1.4, while the faster component (τ = 170 μsec) has only a very slight temperature dependence.

Table 3. Asymmetric capacitance relaxation components of dioleoyllecithin (in decane) bilayers

Temp.	τ_1	$\Delta C_1/C$ (%)	τ_2	$\Delta C_2/C$ (%)
21°C	1.4 ± 0.2 msec	0.8 ± 0.1	170 ± 20 μsec	0.21 ± 0.01
34°C	0.95± 0.5 msec	1.3 ± 1	160 ± 30 μsec	0.26 ± 0.04

From the results with oxidized cholesterol bilayers, and assuming a difference in viscosity of a factor of 30 (judged by the relative effectiveness of valinomycin in inducing cation conductivity in phospholipid and cholesterol membranes [SZABO et al., 1972]), one might anticipate time constants for individual molecular motions of 0.5 to 5 μsec. Because this lies at or below the resolution of our present apparatus, it is not surprising that no corresponding components were detected. In the future, I hope to extend the studies down to 1 μsec resolution, and it will be interesting to see whether another relaxation process can be found. The relaxation components given in Table 3 undoubtedly reflect cooperative motion, and the time constants will depend on the number of units in the cooperating clusters of molecules. Given a faster motion of the individual molecules, the relatively long time constants should imply that the clusters are larger than in the oxidized cholesterol bilayers. Based on current results, I can make no more quantitative statement, but further studies with this and other phospholipid bilayers should bring interesting results.

CONCLUSIONS

Confirmation of Analysis

A crucial question of this work is whether the signal measured is really due to a capacitance change of the bilayer itself and not to some unrelated property of the whole system. As mentioned earlier, control experiments have ruled out the possibility that electrode processes or diffuse double layer effects are involved: the time course of such processes is much slower than those reported here in any case (COSTER, 1973). Further, Magwell-Wagner dispersions, caused by electrical inhomogeneities of the bilayer structure, have also been shown to be orders of magnitude slower than the effects found here (COSTER, SMITH, in press).

A different kind of "artifact" could be caused by charged, lipid-soluble inclusions in the bilayer itself. However, the effect of such impurities would not be expected to be greatly dependent on changes in the aqueous phases, in contrast to what is actually observed (e.g., Figure 7). Thus, this possibility may also be discarded.

We are left with a mechanism of the type proposed: *dielectric* effects caused by molecular reorientations. Whether the different components correspond to the detailed motions suggested remains to be proven. Scattering of laser light may provide a way of detecting the molecular motions--certainly at least the cooperative effects--of the bilayer molecules.

Potential Usefulness

Even without a definite identification of the origin of the dielectric and geometric capacitance relaxations, the fact that these relaxations do reflect molecular dynamics in the bilayer allows their use for comparative studies. Problems that might be suitable for study with this method are:

- phase transitions in bilayers;
- the effect of membrane active material (e.g., valinomycin) on bilayers;
- the detection of membrane/protein interactions (hormone binding, antigen-antibody interactions);
- the effect of the organic solvent, or lack of it, on bilayer structure and dynamics (e.g., MONTAL, MUELLER, 1972);
- similarly, the effects of bilayer stabilizing procedures using cross-linked polypeptides (KING, STEINRAUF, 1972); and
- the influence of asymmetric aqueous phases on the membrane structure.

One interesting sidelight resulting from the capacitance relaxation studies is evident. It has been suggested in the literature that action potential and nerve conduction phenomena may be linked by the ordering and disordering of dipole domains ("electrets"). A model study by WOBSCHALL (1968) indicated that the time constants for such domains must lie between 0.14 and 1.4 msec, which is just where the relaxation components for dioleoyllecithin bilayers are found.

REFERENCES

BABAKOV, A. V., ERMISHKIN, L. M., LIBERMAN, E. A.: Influence of electric field on the capacity of phospholipid membranes. Nature (London) 210, 953-955 (1966).

BAESSLER, H., BEARD, R. B., LABES, M. M.: Dipole relaxation in a liquid crystal. J. Chem. Phys. 52, 2292-2298 (1970).

BAMBERG, E., LAUGER, P.: Channel formation kinetics of gramicidin A in lipid bilayer membranes. J. Membrane Biol. 11, 177-194 (1973).

BOTTCHER, C. J. F.: Theory of electric polarization. Amsterdam-London-New York: Elsevier 1952.

COSTER, H. G. L.: The double fixed charge membrane: Low-frequency dielectric dispersion. Biophys. J. 13, 118-132 (1973).

COSTER, H. G. L., SMITH, J. R.: The molecular organisation of bilayer lipid membranes: A study of the low-frequency Maxwell-Wagner impedance dispersion. Biochem. Biophys. Acta (in press).

DANIEL, V. V.: Dielectric relaxation. London: Academic Press 1967.

GRELL, E., FUNCK, T., EGGERS, F.: Dynamic properties and membrane activity of ion-specific antibiotics. In: Molecular Mechanisms of Antibiotic Action on Protein Biosynthesis and Membranes (ed. E. MUÑOZ, F. GARCIA-FERNANDEZ, D. VAZQUEZ), pp. 646-685. Amsterdam-London-New York: Elsevier 1972.

HANAI, T., HAYDON, D. A., TAYLOR, J.: Polar group orientation and the electrical properties of lecithin bimolecular leaflets. J. Theoret. Biol. 9, 278-298 (1965).

HSU, M., CHAN, S. I.: Nuclear magnetic resonance studies of the interaction of valinomycin with unsonicated lecithin bilayers. Biochemistry 12, 3872-3876 (1973).

KING, T. E., STEINRAUF, L. K.: Stabilization of a lipid bilayer membrane by polylysine. Biochem. Biophys. Res. Commun. 49, 1433-1437 (1972).

MONTAL, M., MUELLER, P.: Formation of bimolecular membranes from lipid monolayers and a study of their electrical properties. Proc. Nat. Acad. Sci. U.S.A. 69, 3561-3566 (1972).

MÖSCHLER, H.: Synthetic models for the biological function of peptides. Ph.D. Thesis No. 5370. ETH, Zürich 1974.

SARGENT, D. F.: An apparatus for the measurement of very small membrane relaxation currents (in preparation [a]).

SARGENT, D. F.: Capacitance relaxation--a new tool for the study of BLM structure and dynamics: Studies on oxidized cholesterol membranes (in preparation [b]).

SCHWYZER, R., TUN-KYI, A., CAVIEZEL, M., MOSER, P.: Homodetic cyclic polypeptides. 15. S,S'-bis(cycloglycyl-L-hemicystylglycylglycyl-L-prolyl), a synthetic bicyclic peptide with cation specificity. Helv. Chim. Acta 53, 15-27 (1970).

SHEETZ, M. P., CHAN, S. I.: Effect of sonication on the structure of lecithin bilayers. Biochemistry 11, 4573-4581 (1972).

SZABO, G., EISENMAN, G., McLAUGHLIN, S. G. A., KRASNE, S.: Ionic probes of membrane structures. Ann. N.Y. Acad. Sci. 195, 273-290 (1972).

TIEN, H. T., CARBONE, S., DAWIDOWICZ, E. A.: Formation of "black" lipid membranes by oxidation products of cholesterol. Nature (London) 212, 718-719 (1966).

WHITE, S. H.: A study of lipid bilayer membrane stability using precise measurements of specific capacitance. Biophys. J. 10, 1127-1148 (1970).

WHITE, S. H., THOMPSON, T. E.: Capacitance, area and thickness variations in thin lipid films. Biochim. Biophys. Acta 332, 7-22 (1973).

WOBSCHALL, D.: An electret model of the nerve membrane. J. Theoret. Biol. 21, 439-448 (1968).

YGUERABIDE, J., STRYER, L.: Fluorescence spectroscopy of an oriented model membrane. Proc. Nat. Acad. Sci. U.S.A. 68, 1217-1221 (1971).

Recognition

Studies on Polypeptide Hormone Receptors*

Robert Schwyzer
Department of Molecular Biology and Biophysics, Swiss Federal Institute of Technology (ETH-Z), CH-8049, Zürich, Switzerland

INTRODUCTION

Transduction of molecular information contained in polypeptide hormones into physiological responses is one of the vital functions of target cell membranes. It is intimately associated with the processes of selective recognition and specific triggering of membrane-bound receptor molecules.

Figure 1 summarizes some of the general notions and termini technici resulting from earlier pharmacological studies. Reversible complexation (binding) of the hormone by cell surface receptors is suggested by the form of the log dose response curve and is presumably the basis of the recognition process. The nature of the stimulus (allosteric?) is still unknown.

Several questions arise from this model: (1) Unity of receptor, or subunits for discriminator (recognition unit), transducer, and amplifier (response generator unit)? (2) Different discriminators for different responses (amplifiers), or only one for a variety? (3) Partial agonists fully activating a part of the receptor population, or partly activating essentially the whole population? (4) Mechanism of receptor reserve and threshold phenomena? (5) Chemical mechanisms of discrimination, transduction, and amplification?

The aim of our studies is to illuminate the discrimination mechanism mentioned in question 5 by chemical synthesis of appropriate analogs and derivatives of polypeptide hormones and the study of their reactions and cross-reactions with various target cells and target cell membranes (see Figure 2). The novel inclusion into this study of crystalline glucose 6-phosphate dehydrogenase (G6PD) was prompted by

---

*The literature references pertinent to this article, if not quoted separately, can be found in reviews by SCHWYZER, 1963, 1964, 1973, 1974; SCHILLER, SCHWYZER, 1973; and in the paper by LANG et al., 1974.

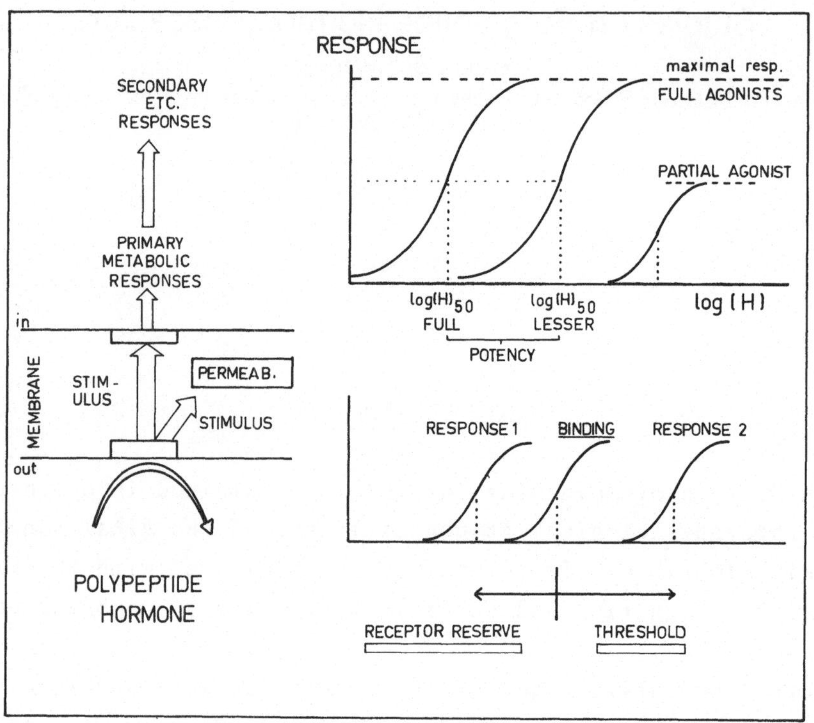

Fig. 1. Current concepts of polypeptide hormone: receptor interactions and some termini technici

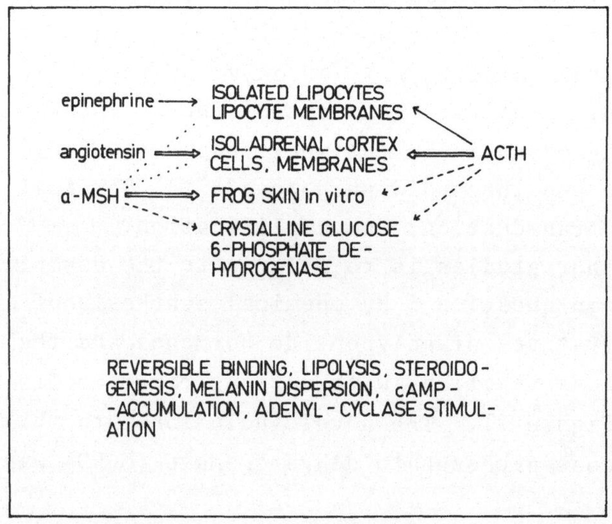

Fig. 2. Main and auxiliary interactions of some hormones with target cells, target cell membranes, and glucose 6-phosphate dehydrogenase (G6PD) used in our studies

our finding that this protein can form specific, reversible complexes with adrenocorticotropin.

FUNCTIONAL ORGANIZATION OF ACTH AND α-MSH DISCRIMINATORS

According to accepted laws of macromolecular interaction, the recognition site of a receptor must be topographically complementary to the active site of its hormone. It is therefore possible to make implications as to its general organization from detailed structure-function studies of the hormone itself. Care must be taken to exclude disturbing effects like transport phenomena and degradation of the hormone which occur in whole animals and perfused target organs. The best systems now available appear to be isolated target cells and plasma membrane preparations.

Adrenocorticotropic hormone (ACTH), alpha-melanocyte-stimulating hormone (α-MSH), and angiotensin are especially amenable to such studies. Their active sites are organized according to the sychnological pattern in which adjacent amino-acid residues are composed into a number of "continuate words," each conveying a part of the total functional information. Chemical synthesis of such relatively short "words"--detached from the rest of the molecule--is feasible and can reveal details about structure-function relationships. (The situation is much more difficult with hormones of the rhegnylogic type, e.g., insulin; cf. SCHWYZER, 1973.)

Many groups over the world have been carrying out such studies over the past 15 years (cf. SCHWYZER, 1963, 1964, 1973, 1974; LAW, 1974; TAGER, STEINER, 1974). An interpretation of the results in terms of ACTH and α-MSH organization is shown in Figure 3.

Adrenal cortex cells appear to have at least partially separate pathways for the stimulation of adenyl cyclase and steroidogenesis: whether they involve different discriminators, or whether one and the same discriminator can trigger either response according to the "message" it encounters, is unknown. In any case, the adenyl cyclase pathway is dependent on its receptor recognizing glutamic acid, E6, and an unsubstituted tryptophan, W9, in the hormone sequence. For the steroidogenic pathway, these requirements are not necessary; however hysine, K21, seems to be recognized specifically only by the "steroidogenic receptor." (An N^{ε} dansyl group at this position diminishes the steroidogenic potency tenfold, but leaves adenyl cyclase activation unimpaired; SCHWYZER, SCHILLER, 1971).

Fat cell receptors appear to be organized in a similar manner. They contain a topographic element that can recognize and be triggered by the hormonal "message + binding" sequence, EHFRWG (Figure 3). The analysis has not been completed with respect to the demonstration of different messages for lipolysis and cyclase stimulation. Again, lipolysis seems to be more affected by dansyl derivatives of K21 than is cyclase stimulation. Other observations strongly suggest at least a partial dissociation of the two pathways with a "safety-catch" mechanism of cyclase stimulation (SCHWYZER, 1974).

The recognition site for the "adrenal address," KKRR, seems also to be present in lipocyte ACTH receptors. The tetradecapeptide, ACTH-(11-24), inhibits responses of adrenal cortex and of fat cells. The binding is about 500 times weaker than that of the complete hormone (K_{diss} ∿2.5 x 10^{-6} ± 0.6 x 10^{-6}M for adrenal cells, and ∿2 x 10^{-6}M for lipocytes), indicating a very similar contribution to hormone interaction by the corresponding sites of both receptors.

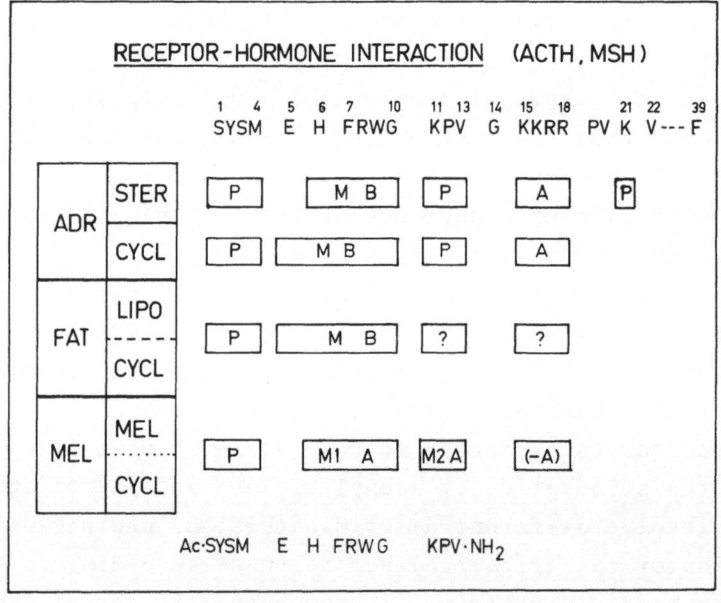

Fig. 3. Functional organization of ACTH and MSH discriminators in adrenal cortex cells (ADR), lipocytes (FAT), and frog melanocytes (MEL) as derived from hormonal structure-function and binding studies. Adenyl cyclase receptors are designated as CYCL, receptors for steroidogenesis, lipolysis, and melanin dispersion as STER, LIPO, and MEL, respectively. "A" stands for "address recognition," "B" for specific binding, "M" for stimulation by hormonal message (triggering), and "P" for effect (binding?) potentiation. Amino acid sequences for ACTH (above) and α-MSH (below) are shown in conventional one-letter symbolism

For many years, it was accepted that melanocyte MSH receptors are similarly organized: that they recognize HFRWG as a "message + binding" sequence, that the "words" Ac•SYSM and KPV•NH_2 mean potentiation, and that the receptors tend to "refuse" the "adrenal address" (G)KKRR (lowering of potency). It appeared well established that the receptor must be triggered by HFRWG, which on its own elicits a potency of about 2×10^4 MSH units/g. Only recently was Dr. Alex Eberle of my laboratory able to show that the sequence KPV•NH_2 displays $\sim 3 \times 10^5$U/g, being enhanced to $\sim 10^7$U/g by N-terminal acetylation. In contrast, PV•NH_2 and Ac•PV•NH_2 are completely inactive.

This finding constitutes the first demonstration of two independent "trigger areas" in a receptor. Their interaction is, in this case, not additive but potentiating. It clearly distinguishes the amphibian MSH receptor from the mammalian ACTH receptors and suggests evolutionary transitions from purely recognitional to functional (or vice versa) sites, or convergent evolution of different proteins to adapt and "read" chemically similar sequences.

DISCRIMINATOR RESERVE MOBILIZATION

With binding studies, it is difficult to distinguish between functional discriminators and binding proteins (without apparent transducing functions). In the case of the lipocyte ACTH receptor, we were fortunate to find a property--discriminator reserve mobilization--that relates the ACTH binding of $K_{ass} = 1.1 \times 10^9 M^{-1}$ to functionality (for a complete discussion, see LANG et al., 1974). A parallel binding site and response increase is brought about very quickly by the phenoxazones actinocine and actinomycin D.

An attempt to correlate membrane fluidity with the phenoxazone effect according to McCONNELL, WRIGHT, and McFARLAND (1972) is not yet complete (Dr. Ursula LANG and Dr. Jean-Luc FAUCHERE). Fatty acid nitroxide spin labels can be dissolved after lowering the pH carefully to 7. Fat cell ghosts (plasma membranes) incorporate the label and can be washed free of all rapidly tumbling molecules that give rise to an "isotropic" spectrum. Addition of epinephrine at concentrations $<10^{-6}$ M appears to cause significant increases of the order parameter (0.4 → 0.6). The ACTH and actinocine influences are as yet less clear cut. A strong influence of ACTH on labeled membranes still containing rapidly tumbling label molecules, which looks like a considerable slowing of the tumbling rate, has been traced to a trivial reaction between the positively charged ACTH and the fatty acid nitroxide anions. Addition of ACTH to clear solutions of the spin label in tris-buffer at pH 7 results in immediate cloudiness and formation of a precipitate. In these cases, the spectra look almost identical to those published by SCHREIER-MUCCILLO et al. (1974) for the interaction between the fatty acid spin label, plasma membranes from smooth muscle of guinea pig ileum, and angiotensin II. In view of the fact that both hormones contain arginine and histidine, I tend to believe that under the conditions of incomplete removal of "free" spin label from the membrane preparation, the spectral changes could result mainly from trivial precipitation effects.

GLUCOSE 6-PHOSPHATE DEHYDROGENASE (G6PD) FROM COW ADRENALS AS SPECIFIC BINDING PROTEIN FOR ACTH AND α-MSH

Following a lead of CRISS and McKERNS (1968) stating that ACTH-(1-24) was able to enhance the activity of highly purified or crystalline G6PD from cow adrenals, SCHILLER and SCHWYZER (1973) investigated a pure, crystalline enzyme preparation obtained from P. G. SQUIRE, Colorado. We were able to detect by fluorescence polarization methods a specific binding of [Lys(dansyl)^{21}]-ACTH-(1-24) to an apparently homogeneous population of binding sites. However, no enhancement of enzyme activity was found, even in the presence of ACTH-(1-24).

The binding studies were extended to [Phe^{2}, Nva(t_2)^{4}]-ACTH-(1-24) and [Ac•Tyr(t_2)^{2}]-α-MSH-(2-13)-dodecapeptide amide by BURGISSER and FAUCHERE in my laboratory, using equilibrium dialysis [the compounds had been prepared by KARLAGANIS and EBERLE (unpublished data), respectively]. It was found that one molecule of a hormone analog is bound for every one of the four subunits of the enzyme, assuming a subunit structure and a molecular weight as determined by SINGH and SQUIRE (1974). The association constants are as follows (in dilute phosphate buffer, <pH 7):

[Lys(dansyl)^{21}]-ACTH-(1-24)	K_{ass}	$\simeq$ 1.5 x 10^{6} $l \cdot mol^{-1}$
[Phe^{2}, Nva(t_2)^{4}]-ACTH-(1-24)	K_{ass}	$\simeq$ 2.3 x 10^{5} $l \cdot mol^{-1}$
[Ac•Tyr(t_2)^{2}]-α-MSH-(2-13)-NH	K_{ass}	$\simeq$ 8.3 x 10^{5} $l \cdot mol^{-1}$

These results could indicate that the main power of recognition by the enzyme is directed to the N-terminal portion of the hormonal amino-acid sequence, because alterations in this part of the molecule tend to lower the binding constant, e.g., the exchange of Tyr^{2} for Phe^{2} and of Met^{4} for Nva^{4}. As expected, replacements of Ac•Ser^{1}-Tyr^{2} by Ac•Tyr^{2} (α-MSH) and of Lys^{21} by $Lys(DNS)^{21}$, which do not influence biological potency to any great extent, give higher binding constants.

G6PD from erythrocytes interacts about 10 times less strongly with $[Lys(DNS)^{21}]$-ACTH-(1-24); other polypeptide hormones, like angiotensin, are not at all bound by G6PD.

We therefore believe that crystalline G6PD is a good and convenient model for studying the chemical details of ACTH and α-MSH interaction. At least it would allow the testing of chemical and physicochemical methods with a readily available, pure compound, before attacking the more difficult problem of interaction with hormone receptors. Even if we assume a large number of receptors, for example about 200 per μ^2 lipocyte surface, an extractive yield of 100%, and a molecular weight of about 10^5 daltons, 1 kg of fat cells would give about 1/10 of 1 mg or about 1 nM of receptor. The yield could, of course, easily drop to only 1 µg, or less, of receptor per kilogram of pure, isolated lipocytes.

Our finding raises a number of fundamental questions. Is there any physiological reason for the specific binding of hormones by this "cytoplasmatic" enzyme? Is the enzyme to be found in membranes? Are membrane receptors related to G6PD by evolution? Do polypeptide hormones, contrary to all belief, enter the target cells (see, e.g., TESSER, FISCH, SCHWYZER, 1972, 1974)? And last, considering the varying specificity of ACTH and MSH receptors for different portions of the hormone amino acid sequence: are receptors possibly related to cell-bound Ig-type antibodies? (These also are not only able to recognize an antigen, but to generate stimuli, e.g., in lymphocytes, where they induce either stimulation or tolerance; they are excreted into the surrounding media [MELCHERS, this volume] much as are receptors for thyroid stimulating hormone [KOHN, this volume].)

CONCLUSION

This chapter has reviewed some of the work that has led to a more detailed understanding of the functional organization of ACTH and α-MSH receptors in terms of recognition and triggering by the hormones. The significant points discussed are as follows: (1) An unexpected second "trigger site" has been found in melanocyte receptors, which is only a recognition (binding) site in the corresponding adrenal cortex cell receptors. (2) Some of the receptors for ACTH in lipocytes are apparently present in a "dormant state" which can be rapidly converted to a "receptive state" by phenoxazones. (Spin label studies of this conversion are only partially complete: they indicate as a side observation the enhancement of the order parameter, S_{10}, in lipocyte ghosts by adrenaline.) (3) Glucose 6-phosphate dehydrogenase from cow adrenal cortex binds ACTH and α-MSH very specifically; this crystalline enzyme is presently being used as a "model" for studying the chemical details of ACTH and α-MSH binding by their discriminators (receptors). (4) Some of the apparent affinity constants derived from binding and log dose-response studies in various systems are compiled and briefly discussed.

ACKNOWLEDGMENTS

The author gratefully acknowledges the experimental contributions of E. Bürgisser, A. Eberle, J.-L. Fauchère, W. Hübscher, G. Karlaganis, Ursula Lang, G.-M. Pelican, Beverly Savage, P. Schiller, W. Schlegel, R. Vogel, and Ursula Walty at different stages of this work, and the financial support by the Swiss National Foundation for the Encouragement of Scientific Research over the past years.

REFERENCES

CRISS, W. E., McKERNS, K. W.: Activation of low adrenal glucose 6-phosphate dehydrogenase by adrenocorticotropin. Biochemistry 7, 2364-2368 (1968).

KOHN, L. D. (this volume). Characterization of the thyrotropin receptor and its involvement in exophthalmos.

LANG, U., KARLAGANIS, G., VOGEL, R., SCHWYZER, R.: Hormone-receptor interactions. Adrenocorticotropic hormone binding site increase in isolated fat cells by phenoxazones. Biochemistry 13, 2626-2633 (1974).

LAW, H. D.: Amino-acids, peptides, and proteins. Specialist Periodical Reports. In: Chemical Structure and Biological Activity, Vol. V, pp. 384-447. London: The Chemical Society 1974.

McCONNELL, H. M., WRIGHT, K. L., McFARLAND, B. G.: The fraction of the lipid in a biological membrane that is in a fluid state: A spin label assay. Biochem. Biophys. Res. Commun. 47, 273-281 (1972).

MELCHERS, F. (this volume). Changes in receptor immunoglobin turnover during B-lymphocyte differentiation.

SCHILLER, P. W., SCHWYZER, R.: (21-N^{ε}-dansyllysine)-corticotropin-(1-24)-tetrakosipeptide, a biologically active derivative of ACTH. A receptor protein and studies with fluorescence depolarization and intramolecular excitation energy transfer. In: Peptides 1971 (ed. H. NESVADBA), pp. 354-366. Amsterdam: North-Holland Publ. Co. 1973.

SCHREIER-MUCCILLO, S., NICULITCHEFF, G. X., OLIVEIRA, M. M., SHIMUTA, S., PAIVA, A. C. M.: Conformational changes at membranes of target cells induced by the peptide hormone angiotensin. A spin label study. FEBS Letters 47, 193-196 (1974).

SCHWYZER, R.: Synthetische polypeptide mit physiologischer wirkung. Ergebn. Physiol. 53, 1-41 (1963).

SCHWYZER, R.: Chemistry and metabolic action of nonsteroid hormones. Ann. Rev. Biochem. 33, 259-286 (1964).

SCHWYZER, R.: Molecular mechanisms of polypeptide hormone action. In: Peptides 1972 (ed. H. HANSON, H.-D. JAKUBKE), pp. 424-436. Amsterdam: North-Holland Publ. Co. 1973.

SCHWYZER, R.: An alternative to the cAMP second messenger concept. Pure Appl. Chem. 37, 299-314 (1974).

SCHWYZER, R., SCHILLER, P. W.: Hormone-rezeptor-beziehungen: synthese und eigenschaften von N^{ε}-dansyllysin^{21}-adrenocorticotropin-(1-24)-tetrakosipeptid. Helv. Chim. Acta 54, 897-904 (1971).

SINGH, D., SQUIRE, P. D.: Molecular weight and subunit structure of bovine adrenal glucose 6-phosphate dehydrogenase. Biochemistry 13, 1819-1825 (1974).

TAGER, H. S., STEINER, D. F. Peptide hormones. Ann. Rev. Biochem. 43, 509-538 (1974).

TESSER, G. I., FISCH, H.-U., SCHWYZER, R.: Limitations of affinity chromatography: Sepharose-bound cyclic 3',5'-adenosine monophosphate. FEBS Letters 23, 56-58 (1972).

TESSER, G. I., FISCH, H.-U., SCHWYZER, R.: Limitations of affinity chromatography: Solvolytic detachment of ligands from polymeric supports. Helv. Chim. Acta 57, 1718-1730 (1974).

Cell Surface Membrane Changes and Growth Control of Cultured Fibroblasts*

Dennis D. Cunningham, Cornelia R. Thrash**, and Tsung-Shang Ho
Department of Medical Microbiology, College of Medicine, University of California, Irvine, California 92664

INTRODUCTION

Two kinds of changes in the cell surface membrane that have been related to growth control processes are: (1) changes measured by increased cell agglutinability with concanavalin A (con A) and certain other plant lectins (reviewed by BURGER, 1973), and (2) changes in the uptake of certain nutrients and/or growth factors (for example, PARDEE, 1964, 1971; WALLACH, 1968; CUNNINGHAM, PARDEE, 1969; SEFTON, RUBIN, 1971; HOLLEY, 1972, 1974).

We have conducted experiments to determine whether there are causal relationships between some of these cell surface membrane changes and control of growth in cultured fibroblasts and have come to the following conclusions: (1) The cell surface change measured by increased con A-specific cell agglutinability that is brought about by brief protease treatment of untransformed fibroblasts is not an event that is sufficient by itself to lead to initiation of cell division and loss of density-dependent growth control (GLYNN, THRASH, CUNNINGHAM, 1973; CUNNINGHAM, HO, in press). (2) The rapid increase in phosphate uptake that takes place when growth is initiated by adding serum to density-inhibited 3T3 mouse cells under usual culture conditions is neither necessary nor sufficient for subsequent initiation of DNA synthesis (CUNNINGHAM, PARDEE, 1969). (3) The similar rapid increase in hexose uptake that takes place after growth initiation (SEFTON, RUBIN, 1971) can be dissociated from initiation of proliferation under certain conditions (THRASH, CUNNINGHAM, in press).

The evidence that supports these conclusions is summarized below.

---

*This work was supported by Grant CA-12306 from the U.S. Public Health Service.

**Currently: Department of Biological Sciences, Stanford University, Stanford, California 92605

PROTEASE-MEDIATED INCREASE IN CON A-SPECIFIC CELL AGGLUTINABILITY

Several observations have led to the proposal that an alteration in the cell surface, detected by increased cell agglutinability with con A and certain other plant lectins, brings about initiation of cell division and escape from density-dependent growth control (reviewed by BURGER, 1973). These include observations that: (1) cells transformed by tumor viruses, chemical carcinogens, and x-irradiation are much more highly agglutinated by con A than their untransformed parental cells (INBAR, SACHS, 1969); (2) con A-specific cell agglutinability correlates with one expression of the transformed phenotype (lack of topo-inhibition) in cells transformed by a temperature-sensitive tumor virus (ECKHART, DULBECCO, BURGER, 1971); (3) density-inhibited cells selected from populations of transformed cells show low agglutinability with the wheat-germ lipase agglutinin (POLLACK, BURGER, 1969); (4) selection for low agglutinability concomitantly selects for cells with density-dependent growth control (INBAR, RABINOWITZ, SACHS, 1969; OZANNE, SAMBROOK, 1971); and (5) there is generally a positive correlation between agglutinability by these lectins and final cell density (POLLACK, BURGER, 1969; WEBER J., 1973). The possibility of a causal relationship between this membrane change and initiation of cell division was suggested by reports that brief protease treatments at the cell surface that brought about the agglutinable state also caused density-inhibited cells to double in number (BURGER, 1970, 1973; BURGER et al., 1972; NOONAN, BURGER, 1973). However, our studies with 3T3 mouse fibroblasts (GLYNN, THRASH, CUNNINGHAM, 1973) as well as with secondary chick embryo fibroblasts and human diploid foreskin fibroblasts (CUNNINGHAM, HO, in press) have shown that brief protease treatments that bring about this surface change do not initiate cell division or lead to a loss of density-dependent growth control.

The amount of con A-specific cell agglutination of quiescent human diploid foreskin and secondary chick embryo fibroblasts after treatment with phosphate-buffered saline (PBS) or various concentrations of pronase or trypsin in PBS for 10 min at 37°C is shown in Figure 1. Con A-specific agglutination of control cells incubated with only PBS ranged from 20% to 30%. As can be seen, treatment with pronase or trypsin increased this agglutination to plateau values of about 80%. Intermediate levels of proteases resulted in graded increases in agglutination. The responses to pronase and trypsin were quite similar; maximal agglutination occurred after a 10-min treatment with 2 to

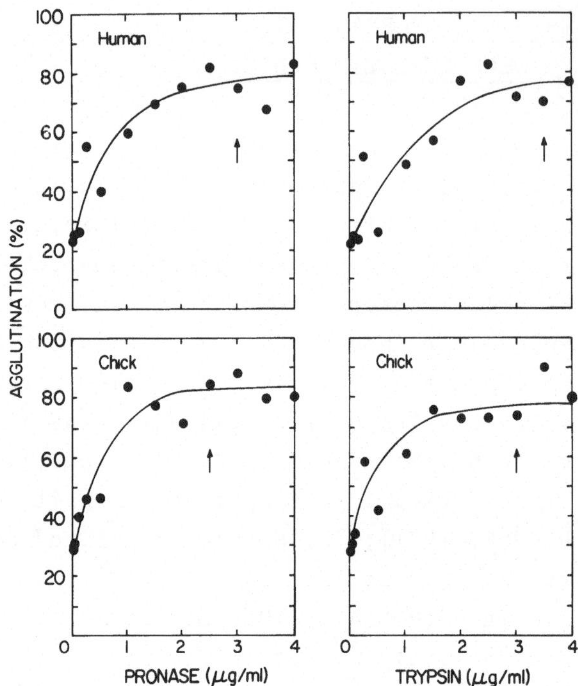

Fig. 1. Effect of 10-min protease treatments on con A-specific agglutination of human and secondary chick embryo fibroblasts. Human fibroblasts were plated at a density of 3.0×10^4 cells/cm^2 and grown to a final density of 3.6×10^4 cells/cm^2 over a 3-day period. Chick fibroblasts were plated at a density of 4.5×10^4 cells/cm^2 and grown to a final density of 1.5×10^5 cells/cm^2 over a 3-day period. Con A-specific agglutination was measured as previously described (OZANNE, SAMBROOK, 1971). The arrows show the protease concentrations at which some rounding up of the cells was detected by phase contract microscopy. (From CUNNINGHAM and HO [in press], with permission from Cold Spring Harbor Laboratory.)

3 μg/ml of either protease. The concentrations of pronase or trypsin that caused some rounding up of the cells (arrows in Figure 1) were in all cases higher than those that brought about maximal con A-specific cell agglutination.

To determine whether this surface change brought about by 10-min protease treatments led to initiation of cell division, we measured cell number on parallel cultures of quiescent human and chick fibroblasts treated in exactly the same way with PBS or the same concentrations of pronase or trypsin in PBS. After these 10-min treatments, the cells were rinsed twice with PBS, and medium that had supported the growth of the cells to the quiescent state ("conditioned" medium) was added back to the cultures. As shown in Figures 2 and 3, there were no significant increases in cell number after treatment with the same

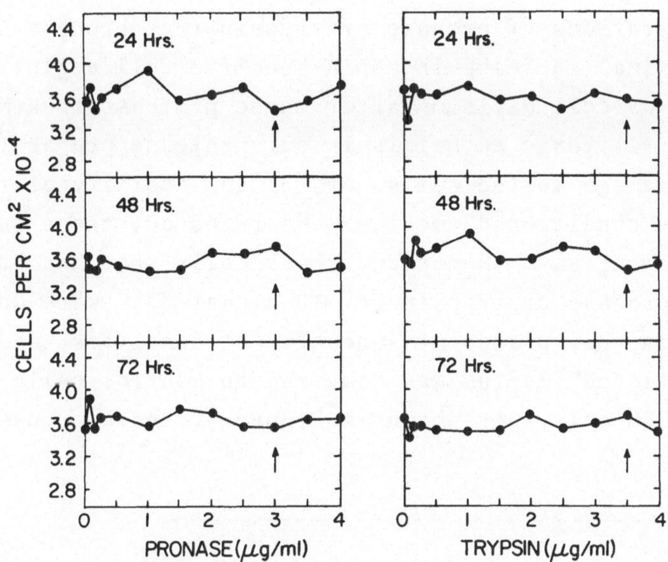

Fig. 2. Effect of 10-min protease treatments on cell density of human fibroblasts. In the same experiment described in Figure 1, parallel cultures of quiescent cells were treated in the same way with PBS or the indicated concentration of pronase or trypsin in PBS for 10 min at 37°C. The cells were then rinsed twice with PBS at 37°C, and the "conditioned" medium that had supported the growth of the cells to the quiescent state was added back to the cultures. Cell number was monitored after 24 hr, 48 hr, and 72 hr. The arrows show the protease concentrations at which some rounding up of the cells occurred. (From CUNNINGHAM and HO [in press], with permission of Cold Spring Harbor Laboratory.)

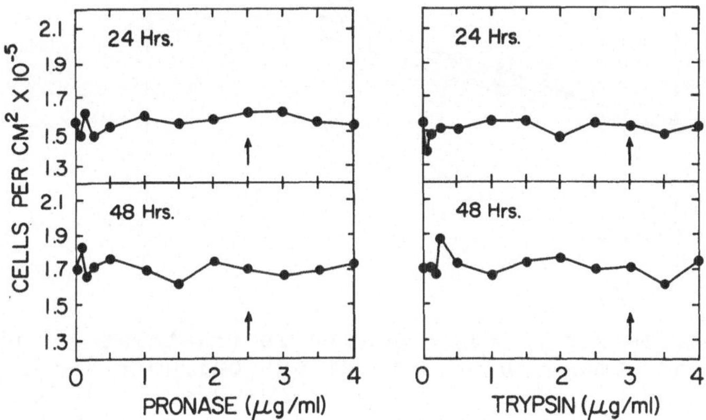

Fig. 3. Effect of 10-min protease treatments on cell density of secondary chick embryo fibroblasts. This experiment was conducted in the same way as described in Figure 2 for human cells. Cell number was measured 24 hr and 48 hr after the protease treatments. (From CUNNINGHAM and HO [in press], with permission from Cold Spring Harbor Laboratory.)

ranges of concentrations of pronase or trypsin that brought about no increase to a maximal increase in con A-specific cell agglutinability.

The absence of cell division after these protease treatments could be a result of: (1) toxic materials in our protease preparations, (2) proteolytic damage to the cells, or (3) an inability of the cells to divide in the "conditioned" medium. We ruled out these possibilities in the following way. Human and chick cells that had been treated with 4 µg/ml of pronase or trypsin for 10 min at 37°C were detached with 0.02% EDTA in PBS, plated at subconfluent densities, and their growth in "conditioned" medium was compared to control cells treated for 10 min with PBS only. As shown in Figure 4, these pronase or

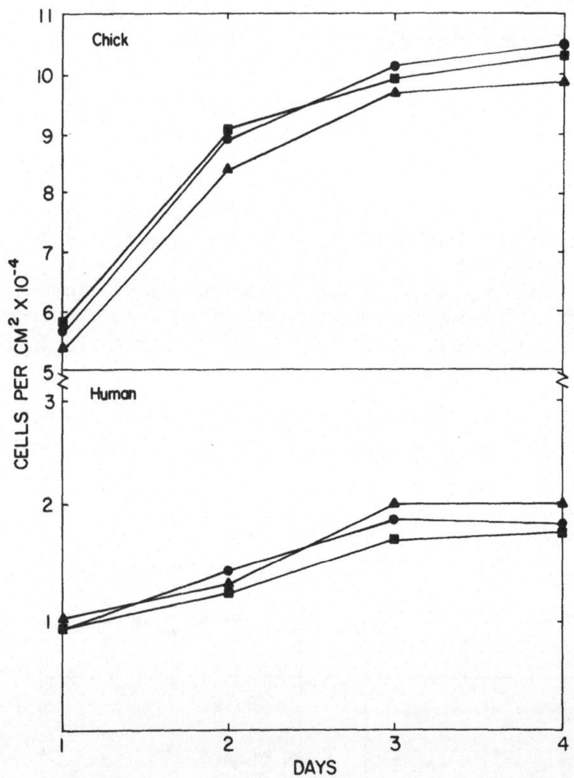

Fig. 4. Effect of 10-min pronase or trypsin treatment on subsequent growth of chick and human fibroblasts at subconfluent densities in "conditioned" medium. In the same experiment described in Figure 1, parallel cultures of chick or human fibroblasts were detached with 0.02% EDTA in PBS after a 10-min treatment with 4 µg/ml pronase (▲———▲, 4 µg/ml trypsin (■———■), or PBS (●———●). The cells were sedimented, resuspended in "conditioned" medium, counted, and plated at subconfluent densities (6.4 x 10^4 cells/cm^2 for chick cells; 1.3 x 10^4 cells/cm^2 for human cells). Cell number was monitored on subsequent days. (From CUNNINGHAM and HO [in press], with permission from Cold Spring Harbor Laboratory.)

trypsin treatments did not reduce the growth of either the chick or human cells. Taken together, these results demonstrate that the protease-mediated surface change measured by increased con A-specific cell agglutinability is not an event sufficient in itself to lead to cell division or escape from density-dependent growth control.

In contrast to these studies using brief protease treatments, we have also added proteases to quiescent fibroblasts and examined the effects of 24-hr to 72-hr treatments on cell number (CUNNINGHAM, HO, in press). We have confirmed the report of SEFTON and RUBIN (1970) that a 24-hr treatment of quiescent secondary chick embryo fibroblasts with trypsin leads to a large increase in cell number (30% to 80%). The trypsin must be present at least 4 hr to 8 hr to bring about a detectable increase in cell number at 24 hr. However, we have detected no significant increase in cell number after growing mouse 3T3, human diploid foreskin, or bovine embryonic trachea fibroblasts to a growth-arrested state under a variety of culture conditions and then adding proteases for periods up to 3 days. Thus, long-term continuous treatment of quiescent fibroblasts with added proteases can lead to initiation of cell division in certain cases, but we do not understand the conditions necessary for the initiation (CUNNINGHAM, HO, in press).

CHANGES IN PHOSPHATE UPTAKE

Another cell surface membrane change that appears to correlate with certain aspects of growth control is changes in levels of phosphate uptake (CUNNINGHAM, PARDEE, 1969; WEBER, EDLIN, 1971; de ASUA, ROSENGURT, DELBECCO, 1974). This finding is illustrated in Figure 5, which shows levels of phosphate uptake measured during 15-min incubation periods into the acid-soluble fractions of 3T3 and polyoma virus-transformed 3T3 (Py3T3) cells in different growth conditions (CUNNINGHAM, PARDEE, 1969). The plain bars show uptake levels measured 3 days after a medium change. The lined bars show levels of phosphate uptake during the 15-min interval immediately following the addition of fresh serum to a final concentration of 10% to parallel cultures. As can be seen, after 3T3 cells grew to confluency and became density-inhibited, the level of phosphate uptake decreased threefold to fourfold. In contrast, Py3T3 cells are not subject to density-dependent growth control, and as Figure 5 shows, the level of phosphate uptake actually increased after the growth of these cells to confluency, indicating that the decrease for 3T3 cells was not simply a result of the confluent state. Addition of fresh serum to confluent 3T3 cells

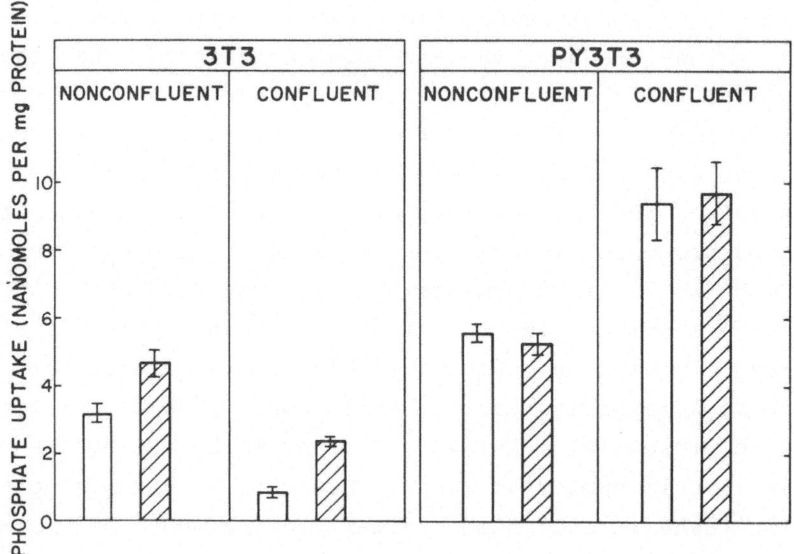

Fig. 5. Levels of phosphate uptake by 3T3 and Py3T3 cells. Levels of (^{32}P) phosphate uptake into the acid-soluble fraction of the cells were measured during 15-min incubation periods. Under these conditions, uptake was linear for at least 30 min. Plain bars: levels of phosphate uptake 3 days after a medium change. Lined bars: levels of phosphate uptake by parallel cultures during the 15-min interval after addition of fresh serum to a final concentration of 10%. (From CUNNINGHAM and PARDEE [1969], with permission from the National Academy of Sciences.)

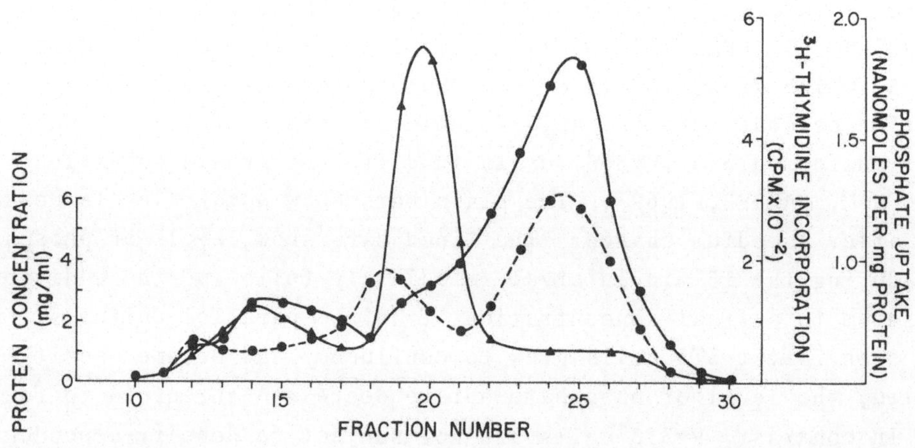

Fig. 6. Fractionation of serum by Sephadex G-200. Fractions were assayed for total protein (•--•) and activities that stimulated phosphate uptake (•——•) and DNA synthesis (▲——▲) by density-inhibited 3T3 cells. (From CUNNINGHAM and PARDEE [1969], with permission from the National Academy of Sciences.)

initiates DNA synthesis and cell division (TODARO, LAZAR, GREEN, 1965), and as shown in Figure 5, this treatment brought about a twofold to threefold increase in phosphate uptake in density-inhibited 3T3 cells within 15 min. The level for growing nonconfluent cells was increased only about 50%, and serum addition did not detectably change the level of phosphate uptake by either nonconfluent nor confluent Py3T3 fibroblasts.

These correlations between levels of phosphate uptake and certain aspects of growth control prompted us to fractionate serum to determine whether the factors that initiate DNA synthesis were similar to or different from factors that bring about the rapid increase in phosphate uptake by density-inhibited 3T3 fibroblasts. These results are shown in Figure 6 (CUNNINGHAM, PARDEE, 1969). As can be seen, most of the activity that stimulated phosphate uptake was present in the smallest molecular weight protein peak which was composed primarily of albumin. A much smaller peak of uptake-stimulating activity of higher molecular weight was also present. Neither of these factors corresponded in size to the factor that was most active in stimulating DNA synthesis, earlier identified by a similar fractionation by TODARO et al. (1967). Thus, even though changes in phosphate uptake correlate with certain aspects of growth control, the rapid increase in phosphate uptake brought about by addition of fresh serum to density-inhibited 3T3 fibroblasts is neither necessary nor sufficient for the subsequent initiation of DNA synthesis under usual culture conditions.

CHANGES IN HEXOSE UPTAKE

Evidence that changes in hexose uptake might participate in growth control comes from reports that uptake of certain hexoses: (1) increases after transformation by tumor viruses (ISSELBACHER, 1972; WEBER, M., 1973; HATANAKA, 1974) and correlates with the expression of the transformed phenotype in chick cells altered by a temperature-sensitive mutant of Rous sarcoma virus (MARTIN et al., 1971); (2) decreases when untransformed fibroblasts reach a density-inhibited monolayer (SEFTON, RUBIN, 1971; WEBER, M., 1973); and (3) rapidly increases after initiating cell division with serum or other agents (SEFTON, RUBIN, 1971; VAHERI, RUOSLAHTI, NORDLING, 1972). We have shown, however, that increased hexose transport and initiation of proliferation can be uncoupled in density-inhibited 3T3 mouse fibroblasts (THRASH, CUNNINGHAM, in press). Levels of cortisol that initiate DNA synthesis and division of the cells (THRASH, CUNNINGHAM, 1973; CUNNINGHAM,

THRASH, GLYNN, 1974) actually decrease hexose transport. In addition, some serum components increase hexose transport without initiating proliferation.

To determine the extent to which initiation of proliferation and increased hexose transport are correlated, we added fresh serum (5.0%) or the glucocorticoid steroid hormone cortisol (3.0×10^{-6}M) and measured DNA synthesis, cell number, and uptake of the hexose analogs, 2-deoxy-D-glucose and 3-0-methyl-D-glucose (Figure 7). These treatments brought about similar increases in the rate of DNA synthesis at 24 hr (sixfold increase) and cell number at 72 hr (30% increase). However, as Figure 7 shows, initiation of DNA synthesis in density-inhibited 3T3 fibroblasts was not always accompanied by increased hexose uptake. Addition of fresh serum to a final concentration of 5.0% brought about an increase in uptake of both 3-0-methyl-D-glucose and 2-deoxy-D-glucose by 1 hr. The increase was maximal by 4 to 6 hr and declined to values characteristic of density-inhibited cells by 24 hr. This result confirms earlier findings of SEFTON and RUBIN (1971). In contrast to

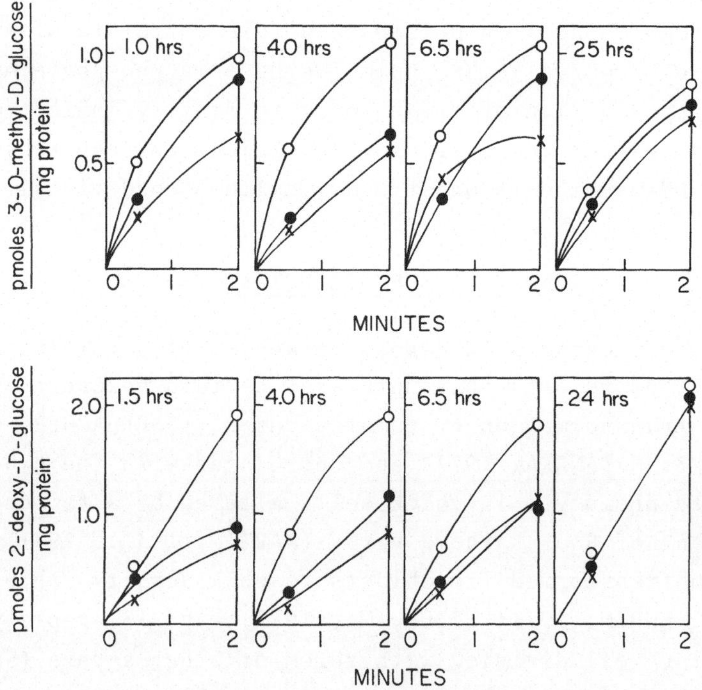

Fig. 7. Time course of (^{3}H) 3-0-methyl-D-glucose and (^{3}H) 2-deoxy-D-glucose uptake by control density-inhibited 3T3 cultures (•——•) and parallel cultures treated with 5% fresh serum (o——o), or 3 μM cortisol (X——X). Times in hours refer to times after serum or cortisol treatment. (From THRASH and CUNNINGHAM [in press], with permission from Nature Magazine.)

serum, cortisol inhibits hexose uptake by certain cells (MUNCK, 1968; ROSEN et al., 1972). Indeed, Figure 7 shows that 3 x 10^{-6}M cortisol, a concentration that initiated DNA synthesis and division similar to 5.0% serum, actually brought about a small decrease in transport of both hexose analogs by density-inhibited 3T3 fibroblasts. Thus, an increase in hexose transport is not necessary for subsequent proliferation under these culture conditions (THRASH, CUNNINGHAM, in press).

To probe further the relationship between initiation of proliferation and increased hexose transport, we added varying levels of fresh serum (0.025% to 25%) to density-inhibited 3T3 fibroblasts and measured 3-0-methyl-D-glucose uptake, DNA synthesis, and cell number. The dose-response curves in Figure 8 show that fresh serum added to 0.5% caused a maximal increase in hexose transport but no significant increase in DNA synthesis or cell number. As Figure 8 shows, higher concentrations of serum brought about increases in DNA synthesis and cell number that were roughly proportional to serum concentration. These higher concentrations caused no additional increase in hexose uptake. Thus, an early increase in hexose uptake is not an event sufficient by itself to initiate proliferation of density-inhibited 3T3 fibroblasts (THRASH, CUNNINGHAM, in press).

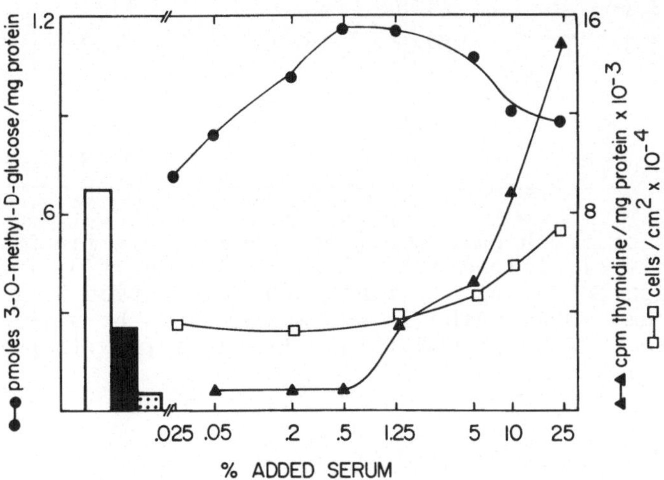

Fig. 8. Effect of concentration of added fresh serum on 3-0-methyl-D-glucose uptake (●——●), DNA synthesis (▲——▲), and cell number (□——□). Hexose uptake was measured during a 1-min incubation 4 hr after adding fresh serum. DNA synthesis was measured at 24 hr, and cell number was monitored at 72 hr. Vertical bars show control values for density-inhibited cultures: plain bar, hexose uptake; solid bar, cell number; stippled bar, DNA synthesis. (From THRASH and CUNNINGHAM [in press], with permission from Nature Magazine.)

Additional experiments indicated that serum might contain multiple factors that affect growth and hexose uptake, and that the relative activities of these factors might vary in different lots of serum. For example, dose-response studies, like those shown in Figure 8, revealed for some lots of serum parallel increases in hexose uptake, DNA synthesis, and cell number. In addition, the relative stimulation of DNA synthesis and hexose uptake varied with different commercial serum lots (Table 1). In fact, the lot that produced the greatest stimulation of hexose uptake had the smallest effect on DNA synthesis. Moreover, various pure or partially purified fractions of serum had different effects on hexose uptake and DNA synthesis. For example, addition of fetuin (0.5 mg/ml) produced a 66% increase of hexose uptake, but had no effect on DNA synthesis (THRASH, CUNNINGHAM, in press).

Table 1. Relative stimulation of DNA synthesis and hexose uptake by different lots of commercial serum[a]

Addition	Thymidine incorporation ($\frac{cpm \times 10^{-4}}{mg\ protein}$)	Thymidine incorporation (% control)	2-deoxy-D-glucose uptake ($\frac{p\ moles}{mg\ protein}$)	2-deoxy-D-glucose uptake (% control)
None	0.11	100	0.90	100
C733308	1.05	950	1.42	159
A9219E	2.66	2420	1.14	128
R2034R	2.12	1930	1.07	119
R832101	2.51	2280	1.02	114
R636017	2.14	1950	1.01	113
W20720	2.47	2240	0.98	110

[a]Different lots of fresh calf serum were added to a final concentration of 5.0% to density-inhibited 3T3 cells. Uptake of (^{3}H) 2-deoxy-D-glucose was measured during a 1-min incubation period 4 hr after the addition of serum. DNA synthesis was measured 24 hr after the addition of serum. (From THRASH and CUNNINGHAM [in press], with permission from Nature Magazine.)

CONCLUDING REMARKS

Our studies on three kinds of growth-arrested fibroblasts in culture have led us to the conclusion that the protease-mediated cell surface alteration measured by increased con A-specific cell agglutinability is not a change that is sufficient to lead to cell division and loss of density-dependent growth control (GLYNN, THRASH, CUNNINGHAM, 1973; CUNNINGHAM, HO, in press). The possibility of a causal relationship between this membrane change and initiation of cell division was suggested by reports that brief protease treatments at the cell surface that brought about the agglutinable state also caused density-inhibited cells to double in number (BURGER, 1970, 1973; BURGER et al., 1972; NOONAN, BURGER, 1973). However, our 10-min protease treatments led to increased con A-specific cell agglutinability, but not to a measurable increase in cell number after "conditioned" medium was returned to protease-treated cells. Although the reasons for this discrepancy in results are not clear, the absence of detectable cell division in our experiments was not the result of toxic materials in our proteases, extensive proteolytic damage to the cells, an inability of the cells to respond to a stimulatory signal, or extensive depletion of the medium such that it could not support further cell divisions.

Although our 10-min protease treatments of growth-arrested fibroblasts led to no detectable increase in cell number, we found that a 24-hr treatment of quiescent secondary chick embryo fibroblasts with trypsin led to a large increase in cell number (CUNNINGHAM, HO, in press), confirming the earlier report of SEFTON and RUBIN (1970). In contrast to these results with quiescent chick embryo fibroblasts, we detected no significant increase in cell number after growing mouse 3T3, human diploid foreskin, or bovine embryonic trachea fibroblasts to a growth-arrested state under a variety of culture conditions and then adding proteases for periods up to 3 days. Thus, long-term treatment of quiescent fibroblasts with added proteases can lead to initiation of cell division in certain cases, but the conditions necessary for the initiation are not well understood. It is possible that initiation by proteases occurs more readily with avian cells, and that mammalian cells are relatively refractory to stimulation by these agents. However, it seems more likely that when different kinds of fibroblasts are grown to quiescence under different nutritional conditions, different factors or conditions eventually limit growth, and long-term protease treatment can initiate division only when certain of these become limiting (CUNNINGHAM, HO, in press).

It should be pointed out that our experiments do not indicate the extent to which the cell surface alteration of *transformed* fibroblasts measured by increased con A-specific cell agglutinability is involved in the loss of density-dependent growth control characteristic of these cells. The molecular changes leading to increased con A-specific agglutinability of transformed fibroblasts might be quite different from those leading to increased agglutinability of untransformed fibroblasts after protease treatment.

Another kind of cell surface membrane change that has been implicated in growth control is changes in levels of uptake of certain nutrients and/or growth factors (PARDEE, 1964, 1971; WALLACH, 1968; CUNNINGHAM, PARDEE, 1969; SEFTON, RUBIN, 1971; HOLLEY, 1972, 1974). For example, uptake of phosphate ion is very rapidly increased after initiating division of density-inhibited 3T3 cells with fresh serum (CUNNINGHAM, PARDEE, 1969; de ASUA, ROSENGURT, DELBECCO, 1974). In addition, phosphate uptake decreases after 3T3 cells grow to a growth-arrested confluent state (CUNNINGHAM, PARDEE, 1969; WEBER, EDLIN, (1971), but does not decrease after growth of Py3T3 cells to confluency (CUNNINGHAM, PARDEE, 1969). However, fractions from serum, obtained by gel filtration on Sephadex G-200, that initiate DNA synthesis in density-inhibited 3T3 cells do not produce an increase in phosphate uptake, and fractions that increase phosphate uptake do not initiate DNA synthesis (CUNNINGHAM, PARDEE, 1969). These results show that under usual culture conditions, the early increase in phosphate uptake is neither necessary nor sufficient for initiation of DNA synthesis by fresh serum. It is noteworthy that 3T3 cells can be arrested in G_1 (or G_0) by limiting phosphate in the medium, and that adding back phosphate leads to reinitiation of DNA synthesis (HOLLEY, 1972, 1974; HOLLEY, KIERNAN, 1974). Furthermore, simultaneous addition of insulin increases the synchrony and extent of initiation of DNA synthesis (HOLLEY, KIERNAN, 1974). Insulin stimulates phosphate uptake, but has little effect on DNA synthesis if added alone (HOLLEY, KIERNAN, 1974). Thus, under conditions in which growth of 3T3 cells is limited by low phosphate, increases in the uptake of this ion by cells appear to participate in initiation of DNA synthesis.

Studies on hexose uptake have also revealed a number of correlations between uptake levels and several aspects of growth control (see preceding section for a summary of these studies). We have examined the rapid increase in hexose uptake that takes place after initiating proliferation of density-inhibited 3T3 cells with fresh serum and have found that under usual culture conditions this increase is neither necessary nor sufficient for subsequent initiation of DNA synthesis and cell division (THRASH, CUNNINGHAM, in press). In these experiments, we measured the uptake of two glucose analogs, 2-deoxy-D-glucose and 3-O-methyl-D-glucose. Levels of cortisol that initiate DNA synthesis and division of density-inhibited 3T3 cells actually bring about small decreases in hexose uptake at various times from 1 to 24 hr after its addition. Furthermore, addition of fresh serum to a final concentration of 0.5% to density-inhibited 3T3 cells brings about a maximal increase in hexose transport at 4 hr, but no detectable increase in DNA synthesis at 24 hr or cell number at 3 days. Higher levels of serum lead to increases in DNA synthesis and cell number that are roughly proportional to serum concentration, but these higher serum concentrations cause no additional increase in hexose uptake. The conclusion that increased hexose uptake and proliferation may be unlinked events is consistent with the findings of GAZDAR et al. (1972) that treatment of mouse sarcoma virus-transformed 3T3 cells with dibutyryl cyclic AMP and theophylline sharply inhibited cell growth yet stimulated glucose uptake. In addition, RUBIN and FODGE (1974) have suggested that 2-deoxy-D-glucose uptake merely reflects the activity of the glycolytic pathway that is activated after quiescent secondary chick embryo fibroblasts are stimulated to proliferate. They found that under certain conditions, external glucose could be entirely omitted from the medium without affecting the initiation of DNA synthesis. However, by using inhibitors of glycolysis, they demonstrated that glucose catabolism is essential for the initiation of DNA synthesis, and discussed the possibility that glycogen could provide an alternate short-term supply of substrate for glycolysis. These results suggest the possibility that the rapid increase in hexose transport after addition of fresh serum might be required for the initiation of DNA synthesis under restricted conditions where glycogen stores are depleted and cells are quiescent in medium containing very low levels of glucose.

Finally, it has been shown that limitation of amino acids in the growth medium can arrest cells in G_1 (or G_0), and that addition of the limiting amino acid(s) leads to initiation of DNA synthesis and cell division (LEY, TOBEY, 1970; HOLLEY, 1974; HOLLEY, KIERNAN, 1974; PARDEE, 1974). This finding strongly indicates that intracellular levels of amino acids can regulate growth. However, initiation of DNA synthesis and cell division by fresh serum brings about no detectable change in amino acid uptake by 3T3 cells (CUNNINGHAM, PARDEE, 1969) and decreases in amino acid uptake by WI-38 cells (WIEBEL, BASERGA, 1969). In addition, growth of 3T3 cells to a density-inhibited state is accompanied by small decreases (30%) in uptake of certain nonmetabolizable amino acids, and no changes in uptake of certain metabolizable amino acids (FOSTER, PARDEE, 1969). Thus, changes in amino acid uptake appear not to be involved in the regulation of growth by serum or cell density.

REFERENCES

BURGER, M. M.: Proteolytic enzymes initiating cell division and escape from contact inhibition of growth. Nature 227, 170-171 (1970).

BURGER, M. M.: Surface changes in transformed cells detected by lectins. Fed. Proc. 32, 91-101 (1973).

BURGER, M. M., BOMBIK, B. M., BRECKENRIDGE, B. M., SHEPPARD, J. R.: Growth control and cyclic alterations of cyclic AMP in the cell cycle. Nature New Biology 239, 161-163 (1972).

CUNNINGHAM, D. D., HO, T. S.: Effects of added proteases on concanavalin A-specific agglutinability and proliferation of quiescent fibroblasts. In: Proteases and Biological Control (ed. E. REICH, D. RIFKIN). New York: Cold Spring Harbor Laboratory (in press).

CUNNINGHAM, D. D., PARDEE, A. B.: Transport changes rapidly initiated by serum addition to "contact inhibited" 3T3 cells. Proc. Nat. Acad. Sci. U.S.A. 64, 1049-1056 (1969).

CUNNINGHAM, D. D., THRASH, C. R., GLYNN, R. D.: Initiation of division of density-inhibited fibroblasts by glucocorticoids. In: Control of Proliferation in Animal Cells (ed. B. CLARKSON, R. BASERGA) Vol. I, pp. 105-113. New York: Cold Spring Harbor Laboratory 1974.

DE ASUA, L., ROSENGURT, E., DULBECCO, R.: Kinetics of early changes in phosphate and uridine transport and cyclic AMP levels stimulated by serum in density-inhibited 3T3 cells. Proc. Nat. Acad. Sci. U.S.A. 71, 96-98 (1974).

ECKHART, W., DULBECCO, R., BURGER, M.: Temperature-dependent surface changes in cells infected or transformed by a thermosensitive mutant of polyoma virus. Proc. Nat. Acad. Sci. U.S.A. 68, 283-286 (1971).

FOSTER, D. O., PARDEE, A. B.: Transport of amino acids by confluent and nonconfluent 3T3 and polyoma virus-transformed 3T3 cells growing on glass cover slips. J. Biol. Chem. 244, 2675-2681 (1969).

GAZDAR, A., HATANAKA, M., HERBERMAN, R., RUSSELL, E., IKAWA, Y.: Effects of dibutyryl cyclic adenosine phosphate plus theophylline on murine sarcoma virus transformed non-producer cells (36930). Proc. Soc. Exp. Biol. Med. 141, 1044-1050 (1972).

GLYNN, R. D., THRASH, C. R., CUNNINGHAM, D. D.: Maximal concanavalin A-specific agglutinability without loss of density-dependent growth control. Proc. Nat. Acad. Sci. U.S.A. 70, 2676-2677 (1973).

HATANAKA, M.: Transport of sugars in tumor cell membranes. Biochim. Biophys. Acta 355, 77-104 (1974).

HOLLEY, R. W.: A unifying hypothesis concerning the nature of malignant growth. Proc. Nat. Acad. Sci. U.S.A. 69, 2840-2841 (1972).

HOLLEY, R. W.: Serum factors and growth control. In: Control of Proliferation in Animal Cells (ed. B. CLARKSON, R. BASERGA) Vol. I, pp. 13-18. New York: Cold Spring Harbor Laboratory 1974.

HOLLEY, R. W., KIERNAN, J. A.: Control of the initiation of DNA synthesis in 3T3 cells: Low molecular-weight nutrients. Proc. Nat. Acad. Sci. U.S.A. 71, 2942-2945 (1974).

INBAR, M., RABINOWITZ, Z., SACHS, L.: The formation of variants with a reversion of properties of transformed cells. III. Reversion of the structure of the cell surface membrane. Int. J. Cancer 4, 690-696 (1969).

INBAR, M., SACHS, L.: Interaction of the carbohydrate-binding protein concanavalin A with normal and transformed cells. Proc. Nat. Acad. Sci. U.S.A. 63, 1418-1425 (1969).

ISSELBACHER, K. L.: Increased uptake of amino acids and 2-deoxy-D-glucose by virus-transformed cells in culture. Proc. Nat. Acad. Sci. U.S.A. 69, 585-589 (1972).

LEY, K. D., TOBEY, R. A.: Regulation of initiation of DNA synthesis in chinese hamster cells. II. Induction of DNA synthesis and cell division by isoleucine and glutamine in G_1-arrested cells in suspension culture. J. Cell Biol. 47, 453-459 (1970).

MARTIN, G. S., VENUTA, S., WEBER, M., RUBIN, H.: Temperature-dependent alterations in sugar transport in cells infected by a temperature-sensitive mutant of Rous sarcoma virus. Proc. Nat. Acad. Sci. U.S.A. 68, 2739-2741 (1971).

MUNCK, A.: Metabolic site and time course of cortisol action on glucose uptake, lactic acid output, and glucose 6-phosphate levels of rat thymus cells in vivo. J. Biol. Chem. 243, 1039-1042 (1968).

NOONAN, K. D., BURGER, M. M.: Induction of 3T3 cell division at the monolayer stage. Exp. Cell Res. 80, 405-414 (1973).

OZANNE, B., SAMBROOK, J.: Isolation of lines of cells resistant to agglutination by concanavalin A from 3T3 cells transformed by SV40. In: The Biology of Oncogenic Viruses (ed. L. G. SILVESTRI) pp. 248-257. New York: American Elsevier Publishing Company, Inc. 1971.

PARDEE, A. B.: Cell division and a hypothesis of cancer. Nat. Cancer Inst. Monog. 14, 7-20 (1964).

PARDEE, A. B.: The surface membrane as a regulator of animal cell division. In Vitro 7, 95-104 (1971).

PARDEE, A. B.: A restriction point for control of normal animal cell proliferation. Proc. Nat. Acad. Sci. U.S.A. 71, 1286-1290 (1974).

POLLACK, R., BURGER, M.: Surface-specific characteristics of a contact-inhibited cell line containing the SV40 viral genome. Proc. Nat. Acad. Sci. U.S.A. 62, 1074-1076 (1969).

ROSEN, J. M., FINA, J. J., MILHOLLAND, R. J., ROSEN, F.: Inhibitory effect of cortisol in vitro on 2-deoxyglucose uptake and RNA and protein metabolism in lymphosarcoma P1798. Cancer Res. 32, 350-355 (1972).

RUBIN, H., FODGE, D.: Interrelationships of glycolysis, sugar transport and the initiation of DNA synthesis in chick embryo cells. In: Control of Proliferation in Animal Cells (ed. B. CLARKSON, R. BASERGA) pp. 801-816. New York: Cold Spring Harbor Laboratory 1974.

SEFTON, B. M., RUBIN, H.: Release from density dependent growth inhibition by proteolytic enzymes. Nature 227, 843-845 (1970).

SEFTON, B. M., RUBIN, H.: Stimulation of glucose transport in cultures of density-inhibited chick embryo cells. Proc. Nat. Acad. Sci. U.S.A. 68, 3154-3157 (1971).

THRASH, C. R., CUNNINGHAM, D. D.: Stimulation of division of density-inhibited fibroblasts by glucocorticoids. Nature 242, 399-401 (1973).

THRASH, C. R., CUNNINGHAM, D. D.: Dissociation of increased hexose transport from initiation of fibroblast proliferation. Nature (in press).

TODARO, G. J., LAZAR, G. K., GREEN, H.: The initiation of cell division in a contact-inhibited mammalian cell line. J. Cell Comp. Physiol. 66, 325-334 (1965).

TODARO, G., MATSUYA, Y., BLOOM, S., ROBBINS, A., GREEN, H.: Stimulation of RNA synthesis and cell division in resting cells by a factor present in serum. In: Growth Regulating Substances for Animal Cells in Culture (ed. V. DEFENDI, M. STOKER) pp. 87-101. Philadelphia: Wistar Institute Press 1967.

VAHERI, A., RUOSLAHTI, E., NORDLING, S.: Neuraminidase stimulates division and sugar uptake in density-inhibited cell cultures. Nature New Biology 238, 211-212 (1972).

WALLACH, D. F. H.: Cellular membranes and tumor behavior: A new hypothesis. Proc. Nat. Acad. Sci. U.S.A. 61, 868-874 (1968).
WEBER, J.: Relationship between cytoagglutination and saturation density of cell growth. J. Cell Physiol. 81, 49-53 (1973).
WEBER, M. J.: Hexose transport in normal and in Rous sarcoma virus-transformed cells. J. Biol. Chem. 248, 2978-2983 (1973).
WEBER, M. J., EDLIN, G.: Phosphate transport, nucleotide pools, and ribonucleic acid synthesis in growing and in density-inhibited 3T3 cells. J. Biol. Chem. 246, 1828-1833 (1971).
WIEBEL, F., BASERGA, R.: Early alterations in amino acid pools and protein synthesis of diploid fibroblasts stimulated to synthesize DNA by addition of serum. J. Cell. Physiol. 74, 191-202 (1969).

Characterization of the Thyrotropin Receptor and Its Involvement in Exophthalmos*

Leonard D. Kohn and Roger J. Winand

Section on Biochemistry of Cell Regulation, Laboratory of Biochemical Pharmacology, National Institute of Arthritis, Metabolism, and Digestive Diseases, National Institutes of Health, Bethesda, Maryland 20014, and Département de Clinique et de Sémiologie Médicales, Institut de Médecine, Université de Liège, B4000 Liège, Belgium

INTRODUCTION

Two theories have been advanced to explain the exophthalmos of Graves' disease. The earlier theory suggested that exophthalmos was caused by the exaggerated production of a pituitary factor whereas the more recent theory implicated an autoimmune mechanism. The first theory was based on the detection of exophthalmogenic activity in anterior pituitary extracts and partially purified thyrotropin (TSH) preparations (DOBYNS, 1966). The second was based on the detection of a "long-acting thyroid stimulator" (LATS) in the sera of exophthalmic patients more frequently than in the sera of hyperthyroid patients without this ophthalmopathy (McKENZIE, 1968). LATS was present in the gamma globulin fraction of these sera and exhibited characteristics of an antibody (KRISS, PHESHAKOV, ROSENBLUM, 1967; SOLOMON, BEALL, 1968).

Unfortunately, neither of these theories was completely satisfactory. In the first case, a factor of pituitary origin, i.e., a high level of TSH, has not been demonstrated in the blood of exophthalmic patients (BATES, 1963). In the second, LATS has not been shown to cause exophthalmos in experimental animals (WINAND, 1970a), and a significant number of patients with malignant exophthalmos have no detectable LATS activity in their sera (McKENZIE, 1968; WINAND, 1970b).

Our own studies of experimental exophthalmos (WINAND, KOHN, 1970; KOHN, WINAND, 1971; WINAND, KOHN, 1972, 1973a, b; AMIR et al., 1973; KOHN, WINAND, in press; KOHN, WINAND, unpublished manuscript; TATE et al., unpublished manuscript [a], [b], [c]; WINAND, KOHN, unpublished

---

*Abbreviations: TSH, thyroid stimulating hormone or thyrotropin; EPF, the exophthalmos producing factor isolated from partial pepsin digests of purified TSH preparations; LH, luteinizing hormone; HCG, human chorionic gonadotropin; ACTH, adrenocorticotropic hormone; LIS, lithium diiodosalicylate; MES, (2-N-morpholino)ethane sulfonic acid; Hepes, N-2-hydroxyethylpiperazine-N'-2-ethane sulfonic acid.

manuscript [a], [b]; WINAND, SALMON, LAMBERT, 1971) suggested a new mechanism for human exophthalmos that incorporates features of both the "pituitary factor" and "autoimmune" theories. Specifically, we suggested that exophthalmos is a two-factor disease caused by the simultaneous presence of a derivative of the TSH molecule with no thyroid-stimulating ability and by the presence of an abnormal or autoimmune gamma globulin that is not LATS (WINAND, KOHN, 1972, 1973b; KOHN, WINAND, in press). The TSH derivative serves as the direct effector and the gamma globulin serves to "guide" this effector by increasing its affinity to a TSH receptor on retro-orbital tissues; i.e., a TSH receptor on a tissue that is not normally considered to be a target for this hormone can be functionally activated by the presence of an autoimmune gamma globulin.

This hypothesis was based on data we have presented earlier and can be summarized as follows:

(1) Purified and homogeneous preparations of bovine TSH have both thyroid stimulating and exophthalmogenic activities in experimental animals (WINAND, KOHN, 1970).

(2) The TSH molecule can be fragmented by partial pepsin digestion to yield a derivative (EPF) that retains its exophthalmogenic activity but has lost its thyroid stimulating activity (KOHN, WINAND, 1971). This derivative is composed of an intact or nearly intact β-TSH subunit but only an amino terminal fragment of the α-TSH subunit (KOHN, WINAND, unpublished manuscript).

(3) This derivative (EPF), as well as TSH, can cause exophthalmos in guinea pigs, a mammalian model of human exophthalmos. In retro-orbital tissues, the exophthalmos is associated with adenylate cyclase stimulating, with increased sulfate transport and activation to PAPS (3'-phosphoadenosine 5' phosphosulfate) and with increased glycosaminoglycan synthesis (WINAND, KOHN, 1973a).

(4) The retro-orbital tissue plasma membranes from guinea pigs contain specific TSH receptors that can bind either TSH or its exophthalmogenic derivative (EPF) (WINAND, KOHN, 1972; TATE et al., unpublished manuscript [a]).

(5) Gamma globulin in the sera of patients with exophthalmos can produce exophthalmos in test animals (KOHN, WINAND, in press; WINAND, SALMON, LAMBERT, 1971), and in vitro this gamma globulin increases the specific binding of both TSH and the TSH derivative (EPF) to TSH receptors on retro-orbital tissue

plasma membranes (WINAND, KOHN, 1972; TATE et al., unpublished manuscript [a]) but not to TSH receptors on thyroid plasma membranes (AMIR et al., 1973; TATE et al., unpublished manuscript [c]).

The idea that exophthalmos might be caused by the activation of an ordinarily latent receptor via an autoimmune agent is an attractive hypothesis that satisfies many of the clinical facets of this disease. More important, however, the data raise several questions pertinent not only to exophthalmos but to receptor recognition and development in general and to the mechanism of receptor message propagation. This report reviews studies (KOHN, WINAND, unpublished manuscript; TATE et al., unpublished manuscript [a], [b], [c]; WINAND, KOHN, unpublished manuscript [a], [b]) that have been directed at an examination of the functional differences between TSH receptors in the thyroid and those in retro-orbital tissue, i.e., receptors in target and nontarget tissues, respectively, that could be correlated with the effect of the autoimmune gamma globulin on binding. The studies describe a structural difference that can be correlated with the gamma globulin effect, the mechanism by which the gamma globulin increases binding, and correlative studies on the solubilization and purification of the TSH receptor. The proposed two-factor mechanism is discussed in terms of its extendability to a therapeutic attack on exophthalmos (WINAND, MAHIEU, 1973) and in terms of its more general applicability to abnormal receptor responses and disease states.

RESULTS

Binding Properties of Thyrotropin Receptors on Thyroid and Retro-Orbital Tissue Membranes

The properties of TSH binding to thyroid and retro-orbital tissue plasma membranes were studied to compare the functional properties of the receptors of a target organ with the receptors of a tissue not usually considered as responsive to TSH. In the studies of binding, both tritiated ([^{3}H]TSH) and iodinated ([^{125}I]TSH) preparations of TSH were used, as well as a tritiated preparation of the exophthalmogenic derivative of the TSH molecule ([^{3}H]EPF), each of which was biologically active (Table 1).

The binding of [^{125}I]TSH, [^{3}H]TSH, and [^{3}H]EPF to guinea pig retro-orbital tissue plasma membranes had the same pH optimum and the same pH dependency (Figure 1) as the binding of [^{125}I]TSH or [^{3}H]TSH

Table 1. Properties of the labeled TSH preparations

Property	Unlabeled TSH	$[^3H]$TSH	$[^{125}I]$TSH Nonenzymatically labeled[a]	$[^{125}I]$TSH Enzymatically labeled[b]
Thyroid stimulating activity (I.U./mg)[c]	24 ± 4	21 ± 4	16 ± 3	19 ± 4
Adenylate cyclase stimulation[d]	Same dose response between 10^{-8} and 10^{-5} M			
Thyroid cell differentiation in tissue culture[e]	+	+	+	+
Specific radioactivity (Ci/mM)[f]	-	0.2-0.5	50-500	50-500

[a]Prepared using the procedure of LESNIAK et al. (1973).

[b]Prepared using a lactoperoxidase procedure (MIYACHI et al., 1972).

[c]Measured in the McKenzie bioassay (McKENZIE, 1958).

[d]Measured as described (KOHN, WINAND, unpublished manuscript; WINAND, KOHN, unpublished manuscript [a]; KRISHNA, WEISS, BRODIE, 1968; WOLFF, JONES, 1971; WOLFF, COOK, 1973); see Figure 1.

[e]Thyroid cells in primary culture form pseudo follicles when exposed to TSH after trypsinization (WINAND, KOHN, unpublished manuscript [b]). Each of the labeled preparations was included in the culture medium in place of unlabeled TSH and at approximately the same concentrations, i.e., 10 mU/ml of media or approximately 0.5 μg/ml of media. Differentiation was monitored microscopically; a plus (+) indicates pseudo follicle formation (WINAND, KOHN, unpublished manuscript [b]).

[f]Measured on 4 to 10 different preparations of each type using a Beckman Model 255 liquid scintillation spectrometer with a tritium counting efficiency of 54% and a ^{125}I counting efficiency of 86%.

to bovine thyroid plasma membranes (WINAND, KOHN, 1972; AMIR et al., 1973; TATE et al., unpublished manuscript [a], [c]). In a like manner, there was a similar sensitivity to the choice and concentration of buffers (Figures 1 and 2A), either by comparison among each of the labeled preparations or by comparison with the TSH receptors of thyroid membranes (WINAND, KOHN, 1972; AMIR et al., 1973; TATE et al., unpublished manuscript [a], [c]). Optimal binding was thus in 0.025 M Tris-acetate or Tris-maleate, independent of the labeled hormone preparation being used and independent of the TSH receptors being studied, either bovine thyroid or guinea pig retro-orbital tissue.

The binding of all three hormone preparations to the guinea pig retro-orbital tissue plasma membranes was sensitive to the concentration of salts in the incubation medium and was especially sensitive if

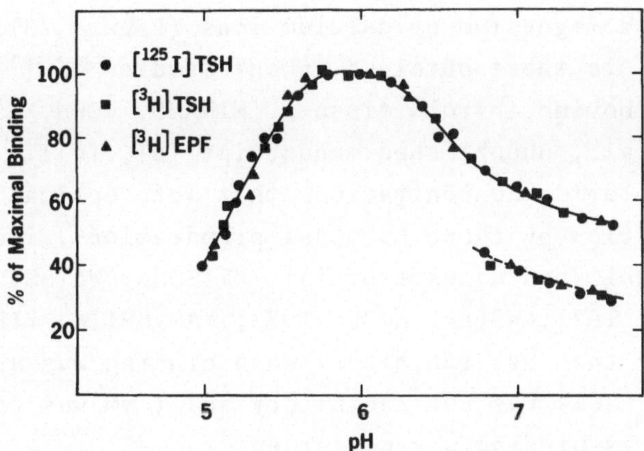

Fig. 1. Binding of [^{125}I]TSH, [^{3}H]TSH, and [^{3}H]EPF to retro-orbital tissue plasma membranes as a function of pH. Buffer in all cases was 0.025 M. For the solid curve (———), the buffers used were acetate, pH 5 to 5.8; Tris-acetate or Tris-maleate, pH 5.5 to 7.6; and Hepes (N-2-hydroxyethylpiperazine-N'-2-ethane sulfonic acid) buffer, pH 7.0 to 7.6. The dashed curve(------) represents the data using Tris-chloride buffers between 6.8 and 7.6, i.e., at the same pH values as Tris-acetate, Tris-maleate, or Hepes buffers. Binding used standard conditions and hormone concentrations of 5 x 10^{-10} M, 5 x 10^{-7} M, and 5 x 10^{-7} M for [^{125}I]TSH (●), [^{3}H]TSH (■), and [^{3}H]EPF (▲) binding. (From TATE et al., unpublished manuscript [a]; by permission of Cold Spring Harbor Laboratory, New York, and Nature Magazine.)

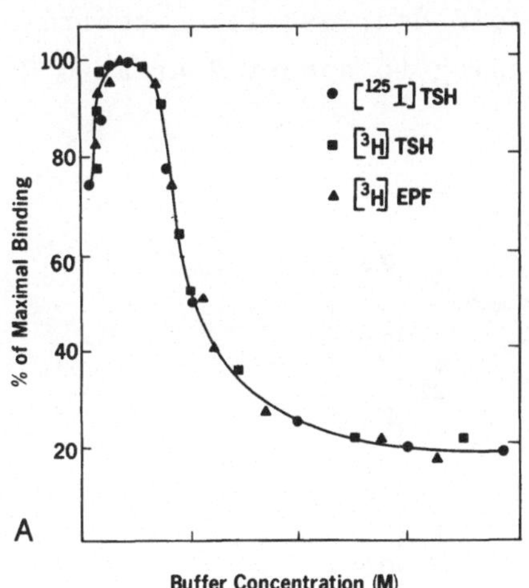

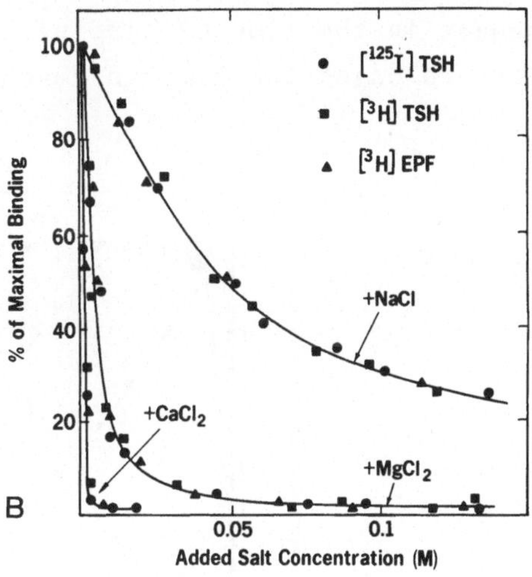

Fig. 2. Binding to retro-orbital tissue plasma membranes. (A) Binding to retro-orbital tissue plasma membranes as a function of buffer concentration. The [^{125}I]TSH (●), [^{3}H]TSH (■), and [^{3}H]EPF (▲) concentrations were the same as in Figure 1; conditions were otherwise standard; the buffer was Tris-acetate, pH 6.0 (TATE et al., unpublished manuscript [a]). (B) Effect of salts on binding to retro-orbital tissue plasma membranes. [^{125}I]TSH (●), [^{3}H]TSH (■), and [^{3}H]EPF (▲) concentrations were the same as in Figures 1 and 2A; conditions were otherwise optimal (TATE et al., unpublished manuscript [a]). The same curve as described for NaCl was obtained with KCl, NH_4Cl, LiCl, and ammonium sulfate. $ZnCl_2$ behaved in a manner between that of $CaCl_2$ and $MgCl_2$. (By permission of Cold Spring Harbor Laboratory, New York.)

these salts included magnesium or calcium ions (Figure 2B). These data are again analogous to those obtained in our studies of [^{125}I]TSH and [^{3}H]TSH binding to bovine thyroid tissues (WINAND, KOHN, 1972; AMIR et al., 1973; TATE et al., unpublished manuscript [a], [c]). In both cases, magnesium chloride concentrations that were optimal for adenylate cyclase stimulation by these hormonal preparations, i.e., 2 to 5 mM (WINAND, KOHN, unpublished manuscript [a]; KRISHNA, WEISS, BRODIE, 1968; WOLFF, JONES, 1971; WOLFF, COOK, 1973; YAMASHITA, FIELD, 1970, 1972), were greater than 90% inhibitory when binding was assayed. In both cases, 0.1 mM $CaCl_2$ was 50% inhibitory and 1 mM was completely inhibitory insofar as binding was concerned.

Binding of [^{3}H]TSH and [^{3}H]EPF to the retro-orbital tissue receptors was both immediate and maximal at either 0°C, 25°C, or 37°C (Figure 3). Continued incubation at 0°C resulted in no decrease in [^{3}H]TSH or [^{3}H]EPF binding, whereas continued incubation at 25°C and 37°C resulted in a progressive decrease in binding activity. Binding of [^{125}I]TSH was essentially immediate at 37°C; at 25°C, it took 5 min to reach the maximal level; and at 0°C it took nearly 30 min (Figure 3). Again, continued incubation at 0°C had no effect, whereas continued incubation at 25°C or 37°C caused progressive decreases in binding. Changes in the [^{125}I]TSH concentration correlated inversely with changes in the time of binding; thus, a 10-fold decrease in concentration prolonged the time to reach maximal binding to almost 2 hr at 0°C

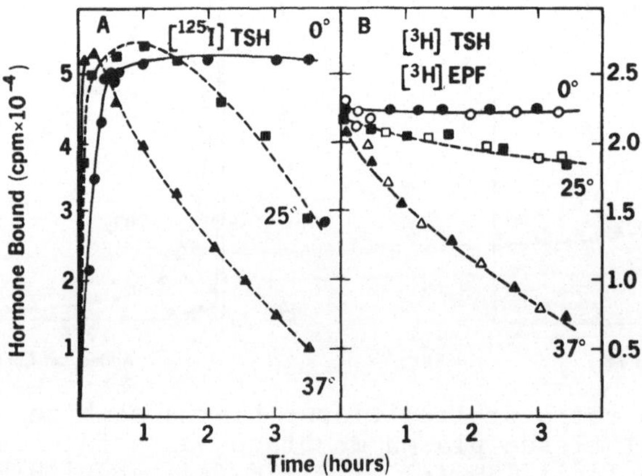

Fig. 3. Binding as a function of temperature for [^{125}I]TSH (A) and [^{3}H]TSH or [^{3}H]EPF (B). Assay conditions were optimal except for binding being performed at the noted time after incubation was initiated (TATE et al., unpublished manuscript [a]). [^{125}I]TSH was at 5×10^{-10} M in (A). In (B), [^{3}H] TSH and [^{3}H] EPF (o,□,Δ; ●,■,▲) were at 5×10^{-7} M. (By permission of Cold Spring Harbor Laboratory, New York, and Nature Magazine.)

and a 10-fold increase in concentration decreased the time to reach optimal binding to less than 10 min at 0°C.

These data are once again analogous to those described for [^{125}I]TSH binding to thyroid plasma membranes, and in both systems the temperature sensitivity appeared to be a problem of membrane stability (Table 2) rather than enzymatic destruction of the hormone (WINAND, KOHN, 1972; AMIR et al., 1973; TATE et al, unpublished manuscript [a], [c]). In both systems, the hormone receptor interaction exhibited time dependency when the concentration of one of its reactants was varied, a phenomenon to be predicted in a bimolecular binding reaction.

When examined as a function of hormone concentration, [^{125}I]TSH and [^{3}H]TSH binding to both thyroid and retro-orbital tissue plasma membranes yielded nonlinear Scatchard plots (Figures 4A and 4B) (SCATCHARD, 1949). In both cases, it was possible to interpret these

Table 2. Effect of preincubation at 37°C on membrane binding and hormone stability

Preincubation conditions	Preincubation components	[^{3}H]TSH bound after 1 min of incubation[a] (cpm)	[^{3}H]EPF bound after 1 min of incubation[a] (cpm)
None	-	24,500	23,100
0°C for 2 hr	Membranes + hormone[b]	24,900	22,800
	Membranes - hormone	23,700	23,400
	Hormone only	24,400	23,200
27°C for 2 hr	Membranes + hormone	12,000	14,200
	Membranes - hormone	6,400	6,200
	Hormone only	24,400	23,000
37°C for 2 hr and 0°C for 2 hr	Membranes + hormone	17,500	19,200
	Membranes - hormone	16,200	15,400
	Hormone only	24,900	23,600

[a]Preincubations were performed using the noted conditions of time and temperature. In a volume of 90 µl, the preincubation mixture contained the noted components as well as the albumin and buffer contained in a standard binding assay (TATE et al., unpublished manuscript [a], [c]). The binding assay was initiated by adding in a 10 µl volume either an appropriate amount of enzyme or sufficient [^{3}H]TSH and [^{3}H]EPF to yield a 5×10^{-7} M concentration. Assay conditions were optimal (TATE et al., unpublished manuscript [a], [c]) except for filtration after only 1 min of incubation rather than 1 hr.

[b]Hormone is either [^{3}H]TSH or [^{3}H]EPF; membranes are from retro-orbital tissue, but thyroid plasma membranes yield analogous data.

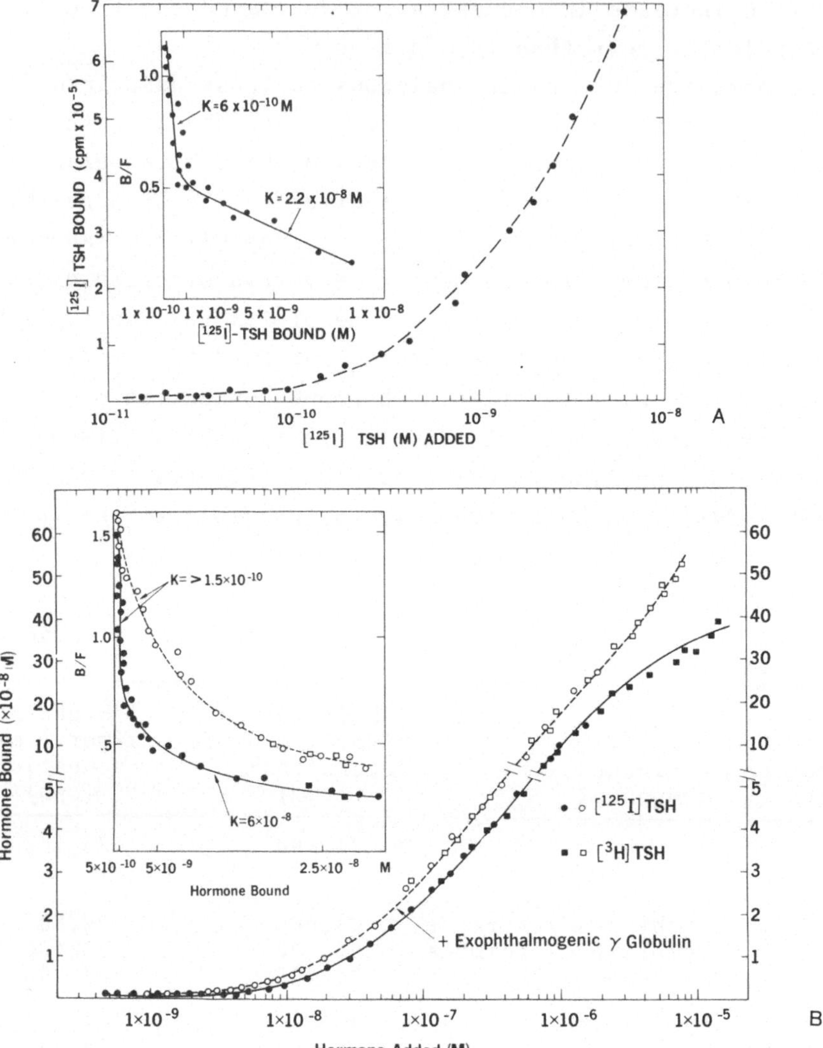

Fig. 4. Dependence of binding on hormone concentration. (A) Binding of [^{125}I]TSH to thyroid plasma membranes as a function of hormone concentration. Inserted is a Scatchard plot (SCATCHARD, 1949) of the data. Optimal assay conditions were used (TATE et al., unpublished manuscript [c]). (B) Binding of TSH to retro-orbital tissue plasma membranes as a function of hormone concentration and the effect of gamma globulin from a patient with malignant exophthalmos on [^{125}I]TSH (●,○) and [^{3}H]TSH (■,□) binding to these retro-orbital tissue plasma membranes (TATE et al., unpublished manuscript [a]). The solid curve is that obtained either in the absence of gamma globulin or in the presence of 15 µg of gamma globulin from a normal subject or a patient with Graves' disease who had no exophthalmos and was LATS positive. The dashed curve (------) represents the binding in the presence of 15 µg of gamma globulin from a patient with malignant exophthalmos. This amount of gamma globulin gave approximately 55% to 60% of the maximal stimulation of hormone binding at a TSH concentration of 5×10^{-8} M added hormone. The insert in Figure 4B displays part of these data as a Scatchard plot (SCATCHARD, 1949). (By permission of Cold Spring Harbor Laboratory, New York.)

curves as representing a heterogeneous population of receptor sites that could be grouped into two major classes--those with high affinity and those with low affinity. The apparent binding constant of high affinity groups was 6 x 10^{-10} M for the thyroid plasma membranes and a nearly identical 15 x 10^{-10} M for the retro-orbital tissue membranes; the apparent binding constants for the low-affinity groups were also similar, 2.2 x 10^{-8} M and 6 x 10^{-8} M for the two-membrane preparations, respectively. Hill plots (HILL, 1913) of the data from both membrane preparations indicated that these sites were not independent but rather exhibited a negatively cooperative (LEVITZKI, KOSHLAND, 1969) relationship since the slope of the plot was less than unity (Figure 5) (TATE et al., unpublished manuscript [a], [c]).

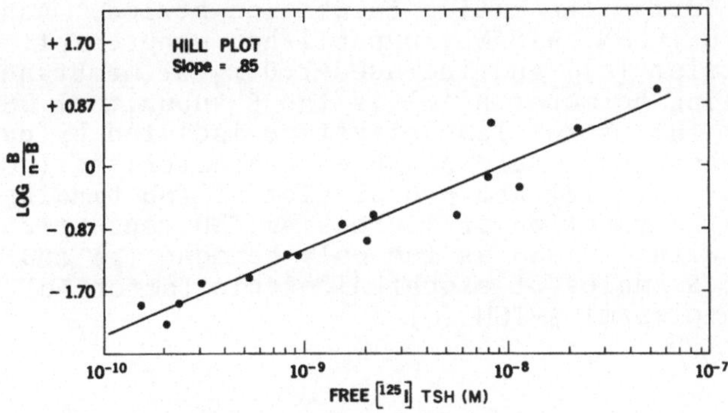

Fig. 5. Hill plot (HILL, 1913) of the data in Figure 4A. An analogous plot was obtained with data from Figure 4B

Inhibition of binding using [^{3}H]TSH or [^{3}H]EPF as the labeled hormone is relatively simple to interpret since only 2% to 3% of the hormone added is bound at a 10^{-6} M concentration of hormone. Inhibition kinetics are thus performed at saturating hormone levels where effectively all the unlabeled hormone will act as a competing agent. Under these conditions, unlabeled TSH inhibits [^{3}H]TSH binding in a clearly theoretical fashion, i.e., the addition of equal concentrations of labeled and unlabeled hormone to an incubation mixture resulted in approximately a 50% inhibition of binding (Figure 6A) (WINAND, KOHN, 1972; AMIR et al., 1973; TATE et al., unpublished manuscript [a], [c]; WINAND, SALMON, LAMBERT, 1971). To achieve the same level of inhibition required 4 to 5 times higher concentrations of the unlabeled exophthalmogenic derivative of the TSH molecule (EPF), 10 times higher concentrations of LH, 40 times greater concentrations of the β-TSH

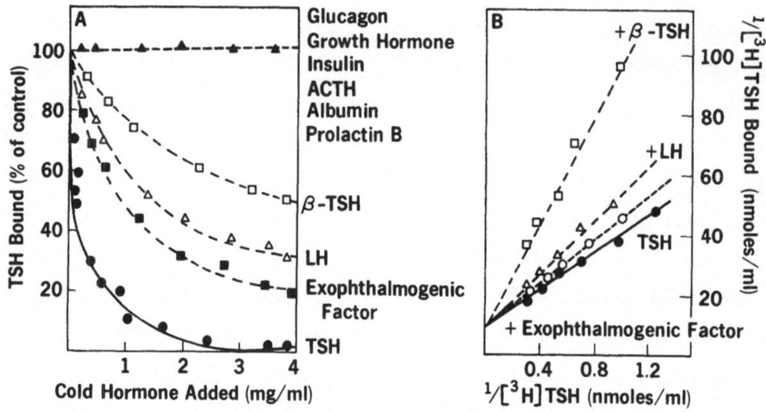

Fig. 6. The effect of unlabeled TSH (•), TSH subunits, and exophthalmogenic factor (EPF) on [^{3}H]TSH binding to thyroid plasma membranes. (A) Results are expressed as the percent of [^{3}H]TSH remaining bound when incubation mixtures contained increasing amounts of the unlabeled hormone during the entire incubation period. Assay conditions were standard (KOHN, WINAND, unpublished manuscript; TATE et al., unpublished manuscript [c]) and included 1.0 mg of membrane protein. LH (Δ) is luteinizing hormone; β (□) is the β subunit of TSH. Exophthalmogenic factor (■) is the TSH derivative isolated by partial pepsin digestion of TSH whose structure has been characterized (KOHN, WINAND, unpublished manuscript. (B) Reciprocal plot of TSH binding to thyroid plasma membranes as a function of increasing TSH concentration. Binding was performed with [^{3}H]TSH as the only hormone (•) and with [^{3}H]TSH in the presence of 5 nmoles of exophthalmogenic factor (o), 4 nmoles/ml of LH (Δ), or 60 nmoles/ml β-TSH (□)

subunit, and over 200 times the concentration of the α-TSH subunit. Double reciprocal plots of the [^{3}H]TSH binding data in the range of 5×10^{-7} M to 5×10^{-6} M indicated that the exophthalmogenic factor, LH, the β-TSH subunits, and the α-TSH subunits were competitive inhibitors of [^{3}H]TSH binding (Figure 6B).

In the case of [^{125}I]TSH, however, inhibition is performed under nonsaturating conditions, where as much as 40% to 50% of the total hormone added to the incubation can be bound even after the addition of unlabeled hormone; accordingly, nonlinear inhibition curves were obtained. This is evident in Figure 7, where [^{125}I]TSH binding is measured in the presence of unlabeled TSH, unlabeled luteinizing (LH) hormone, and the unlabeled subunits of TSH (TATE et al., unpublished manuscript [c]). When, however, corrections are made for the [^{125}I]TSH bound by considering the total of the labeled and unlabeled TSH concentration and the percent of this total that should be bound, a theoretic curve can be derived. As noted in Figure 7, the theoretic curve (dashed line) derived for unlabeled TSH inhibition of [^{125}I]TSH binding encompassed the actual data points reasonably well.

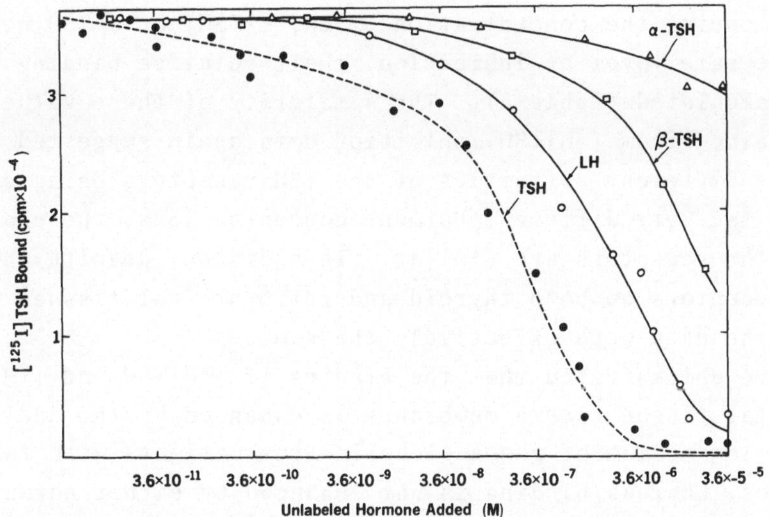

Fig. 7. Inhibition of [^{125}I]TSH binding to thyroid plasma membranes by unlabeled TSH (•), LH (○), β-TSH (□), α-TSH (Δ), and either albumin, prolactin, ACTH (adrenocorticotropic hormone), insulin, glucagon, or growth hormone (□) (TATE et al., unpublished manuscript [c]). Assay conditions were optimal; [^{125}I]TSH was at 1.5 x 10^{-10} M. The dashed line drawn through the binding values obtained in the presence of unlabeled TSH is a theoretical curve, derived by compensating for the amount of labeled hormone that would be bound at the concentrations of total hormone, labeled and unlabeled, present in the assay mixture. This is an important correction, since in the 1 x 10^{-9} M total concentration range, over 40% of the total hormone added continues to be bound under these conditions

Table 3. Relative affinities of EPF, LH, and the subunits of TSH to retro-orbital tissue and thyroid TSH receptors

Hormonal factor	Relative affinity to retro-orbital tissue TSH receptors		Relative affinity to bovine thyroid TSH receptors	
	[^{3}H]TSH data[a]	[^{125}I]TSH data[b]	[^{3}H]TSH data[c]	[^{125}I]TSH data[c]
TSH	100	100	100	100
EPF	25	17	16	-
LH	11	9	10	12.5
β-TSH	2.5	2.6	2.1	2.5
α-TSH	<0.5	<0.5	<0.5	0.4

[a]Data are derived from plots analogous to those in Figure 6 by comparing the concentrations necessary to yield 50% inhibition.

[b]Data are derived from [^{125}I]TSH inhibition curves analogous to those in Figure 7. The values were obtained by comparing 50% inhibition concentrations.

[c]Data taken from Figures 6 and 7 by comparing concentrations necessary to yield 50% inhibition.

By comparing the concentration of LH, β-TSH, or α-TSH necessary to achieve the same level of inhibition, their relative binding affinities could be calculated (Table 3). The similarity of these values to the values obtained from [^{3}H]TSH inhibition data again suggested that, despite the different affinities of the TSH receptors being examined with these two very different hormone concentrations, the characteristics of the receptors are similar. In addition, despite the fact that TSH receptors on both thyroid and retro-orbital tissues were examined, the data were effectively the same.

We have demonstrated that the binding of [^{3}H]TSH and [^{3}H]EPF to retro-orbital tissue plasma membranes is enhanced by the addition to the incubation medium of gamma globulin from patients with malignant exophthalmos, whereas binding is not enhanced by either normal gamma globulin or gamma globulin from patients with Graves' disease but no exophthalmos (WINAND, KOHN, 1972; AMIR et al., 1973; TATE et al., unpublished manuscript [a], [c]). Double reciprocal plots of such data indicate that there is both an increase in affinity and an increase in the number of sites binding [^{3}H]TSH or [^{3}H]EPF at these high concentrations of hormone (>1 x 10^{-6} M), i.e., there is both an increase in K_m values and a twofold increase in TSH maximal binding and a fourfold to fivefold increase in exophthalmogenic factor binding when those data are extrapolated toward infinite concentrations of hormone. In contrast to these results, the exophthalmogenic gamma globulin preparations had no effect or actually inhibited [^{3}H]TSH binding to thyroid plasma membranes (WINAND, KOHN, 1972; AMIR et al., 1973; TATE et al., unpublished manuscript [a], [c]).

Studies of the effect of the gamma globulin from malignant exophthalmos on the binding of [^{125}I]TSH and [^{3}H]TSH to the retro-orbital membrane receptors demonstrated the effect of the abnormal globulin even at hormone concentrations in the 10^{-8} M to 10^{-9} M range (Figure 4B). As can be seen in the Scatchard plot (SCATCHARD, 1949) of these data, there appears to be a general shift of binding sites toward higher affinities. Thus, whereas a group of high affinity sites with association constants of 15 x 10^{-10} M or lower can still be recognized, the lower affinity sites that could be grouped into a class with an association constant of 6 x 10^{-8} M is no longer recognizable. Instead, there is the appearance of a large number of sites with progressively increasing affinities in this area of the curve. In summary, these data indicate that the presence of the gamma globulin from malignant exophthalmic patients in these incubation mixtures results in an increase in the number of sites that can recognize the hormone and an

Table 4. Binding of [^{125}I]gamma globulin to retro-orbital tissue membranes

Unlabeled additions to the incubation containing [^{125}I]gamma globulin	^{125}I labeled gamma globulin added	
	[^{125}I]gamma globulin from a patient with malignant exophthalmos[a] (cpm)	[^{125}I]gamma globulin from an LATS positive patient with Graves' disease but *no* exophthalmos[b] (cpm)
None	890	920
+ exophthalmogenic gamma globulin, 30 µg	450	530
+ normal gamma globulin, 30 µg	480	515
+ TSH[c]	12,400	840
+ EPF[d]	15,200	950
+ TSH and exophthalmogenic gamma globulin, 30 µg .	6,800	490
+ EPF and exophthalmogenic gamma globulin, 30 µg .	7,400	520
+ TSH and normal gamma globulin, 30 µg	11,600	590
+ EPF and normal gamma globulin, 30 µg	14,500	510

[a]The [^{125}I]gamma globulin from a malignant exophthalmos patient was biologically active in the fish bioassay (BROUHON-MASSILLON, 1960; DEDMAN, FAWCETT, MORRIS, 1967). It was incubated with retro-orbital tissue plasma membranes using the same optimal conditions as described for TSH binding assays with the exception that no labeled TSH was included in the incubation (TATE et al., unpublished manuscript [a], [c]). Thirty micrograms of the labeled gamma globulin were present in the 100 µl incubation mixture; i.e., it was twice the concentration in Figure 4B and was at its maximal point of stimulation. The [^{125}I]gamma globulin had 230,000 cpm/30 µg.

[b]The [^{125}I]gamma globulin from an LATS positive Graves' disease patient without exophthalmos was prepared as described (TATE et al., unpublished manuscript [a]) and was biologically active in a LATS assay (McKENZIE, 1968). It was incubated in the same way and at the same concentration as the malignant exophthalmos gamma globulin. Thirty micrograms had 290,000 cpm. Analogous data to those presented in this column were obtained if ^{125}I normal gamma globulin was used at the same concentration and similar specific activity.

[c]TSH was added at 5×10^{-7} M.

[d]EPF was added at 5×10^{-7} M.

increase in their affinity to the hormone or an alteration of the negatively cooperative relationship between the sites.

The mechanism by which the gamma globulin acted in this process was investigated (TATE et al., unpublished manuscript [a], [c]), using [^{125}I]gamma globulin as the binding agent rather than labeled hormone (Table 4). Alone, the [^{125}I]gamma globulin from patients with malignant exophthalmos did not bind better than either gamma globulin from normals or from Graves' disease patients without exophthalmos, and whatever did bind could be prevented by the addition of unlabeled normal gamma globulin to the incubation mixture (Table 4). In contrast, the addition of either unlabeled TSH or its unlabeled exophthalmogenic derivative to the incubation resulted in significant [^{125}I]gamma globulin binding in incubations containing the malignant exophthalmogenic gamma globulin preparations but not in incubations containing the other gamma globulin preparations. Normal gamma globulin and gamma globulin from Graves' disease patients without exophthalmos could not significantly inhibit this binding, whereas unlabeled gamma globulin from either the same or a different patient with malignant exophthalmos was able to compete.

In sum, then, with the exception of this gamma globulin effect, the TSH receptors on bovine thyroid plasma membranes and those on retro-orbital tissue plasma membranes functioned effectively identically in vitro independent of the labeled hormone preparation used, [^{3}H]TSH or [^{125}I]TSH, and independent of the hormone concentration used to test these properties, 10^{-6} M or 10^{-10} M, respectively. Both receptor preparations had effectively the same pH and buffer optima, the same salt inhibition phenomena, the same temperature optima, the same nonlinearity of binding as measured in Scatchard plots, and the same specificities toward hormone analogs. Functionally they could be distinguished only by the stimulatory effect of gamma globulin from exophthalmos patients on TSH or EPF binding to retro-orbital tissue TSH receptors.

Structural Differences between the TSH Receptors from Retro-Orbital and Thyroid Plasma Membranes

During the course of our studies on the solubilization of the TSH receptor from thyroid membranes (TATE et al., unpublished manuscript [b]), it was noted that trypsin did not destroy binding activity whereas other proteases or protease mixtures did. In contrast, trypsin at the same concentration did destroy the binding of TSH to receptors on intact thyroid plasma membranes (Figure 8A). This discrepancy was

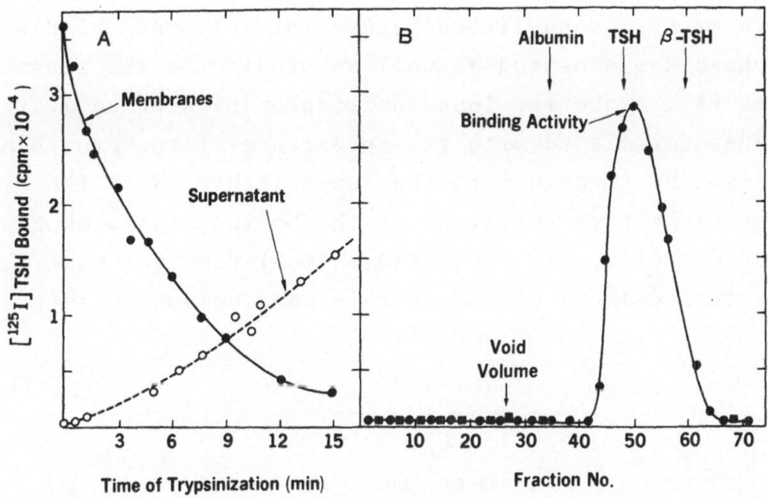

Fig. 8A. Loss of binding activity in thyroid plasma membranes (•) after exposure to trypsin and the release of "solubilized" receptor activity (o) into the supernatant solution (TATE et al., unpublished manuscript [c]). Membranes were exposed to TPCK trypsin at room temperature, at a 20:1 weight ratio of membranes to trypsin, and in an 0.025 M Tris-maleate buffer, pH 7.8. At the noted times, aliquots were removed and mixed with an amount of cold soybean trypsin inhibitor (Calbiochem) fivefold in excess of the trypsin concentration, and the suspensions were chilled to 2°C to 4°C. One aliquot from this suspension was immediately assayed for binding activity using optimal conditions and a 1.5×10^{-10} M [^{125}I]TSH concentration (•); a second aliquot was centrifuged using a Beckman microfuge, and the supernatant was assayed (o) using a modified binding assay developed for our studies of solubilized TSH receptors (TATE et al., unpublished manuscript [b]). Controls included duplicate incubations and assays containing trypsin or trypsin plus trypsin inhibitor at identical concentrations but either in the absence of membranes or with membranes added at the time of assay

Fig. 8B. Chromatography on Sephadex G100 of the TSH receptor activity released into the supernatant by trypsinization of bovine thyroid plasma membranes (TATE et al., unpublished manuscript [c]). Binding activity (•) was measured on aliquots of the noted fractions using optimal conditions for the solubilized receptor assay (TATE et al., unpublished manuscript [b]) and 1.5×10^{-10} M [^{125}I]TSH. Solubilized receptor activity from bovine plasma membranes was prepared as described in Figure 8A; the supernatant after 15 min of a large-scale incubation was used. The elution peaks of the three markers, albumin, TSH, and β-TSH, are noted by the arrows; their molecular weights are 67,000, 27,000 to 28,000, and 13,000 to 14,000, respectively. The column was 50 cm x 0.9 cm in size and was equilibrated and eluted with 0.025 M Tris-acetate, pH 6.0. The elution rate was 20 ml/hr. Fractions of 0.4 ml were collected. The void volume is the peak of elution of a dextran blue dye marker of >200,000 mol wt resolved (TATE et al., unpublished manuscript [c]) when binding in the supernatant phase was measured as well as binding to the plasma membranes (Figure 8A), since the loss in receptor binding activity on the plasma membranes correlated with the appearance of receptor binding activity released by trypsin into the supernatant. When the binding activity released by trypsinization of the bovine plasma membranes was eluted on Sephadex G100, its estimated molecular weight was 15,000 to 30,000, based on standards eluted on this same column (Figure 8B).

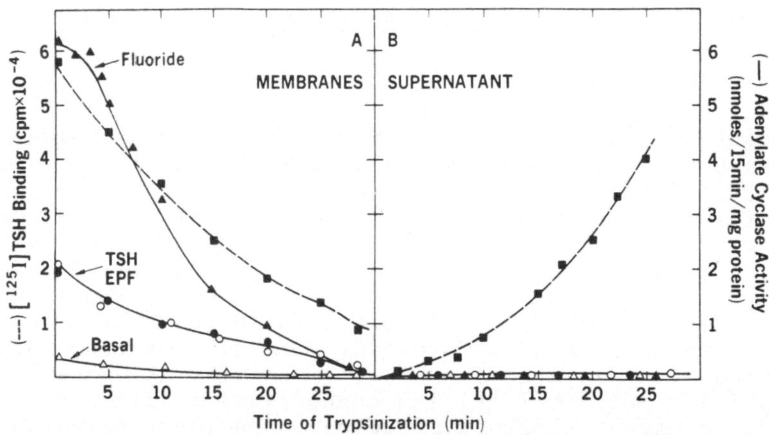

Fig. 9A. Effect of trypsin on TSH (o), EPF (•), and fluoride (▲), stimulated adenylate cyclase activity (———) of retro-orbital tissue plasma membranes compared to the effect of trypsin on TSH binding (■-----■) to these membranes (WINAND, KOHN, unpublished manuscript [a]). Membranes were exposed to TPCK trypsin at room temperature at a 20:1 ratio of membranes to trypsin and in 0.025 M Tris-maleate, pH 7.8. At the noted times, aliquots were removed and mixed with an amount of cold soybean trypsin inhibitor (Calbiochem) that was fivefold in excess of the trypsin concentration and the suspensions were chilled from 2°C to 4°C. Aliquots from this suspension were assayed for [^{125}I]TSH binding (■), for their TSH (o), EPF (•), and fluoride (▲) stimulated adenylate cyclase activities, and for their basal (Δ) adenylate cyclase activities. The binding and adenylate cyclase assays were the same as have been described (TATE et al., unpublished manuscript [a], [c]; WINAND, KOHN, unpublished manuscript [a]). Fluoride was at a 10 mM concentration

Fig. 9B. Recovery of TSH binding activity (■) in the supernatant solutions from tryptic digests of retro-orbital tissue membranes in the absence of recoverable TSH or EPF stimulable adenylate cyclase activity. Aliquots of the tryptic digests described in Figure 9A were centrifuged in a Beckman microfuge and the supernatant solutions were assayed for [^{125}I]TSH binding (■), for TSH (o), EPF (•), and fluoride (▲) stimulated adenylate cyclase activities, and for basal (Δ) adenylate cyclase activity. Assays were the same as in Figure 9A with the exception of the binding assay; this used a solubilized receptor assay described in an accompanying report (TATE et al., unpublished manuscript [b])

Analogous trypsinization of retro-orbital tissue plasma membranes (TATE et al., unpublished manuscript [a]) also destroys their TSH and exophthalmogenic derivative (EPF) binding activities and releases specific binding activity into the media (Figs. 9A,9B). In contrast to the thyroid receptor released, however, the receptor activity released by trypsinization of retro-orbital tissue membranes eluted at the front of a G100 column (Figure 10) (mol wt >75,000) and on sucrose gradients had a molecular weight of between 100,000 and 150,000. In both cases, the trypsin-solubilized receptor activity eluted from the columns was specific for [^{125}I]TSH binding, in that analogous results to those in Figure 6 were obtained when binding activity was measured in the presence of unlabeled TSH. Albumin, ACTH, prolactin, human growth hormone, and insulin did not displace [^{125}I]TSH or [^{3}H]TSH bound to the trypsin-released receptor. Muscle cell and adrenal cell plasma membranes (MIYACHI et al., 1972) similarly trypsinized yielded no specific TSH binding component in the supernatant solutions.

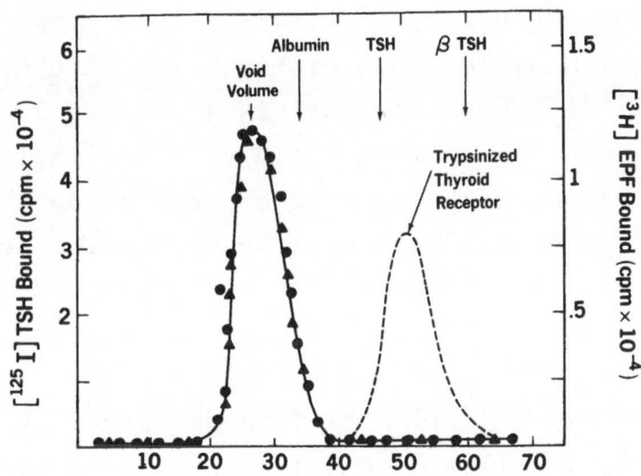

Fig. 10. Chromatography on Sephadex G100 of the TSH receptor activity released into the supernatant and by trypsinization of retro-orbital tissue plasma membranes (TATE et al., unpublished manuscript [a]). Binding activity was measured under optimal conditions for the solubilized receptor assay (TATE et al., unpublished manuscript [b]) and using both [^{125}I]TSH (•) and [^{3}H]EPF (▲) at 5×10^{-10} M and 5×10^{-7} M concentrations. Solubilized receptor activity was prepared as described in Figure 9; the supernatant after 20 min of a large-scale incubation was used. The column was 50 cm x 0.9 cm in size and was equilibrated with 0.025 M Tris-acetate, pH 6.0. The elution rate was 20 ml/hr. Fractions of 0.4 ml were collected. The void volume is the peak of elution of a dextran blue dye marker >200,000 in mol wt; albumin (67,000), TSH (27,000 to 28,000), and β-TSH (13,000 to 14,000) were markers used to calibrate the column. The elution area of the trypsin solubilized TSH receptor fragment described in Figure 8B is noted by the dashed line (------). The same data were obtained if the column was eluted in 0.001 M bicarbonate, pH 9.5, containing 0.1% Triton X100

Solubilization and Purification of the 15,000 to 30,000 TSH Receptor Fragment

The TSH receptor could be solubilized (TATE et al., unpublished manuscript [c]) by treatment with 0.1 M lithium diodosalicylate. The pH optimum of TSH binding to solubilized TSH receptors, pH 5.5 to 6.0, was similar to the optimum described for TSH binding to TSH receptors on thyroid or retro-orbital tissue (Figure 1) (DOBYNS, 1966; WINAND, KOHN, 1972). The pH curve was, however, slightly different, in that binding was slightly less sensitive to lower pH buffers and more sensitive to higher pH buffers. Optimal binding was in either MES, Tris-acetate, or Tris-maleate, pH 5.5 to 6.0.

Optimal buffer concentration was broader in range and salt inhibition phenomena were less stringent for TSH binding to the solubilized receptors than for TSH binding to the receptors in thyroid and retro-orbital tissue membranes (Figures 2A and 3) (WINAND, 1970a; WINAND, KOHN, 1972). Thus optimal buffer concentration was between 0.01 M and 0.075 M rather than between 0.01 M and 0.03 M; 50% inhibition by sodium chloride occurred at 0.5 M rather than 0.05 M; and 50% inhibition by magnesium chloride occurred at 2.5 mM rather than 1 mM.

Binding of $[^3H]$TSH was maximal within 1 min at 0°C, 25°C, and 37°C. Binding of $[^{125}I]$TSH was maximal in 45 min to 1 hr at 0°C, in 5 to 10 min at 25°C, and in less than 1 min at 37°C. Both $[^{125}I]$TSH and $[^3H]$TSH binding decreased with the time of incubation at 37°C, but these effects were again much less prominent than the decreases in binding activity previously described in thyroid and retro-orbital tissue plasma membranes incubated at these same temperatures (Figure 5) (WINAND, KOHN, 1972).

Whereas $[^3H]$TSH and $[^{125}I]$TSH binding was linear with respect to the concentration of solubilized membrane protein in the assay, $[^3H]$TSH and $[^{125}I]$TSH binding as a function of hormone concentration yielded a nonlinear Scatchard plot (Figure 11) (SCATCHARD, 1949) analogous to that obtained when binding was measured with thyroid plasma membrane TSH receptors (WINAND, KOHN, 1972; AMIR et al., 1973; TATE et al., unpublished manuscript [a], [c]). Again the receptor sites could be grouped into two general classes having high affinities: ($K = 5 \times 10^{-10}$ M) and low affinities ($K = 2 \times 10^{-8}$ M).

Analogous LIS solubilization of bovine muscle plasma membranes yielded no specific binding in the LIS supernatant. Binding was specific for the solubilized TSH receptor, in that cold TSH could compete for labeled TSH binding or chase-labeled TSH from the receptor, whereas albumin, ACTH, glucagon, insulin, prolactin, parathyroid hormone, and

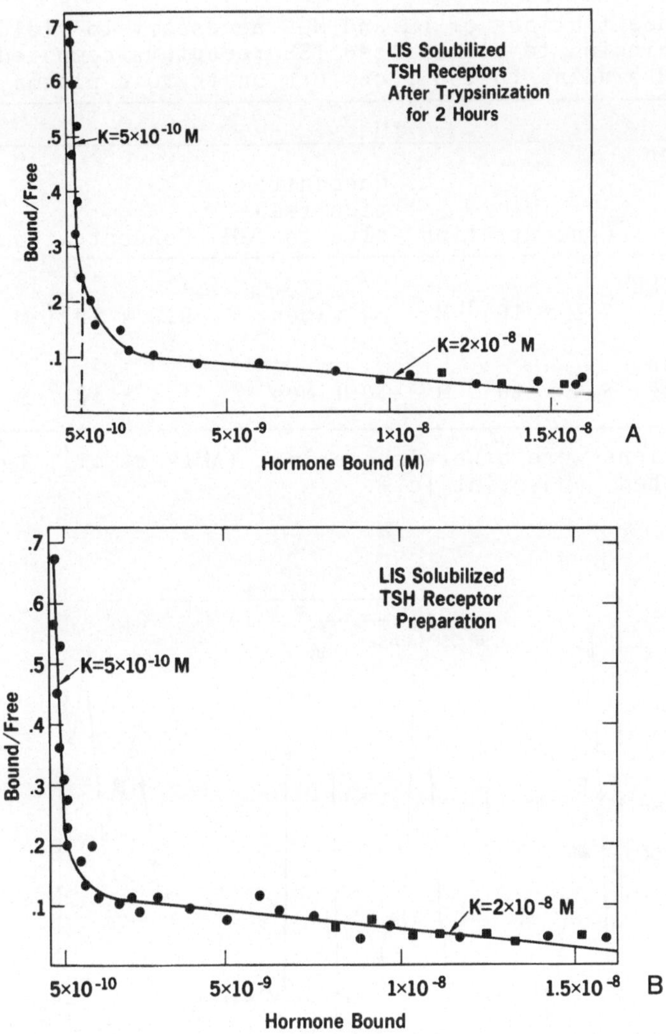

Fig. 11A. Scatchard plot (SCATCHARD, 1949) of [^{125}I]TSH (●) and [^{3}H]TSH (■) binding to the LIS solubilized TSH receptor preparation

Fig. 11B. Scatchard plot (SCATCHARD, 1949) of [^{125}I]TSH (●) and [^{3}H]TSH (■) binding to the 15,000 to 30,000 mol wt receptor fragment isolated from Sepharose chromatography of the LIS solubilized TSH receptors after 2 hr of trypsin treatment

human growth hormone could not. Of interest was the effect of luteinizing hormone (LH) and human chorionic gonadotropin (HCG) on TSH binding to the solubilized TSH receptor as opposed to the TSH receptor on plasma membranes (Table 5). Whereas 10 times more LH than TSH was necessary to yield 50% inhibition of TSH binding to the plasma membranes, only 4 to 5 times more was necessary to yield 50% inhibition of TSH binding to the solubilized receptor. Nearly 20 times more HCG than

Table 5. Concentrations of LH and HCG necessary to yield 50% inhibition of TSH binding to solubilized TSH receptors compared to their effect on TSH binding to TSH receptors on thyroid plasma membranes

50% Inhibition of binding 0.5×10^{-7} M [^{3}H]TSH[a]	LH		HCG	
	Concentration	Concentration relative to TSH	Concentration	Concentration relative to TSH
Solubilized TSH receptors[a]	2.3×10^{-7} M	4 times	9.8×10^{-7} M	20 times
Plasma membrane TSH receptors[a]	5.4×10^{-7} M	10 times	1.3×10^{-7} M	2-3 times

[a]Assay conditions were otherwise optimal (AMIR et al., 1973; TATE et al., unpublished manuscript [c]).

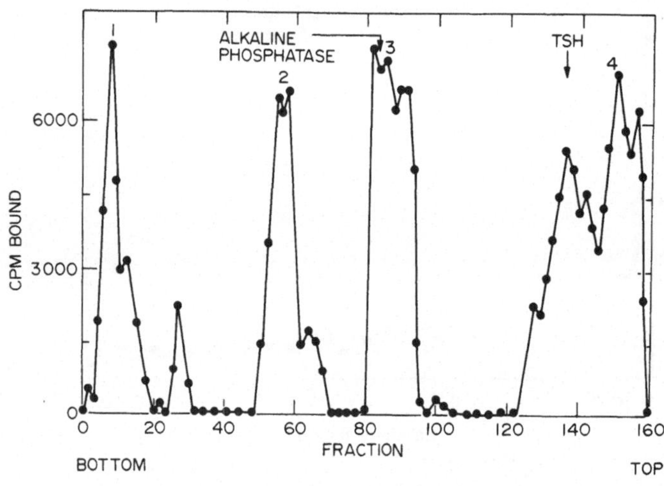

Fig. 12. Sucrose gradient centrifugation of the LIS solubilized TSH receptor preparation. The gradient was 5% to 20% sucrose, its formation and centrifugation in an SW 50.1 rotor for 8 hr at 50,000 rpm was the same as described by MARTIN and AMES (1961). The location of TSH and alkaline phosphatase in the same and duplicate gradients are noted with arrows; their molecular weights are 27,500 and 77,500, respectively. Fractions of 0.087 ml were collected and assayed for soluble binding activity (TATE et al., unpublished manuscript [b]). The background counts from duplicate gradients without added receptors were 2,400 ± 500 cpm, and this value was subtracted from all values to yield the values presented

TSH was necessary to yield 50% inhibition of TSH binding to plasma membranes but only 2 to 3 times more HCG was necessary to yield 50% inhibition of TSH binding to the solubilized receptors. Thus, the specificity for two structural analogs of TSH was markedly decreased in the solubilized receptor despite similar affinities for TSH in the two preparations as measured by Scatchard plots of TSH binding.

When LIS solubilized receptor preparations were applied to sucrose gradients (Figure 12), a heterogeneous population of receptors was detected. There were four major peaks of binding activity with estimated molecular weights of 280,000 (peak 1), 160,000 (peak 2), 75,000 (peak 3), and 15,000 to 30,000 (peak 4) as determined by standards run on the same or duplicate gradients. Experiments to date have not been able to exclude the possibility that the 280,000 and 160,000 molecular weight components are aggregates of the 75,000 mol wt component, although their location on the gradient is not changed when gradients are run in the presence of 0.1% to 1% Triton X100.

The 15,000 to 30,000 mol wt component appears to be a fragment of the higher molecular weight structures. This conclusion was first suggested by its variable presence in different LIS solubilized receptor preparations. This was confirmed by the results of Sepharose 4B chromatography of LIS solubilized receptor preparations before and after trypsin digestion (Figure 13).

As noted in this experiment, the high molecular weight binding components are completely converted to a low molecular weight component after 2 hr of trypsinization. This low molecular weight component comigrates with peak 4 material in sucrose gradients and comigrates with the 15,000 to 30,000 mol wt receptor fragments that are released by trypsinization of intact thyroid plasma membranes, as described in Figure 8.

The 15,000 to 30,000 mol wt receptor component formed by trypsinization of LIS solubilized receptor preparations is specific in its binding of [^{125}I]TSH since binding was inhibited by cold TSH but not by albumin or several other polypeptide hormones; these data were analogous to those presented in Figure 6. When binding was evaluated as a function of hormone concentration (Figure 11B), a nonlinear Scatchard plot, effectively identical to that of the heterogeneous TSH receptor produced by LIS solubilization, was found.

The LIS solubilized TSH receptor preparation and the trypsinized LIS solubilized TSH preparation were applied to TSH-Sepharose columns (0.9 cm x 5 cm) (TATE, WINAND, KOHN, in press) in an effort to purify the receptor. In both cases, elution with TSH was unable to release the receptor activity. Elution with 0.2 M acetate, pH 2.3, released 10% to 30% of the LIS solubilized receptor preparation and about 50% of the trypsinized LIS solubilized receptor preparation (Table 6). The trypsin-treated LIS solubilized TSH receptor preparation was approx-

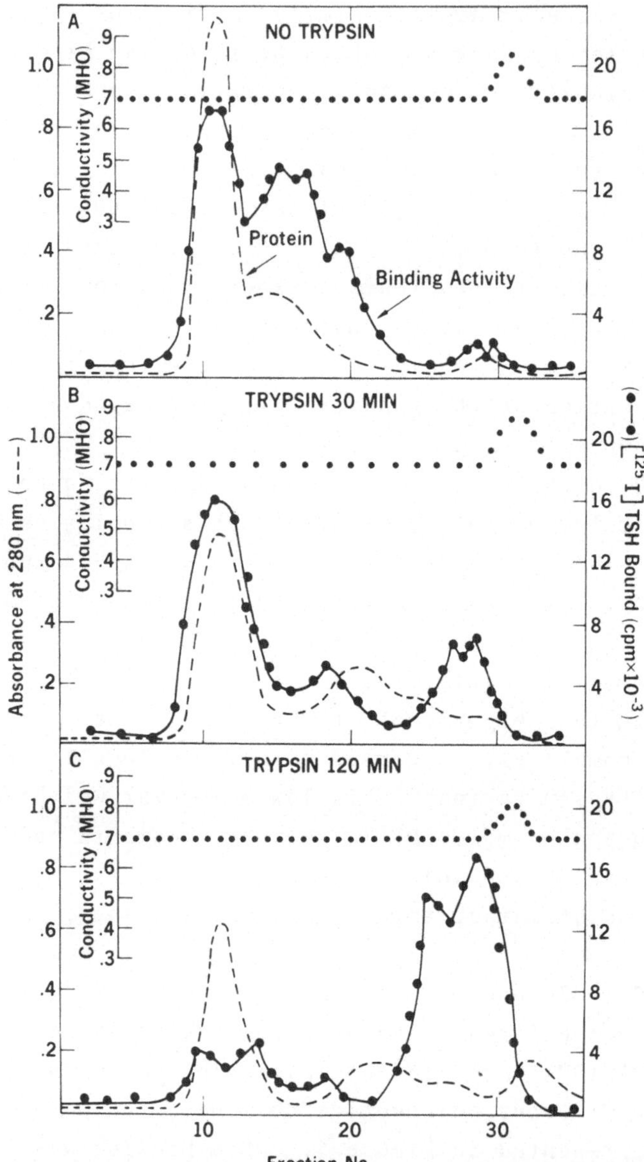

Fig. 13. Chromatography on Sepharose 6B of the LIS solubilized TSH receptor preparation before trypsinization (A), 30 min after trypsinization (B), and 120 min after trypsinization (C). The column was 0.9 cm x 50 cm; it was equilibrated and eluted with 0.02 M sodium bicarbonate, pH 9.4. Trypsinization was with TPCK trypsin. Aliquots of the LIS solubilized TSH receptor preparation were exposed to a 20:1 weight ratio of trypsin for the times noted; the pH was 8 and the temperature 25°C. At that point, a fivefold excess of soybean trypsin inhibitor (to trypsin) was added and the aliquots were loaded onto the column and eluted at a 10 ml/hr rate. Each fraction contained 0.4 ml. Triplicate 50-µl aliquots were assayed for binding activity (TATE et al., unpublished manuscript [b]) (●———●). Protein was measured as 280 nm absorbance (------). Conductivity measurements were used to locate the salt peak, which served both to insure that the elution

imately 120-fold purified over the original LIS solubilized TSH receptor preparation, approximately 250-fold purified over the membrane preparations, and approximately 5,000- to 25,000-fold purified over whole cell preparations. It demonstrated the same hormone specificity and salt-inhibition phenomena as the LIS solubilized preparation untreated with trypsin, and yielded the same curvilinear Scatchard plot as detailed in Figure 11 when binding was studied as a function of hormone concentration.

As noted in Table 6, the binding activity of the LIS solubilized TSH receptor preparation could be completely absorbed by TSH-Sepharose preparations formed by coupling TSH to cyanogen bromide-activated Sepharose 2B or 4B. Sepharose 2B or 4B alone did not absorb the binding activity. Procollagen or collagen-Sepharose preparations formed by coupling procollagen or collagen to cyanogen bromide-activated Sepharose also did not absorb binding activity. In contrast, neuraminidase-Sepharose and conconavalin A-Sepharose did decrease the binding activity of the LIS solubilized preparations exposed to them (Table 7).

The 15,000 to 30,000 mol wt receptor fragment has been isolated as a labeled glycoprotein from thyroid cell cultures (WINAND, KOHN, unpublished manuscript [b]). Cultures of either monolayer growing thyroid cells or of follicular thyroid cells were incubated with 0.2 µCi/ml of $[^{14}C]$glucosamine (3 mCi/mM, Commission de l'Energie Atomique, France) for 24 hr. After labeling, they were exposed to 2.5% trypsin (Difco Laboratories) for 20 min. The tryptic digestion was stopped by the addition of a fivefold excess of purified soybean trypsin inhibitor (Sigma Chemical Company), and the cells and media were separated by centrifugation. When the supernatant was concentrated and chromatographed on Sephadex G100, a major peak of radioactivity was detected, which moved just after the TSH marker (Figure 14A). When rechromatographed on Sephadex G100 in the presence of TSH, its position was shifted toward the void volume (Figure 14B), indicating that TSH was able to bind to this material and create a higher molecular weight complex. When $[^{3}H]$TSH was mixed with this $[^{14}C]$glucosamine-labeled

---

column procedures and volumes were the same and to define the area of the low molecular weight binding activity. The control with no trypsin had TPCK trypsin and soybean trypsin inhibitor added just before it was loaded onto the column; control aliquots chromatographed at zero time and after being allowed to stand 2 hr at room temperature had the same elution patterns

Table 6. Results of affinity chromatography of LIS solubilized or trypsin treated LIS solubilized receptor preparations using TSH-Sepharose[a]

	Total binding activity (cpm bound x 10^{-6}/mg protein)	Total protein (mg)	Specific activity (cpm bound x 10^{-6}/mg protein)
Starting membranes	270	50.2	5.4
LIS solubilized TSH receptor preparation . . .	125	12.5	10
Trypsin treated LIS solubilized TSH receptor preparation	118	1.9	62
LIS solubilized TSH receptor activity eluted from TSH Sepharose	20[b]	0.12	170
Trypsin treated LIS solubilized receptor preparation eluted from TSH-Sepharose	62	<0.05	>1,240

[a]TSH-Sepharose was prepared as described and columns were washed as described (TATE, WINAND, KOHN, in press) prior to loading of the samples. After loading, they were washed with equilibrating buffer and eluted with 0.2 M sodium acetate, pH 2.5. These fractions were brought to pH 9.4 with NaOH and 0.02 M sodium bicarbonate buffer.

[b]This result was variable and could be as low as 10 or as high as 32. It appeared to be worse when higher protein concentrations of the LIS solubilized receptor preparation were applied to the columns.

Table 7. Effects of Sepharose derivatives on the [^{3}H]TSH binding activity of the LIS solubilized TSH receptor preparation[a]

Additions	Net cpm bound
None	11,773
TSH Sepharose	0
Sepharose 2B or 4B	10,998
Neuraminidase-Sepharose	6,334
Concanavalin A-Sepharose	4,115
Procollagen-Sepharose	11,200
Collagen-Sepharose	11,800

[a]The standard 150 λ binding assay was carried out with 40 λ of a membrane extract that had been pretreated with the Sepharose derivatives indicated. Pretreatment consisted of incubating with shaking 100 μl of membrane extract with 100 μl of the indicated packed volume of swollen, buffer-washed Sepharose derivative beads for 30 minutes at 37°C. The beads were then pelleted by centrifugation and the supernatant fluid was assayed for binding activity.

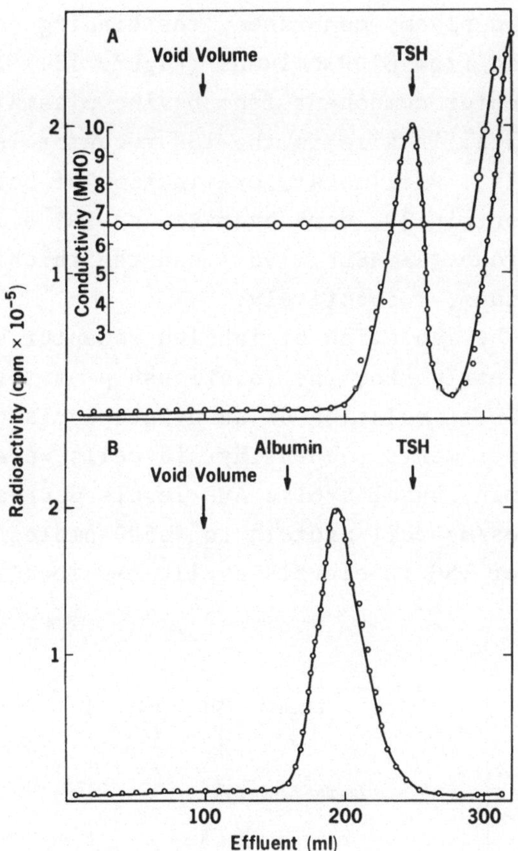

Fig. 14. Chromatography on Sephadex G100 (4 cm x 75 cm column) of ^{14}C labeled glycopeptides (o———o) obtained after trypsinization of thyroid cells (A). A major peak of radioactivity has a slightly lower mobility than TSH while a large amount of radioactivity is eluted with the salt peak as indicated by conductivity measurements (o———o). The first peak was lyophilized, taken up in 1 ml of 0.002 M Tris-HCl, pH 7.5, and incubated with 5 μg of purified TSH for 1/2 hr. It was then chromatographed on the same column (B)

material, the [^{3}H]- and [^{14}C]-labels both coeluted in this more advanced position of the column. When [^{3}H]TSH and a 10-fold excess of cold TSH were incubated with the [^{14}C]glucosamine labeled material, the [^{3}H]TSH migrated in the position of marker TSH, whereas the [^{14}C]glucosamine labeled material remained in the higher molecular weight position, i.e., [^{3}H]TSH binding was displaced by cold TSH. Analogous displacement was not obtained with albumin.

When the [^{14}C]-glucosamine-labeled material released by trypsinization of the cultured thyroid cells was mixed with the 15,000 to 30,000 TSH receptor component that could be obtained by trypsinization of either intact thyroid plasma membranes or solubilized TSH receptors from thyroid plasma membranes, the binding activity and [^{14}C] radioactivity coeluted from G100 columns (Figure 15). Thus, the 15,000 to 30,000 mol wt receptor component from bovine plasma membranes appeared to be identical in size to the TSH receptor isolated from cultured thyroid cells. Preliminary examination of both materials revealed that they contain 30% carbohydrate and 10% sialic acid, as measured by the anthrone (ASHWELL, 1957) and the thiobarbituric acid (WARREN, 1963) reactions, respectively.

In addition to the isolation of labeled receptor, culture studies have allowed us to clearly show the relationship of the TSH receptor interaction and of TSH stimulation of adenylate cyclase activity in tryptic digestion experiments. When thyroid cells were treated with 2.5% trypsin for 30 min, basal cyclic AMP levels decreased only slightly, 6,000 pmoles/mg cell protein to 4,500 pmoles/mg cell protein; whereas the ability of TSH to elevate cyclic AMP levels was eliminated,

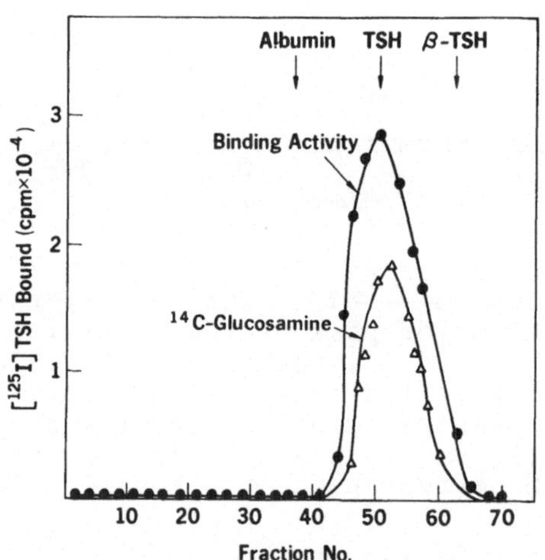

Fig. 15. Cochromatography on Sephadex G100 of the TSH receptor activity (●) released into the supernatant by trypsinization of bovine thyroid plasma membranes (Figure 8) and the ^{14}C labeled receptor activity released by trypsinization of [^{14}C]glucosamine labeled thyroid cells (Δ) in culture (Figure 14). Binding activity was measured on aliquots of the noted fractions using optimal conditions for the solubilized receptor assay (TATE et al., unpublished manuscript [b]) and 1.5 x 10^{-10} M [^{125}I]TSH. Solubilized receptor activity from bovine plasma membranes was prepared as previously described (Figure 8). The elution peaks of the three markers, albumin, TSH, and β-TSH, are noted by the arrows; their molecular weights are 67,000, 27,000 to 28,000, and 13,000 to 14,000, respectively. The column was 50 cm x 0.9 cm in size and was equilibrated and eluted with 0.025 M Tris-acetate, pH 6.0. The elution rate was 20 ml/hr. Fractions of 0.4 ml were collected. The void volume is the peak of elution of a dextran blue dye marker of >200,000 mol wt

18,700 pmoles/mg cell protein to the basal level of 4,600 pmoles/mg cell protein (Table 8). Simultaneous with this loss of TSH stimulable adenylate cyclase activity, in vitro TSH binding measured in both follicle forming and monolayer thyroid cell plasma membranes became unmeasurable (Table 8). Most important, during the next 48 hr, as the trypsinized cells regrew their TSH receptors and regained their binding ability, the ability of TSH to stimulate adenylate cyclase activity returned in a parallel fashion as measured by the ability of TSH to elevate cyclic AMP levels (Table 8).

Table 8. TSH binding to trypsinized thyroid cells correlated with the ability of TSH to stimulate adenylate cyclase activity[a]

	Cyclic AMP content[b]		[^{3}H]TSH binding[b]	
	Monolayer cells[c]	Follicular cells[d]	Monolayer cells	Follicular cells
	(pmoles/mg cell protein)		(nmoles/ml 20 μg membrane protein)	
Intact cells	6,080 ± 670[f]	18,700 ± 1,240	0.5	0.9
Trypsinized cells	4,570 ± 840	4,684 ± 915	0	0
Trypsinized cells after 24 hr[e]	5,120 ± 630	8,145 ± 575	0.22	0.33
Trypsinized cells after 48 hr[e]	5,490 ± 545	16,650 ± 1,035	0.42	0.78

[a]Adenylate cyclase activity was measured as the cyclic AMP levels stimulable by TSH. The possibility that this was a phosphodiesterase effect was tested for and eliminated by analysis of the cyclic nucleotide phosphodiesterase activity (KRISHNA, WEISS, BRODIE, 1968; WOLFF, JONES, 1971; WOLFF, COOK, 1973) in the cultures, i.e., no difference in this activity could be measured in the different cultures and the different conditions.

[b]Cyclic AMP levels were measured as described (WINAND, KOHN, unpublished manuscript [b]).

[c]Monolayer cells are those thyroid cells grown in the absence of TSH after trypsinization (WINAND, KOHN, unpublished manuscript [b]).

[d]Follicular cells are those thyroid cells that differentiate and form follicles when grown in the presence of TSH after trypsinization (WINAND, KOHN, unpublished manuscript [b]).

[e]Times after distribution into Petri dishes containing fresh medium.

[f]Values are the average of 6 experiments ± 2 S.D.

DISCUSSION

This report summarizes data that show from a functional point of view that the TSH receptor on bovine thyroid plasma membranes has many similarities to the TSH receptor on retro-orbital tissue plasma membranes, despite the fact that the former are the nominal target whereas the latter are apparently expressed only in a disease state. The receptors have the same optimal pH, the same salt-inhibition effects, the same temperature effects, and similar inhibition by LH and the subunits of TSH. Both receptors are heterogeneous on Scatchard plots of binding activity and have negative slopes in Hill plots of these data, i.e., both appear to exhibit some form of negative cooperativity. The only differences in these respects were the association constants calculated for the two major classes of sites that could be visually detected in the Scatchard plots, the so-called high affinity class and low affinity class. For bovine thyroid TSH receptors, the association constants were 6 x 10^{-10} M and 2 x 10^{-8} M, respectively; for guinea pig retro-orbital tissue TSH receptors these values were 15 x 10^{-10} M and 6 x 10^{-8} M, respectively, using the same [^{125}I]TSH preparations.

These similarities are in contrast to the significant functional difference that was detected when gamma globulin preparations from patients with malignant exophthalmos were shown to enhance TSH and EPF binding to retro-orbital but not thyroid plasma membranes. The gamma globulin from malignant exophthalmos patients appears to enhance TSH binding or the binding to its exophthalmogenic derivative by both increasing the number of TSH receptors and by shifting the association constants of receptors to much lower levels, i.e., by increasing the affinities of receptors for TSH such that greater numbers of receptors have high affinity association constants (15 x 10^{-10} M) or values near that. Although the mechanism by which the gamma globulin enhances binding is not completely clarified, the data summarized in Table 4 show that the gamma globulin does not modify the binding site or the receptor directly. Its action requires the presence of TSH or its exophthalmogenic derivative; hence, it must function to modify the TSH or EPF molecule itself or to interact with the membrane only after TSH has been bound, perhaps to alter the negative cooperativity effect, or both. Either of these mechanisms presupposes a structurally different TSH receptor on retro-orbital tissue and thyroid tissue plasma membranes. In the first case, there must be a different receptor-site topography; in the second case, there is some unique topographical feature near the receptor site which allows the gamma globulin to prevent the transmission of some structural or conformational information from the first to the second receptor. In this regard, tryptic release of different sized receptor fragments from retro-orbital and thyroid plasma membranes is highly suggestive, although the different animal species from whence the receptor came must be considered. Certainly, the strong conservation of pH, salt, and temperature properties between the receptors suggests a strong similarity in at least some portion of the receptor topography or structure despite their different sources and argues for the relationship of this structural difference to the gamma globulin effect; final proof will require further experimental efforts.

In regard to the effect of trypsin, these data have shown that tryptic digestion can release from thyroid plasma membranes a 15,000 to 30,000 mol wt component that has specific TSH receptor activity and all of the binding properties of the receptor in situ. Work by LEVEY et al. (1974) has indicated that this may not be a unique phenomenon, i.e., these workers have also recovered glucagon receptor activity in similarly sized units released from cat myocardium. Although the final purity of the material has not been established, the presence of 30% carbohydrate and 10% sialic acid is compatible with the effects of Con A and neuraminidase on binding activity (Table 7).

The relationship of in vitro binding studies to physiological hormone effects has always been a difficult question to answer. In the present report an important correlation is presented. In vitro binding is lost on the plasma membranes isolated from trypsinized whole cells, whereas binding gradually returns over the first 48 hr of cell growth in fresh media. TSH stimulation of adenylate cyclase activity is coincidentally lost and regained parallel to the receptor changes measured using these in vitro binding measurements. These data not only offer one of the most direct correlations of TSH binding in vitro with TSH stimulation of cyclase activity; they also clearly show that in vitro binding measurements are a reflection of in vivo action, despite the unusual conditions of the assay that optimize binding results. These data and the data of MACCHIA and MELDOLESI (1974), who show the absence of in vitro TSH receptor activity in tumors which cannot respond to TSH functionally, are clear proofs of the relevance of in vitro binding studies and of the close in vivo coupling of receptor and hormone-stimulated cyclase activity.

In sum, our initial observation that the gamma globulin from malignant exophthalmic patients enhances TSH binding and the binding of a modified TSH molecule to TSH receptors of retro-orbital tissue plasma membranes has been documented at much lower in vitro concentrations of hormone, i.e., at levels that are at least much closer to physiologic reality. The mechanism of the gamma globulin effect has been further clarified and a structural difference between the receptors has been detected, which should allow a direct structural attack on one aspect of exophthalmos. Comparative binding studies have thus succeeded in giving us a picture of TSH receptor in a tissue not normally presumed to be a target of TSH action, a receptor with strong functional and therefore perhaps many structural similarities to the thyroid TSH receptor, but a receptor with at least one structural difference that may explain its ascendancy to pathologic significance in the presence of one of the autoimmune complications of Graves' disease.

The importance of the autoimmune influence on receptors has been emphasized by the success of preventive therapeutic measures for exophthalmos that have been developed in the course of and as a consequence of our investigations correlating the development of the serum gamma globulin with the existence of active eye disease (WINAND, MAHIEU, 1973). In this regard, we have shown that cellular immunity, as measured by the inhibition of leukocyte migration by retro-orbital tissue antigens (the MIF assay), appears to precede the appearance of measurable serum gamma globulin concentrations and active eye disease. Thus, patients who have Graves' disease without exophthalmos but who have a positive MIF assay appear to be susceptible to the development of exophthalmos and can be "protected" by treatment with low doses of azaothioprine.

We previously suggested a wider prevalence of the autoimmune-receptor interaction, based on our correlation of positive MIF assays using skin antigens with several patients with pretibial myxedema (MAHIEU, WINAND, 1972). More recently, KONISHI, HERMAN, and KRISS (1974) have supported the two-factor mechanism further by their finding that thyroglobulin-antithyroglobulin complexes bind significantly better to eye muscle than heart or skeletal muscle and better than thyroglobulin alone. These data suggest the muscle involvement in the ophthalmopathy of Graves' disease is another example of two-factor or autoimmune-receptor mechanism invoking a pathological state.

SUMMARY

This report summarizes studies that have characterized the in vitro binding properties of thyrotropin receptors on retro-orbital tissue and thyroid plasma membranes. These properties are effectively identical in pH optima, temperature effects, salt inhibition, and inhibition by other hormones, thyrotropin analogs, or thyrotropin subunits. Both receptors have a similarly heterogeneous population of receptors in regard to their affinity for thyrotropin, and both have negative Hill plots, suggesting a negatively cooperative relationship among the receptors. Despite these functional similarities, gamma globulin from patients with malignant exophthalmos enhances thyrotropin binding to retro-orbital tissue plasma membranes but not to thyroid plasma membranes. Normal gamma globulin and gamma globulin from Graves' disease patients without exophthalmos do not have this effect. Tryptic digestion of thyroid plasma membranes releases a specific thyrotropin receptor 15,000 to 30,000 in mol wt, whereas tryptic digestion of retro-orbital tissue membranes releases a 75,000 to 150,000 mol wt thyrotropin receptor. This structural difference may correlate with the different effects of exophthalmic gamma globulin on binding in these two membrane preparations.

The thyrotropin receptor has been solubilized from thyroid plasma membranes. Although the solubilized preparation has a heterogeneous population of receptors on sucrose gradients, tryptic digestion creates a 15,000 to 30,000 mol wt receptor fragment with all the binding properties of the solubilized receptor. This fragment has been partially purified. Preparations of the fragment contain 30% carbohydrate and 10% sialic acid, and receptor activity can be lost by incubation with Con A-Sepharose or neuraminidase-Sepharose.

In vitro binding has been shown to correlate with thyrotropin stimulation of adenylate cyclase activity in thyroid cell cultures, i.e., in vitro binding assays actually measure the functional receptor population.

ACKNOWLEDGMENTS

We are indebted to Drs. J. E. Rall and H. Tabor, National Institute of Arthritis, Metabolism, and Digestive Diseases, U.S. National Institutes of Health, and Professor A. Nizet, Department of Medicine, University of Liège, for their support and encouragement throughout these investigations.

REFERENCES

AMIR, S. M., CARRAWAY, T. F., Jr., KOHN, L. D., WINAND, R. J.: J. Biol. Chem. 248, 4092-4100 (1973).
ASHWELL, G.: Methods in Enzymology 3, 84 (1957).
BATES, R. W.: In: Thyrotropin (ed. S. C. WERNER), p. 290. Springfield, Ill.: Charles C. Thomas 1963.
BROUHON-MASSILLON, L.: Doc. Ophthalmol. 17, 249-302 (1960).
DEDMAN, M. L., FAWCETT, Y. S., MORRIS, C. J. O. R.: J. Endocrinol. 39, 197-202 (1967).
DOBYNS, B. M.: In: The Pituitary Gland (eds. G. HARRIS, B. DONOVAN) Vol. I, p. 411. London: Butterworths 1966.
HILL, A. J.: Biochem. J. 7, 471 (1913).
KOHN, L. D., WINAND, R. J.: J. Biol. Chem. 246, 6570-6575 (1971).
KOHN, L. D., WINAND, R. J.: Israel J. Med. Sci. (in press).
KOHN, L. D., WINAND, R. J.: Unpublished manuscript.
KONISHI, J., HERMAN, M. M., KRISS, J. P.: Endocrinology 96, 434 (1974).
KRISHNA, G., WEISS, B., BRODIE, B. B.: J. Pharmacol. Exp. Ther. 163, 379-385 (1968).
KRISS, J. P., PHESHAKOV, V., ROSENBLUM, F.: J. Clin. Endocrinol. Metab. 27, 582 (1967).
LESNIAK, M. A., ROTH, J., GORDEN, P., GAVIN, J. R.: Nature New Biol. 241, 20-21 (1973).
LEVEY, G. S., FLETCHER, M. A., KLEIN, I., RUIZ, E., SCHENK, A.: J. Biol. Chem. 249, 2665-2673 (1974).
LEVITZKI, A., KOSHLAND, D. E., Jr.: Proc. Nat. Acad. Sci. U.S.A. 62, 1121-1128 (1969).
MACCHIA, V., MELDOLESI, M-F.: In: Advances in Cytopharmacology II (eds. B. CECCARELLI, F. CLEMENTI, J. MELDOLECI), pp. 33-37. New York: Raven Press 1974.
MAHIEU, P., WINAND, R. J.: J. Clin. Endocrinol. Metab. 34, 1090 (1972).
MARTIN, R. G., AMES, B. N.: J. Biol. Chem. 236, 1372-1379 (1961).
McKENZIE, J. M.: Endocrinology 63, 372 (1958).
McKENZIE, J. M.: Physiol. Rev. 48, 252 (1968).
MIYACHI, Y., VAITUKAITIS, J. L., NIESCHLAG, E., LIPSETT, M.: J. Clin. Endocrinol. Metab. 34, 23-28 (1972).
SCATCHARD, G.: Ann. N.Y. Acad. Sci. 51, 660 (1949).
SOLOMON, D. H., BEALL, G. N.: J. Clin. Endocrinol. Metab. 28, 1496 (1968).
TATE, R. L., BOLONKIN, D., LUBER, J. H., WINAND, R. J., KOHN, L. D.: Unpublished manuscript (a).
TATE, R. L., HOLMES, J. M., WINAND, R. J., KOHN, L. D.: Unpublished manuscript (b).
TATE, R. L., SCHWARTZ, H. I., HOLMES, J. M., WINAND, R. J., KOHN, L. D.: Unpublished manuscript (c).
TATE, R. L., WINAND, R.J., KOHN, L.D.: Methods in Enzymology (in press).
WARREN, L.: Methods in Enzymology 6, 463-465 (1963).
WINAND, R. J.: In: Rapports de la XIeme Reunion des endocrinologistes de langue Francais, p. 309. Paris: Masson et Cie 1970a.
WINAND, R. J.: In: Contributions a l'etude de l'exophthalmie endocrinienne, p. 67. Liege: Vaillant Carmanne 1970b.
WINAND, R. J., KOHN, L. D.: J. Biol. Chem. 245, 967-975 (1970).
WINAND, R. J., KOHN, L. D.: Proc. Nat. Acad. Sci. U.S.A. 69, 1711-1716 (1972).
WINAND, R. J., KOHN, L. D.: Endocrinology 93, 670-680 (1973a).
WINAND, R. J., KOHN, L. D.: In: Endocrinology: Proceedings of the Fourth International Congress of Endocrinology, Washington, D.C., June 18-24, 1972 (ed. R. O. SCOW; co-eds. F. J. G. EBLING, I. W. HENDERSON), pp. 1150-1155, International Congress Series No. 273, Excerpta Med. Int. Congr. Ser. Amsterdam 1973b.

WINAND, R. J., KOHN, L. D.: Unpublished manuscript (a).
WINAND, R. J., KOHN, L. D.: Unpublished manuscript (b).
WINAND, R. J., MAHIEU, P.: Lancet 1, 1196 (1973).
WINAND, R. J., SALMON, J., LAMBERT, P. H.: In: Further Advances in Thyroid Research, pp. 583-593. Vienna: Verlag der Wiener Medizinischen Akademie 1971.
WOLFF, J., COOK, G. H.: J. Biol. Chem. 248, 350-355 (1973).
WOLFF, J., JONES, A. B.: J. Biol. Chem. 246, 3939-3947 (1971).
YAMASHITA, K., FIELD, J. B.: Biochem. Biophys. Res. Commun. 40, 171-178 (1970).
YAMASHITA, K., FIELD, J. B.: J. Clin. Invest. 51, 463-472 (1972).

Binding and Gating by the Acetylcholine Receptor*

Arthur Karlin, Mark G. McNamee, and Cheryl L. Weill
Department of Neurology, College of Physicians and Surgeons, Columbia University, New York, New York 10032

INTRODUCTION

The receptors for acetylcholine transduce the binding of acetylcholine into permeability changes of postsynaptic membranes. Although there has been much progress in the development of means of identification of the receptors in isolation and in their purification and gross characterization (see review by KARLIN, 1974), the mechanisms of the transductions and of the ionic permeation processes remain undetermined.

The sources for the purification of receptors have been the electric organs of the fish *Electrophorus electricus* and of various species of *Torpedo*. These specialized tissues, derived embryologically from muscle (BENNETT, 1971), are highly enriched in acetylcholine receptors as well as in the other components (NACHMANSOHN, 1959) involved in cholinergic transmission. Pharmacologically and physiologically, the receptors in these tissues have specificities for agonists and antagonists and for permeating cations similar to the receptors present in the end-plates of vertebrate striated muscle.

The assays for isolated receptors that have been developed are based on the pharmacological properties of the acetylcholine binding site. Three inferred properties of the binding site being used as bases for assays are: (1) the rapidly reversible binding of acetylcholine and similarly acting agonists and of competitive inhibitors such as (+)-tubocurarine (ELDEFRAWI, ELDEFRAWI, 1973); (2) the slowly reversible binding of the curarimimetic α-neurotoxins of snake venoms (LEE, 1972); and (3) the susceptibility of the binding site to covalent modification by reduction and affinity alkylation (KARLIN, COWBURN,

*This work was supported by NIH Research Grant NS 07065, by NSF Research Grant GB 15906, and by a gift from the New York Heart Association, Inc.

1973). No assay has yet been devised to measure the ability of the isolated receptor to control cation permeability, although promising steps in this direction have been taken (HAZELBAUER, CHANGEUX, 1974; RAFTERY et al., 1974).

CHARACTERISTICS OF THE BINDING SITE

Potent agonists and antagonists of the receptor have at least one quaternary ammonium moiety, and it is highly likely that the acetylcholine binding site contains one or more negatively charged amino acid residues that interact with this moiety. In addition, potent ligands are fairly hydrophobic (PODLESKI, 1969; MAUTNER, BARTELS, WEBB, 1966); hydrophilic substituents appear to interfere with binding (choline, for example, binds very weakly). An area of hydrophobic interaction in the binding site thus seems likely. Furthermore, some (but not all) potent ligands can accept a hydrogen bond, and this may contribute to their binding (BEERS, REICH, 1970).

There is considerable evidence, obtained primarily in *Electrophorus* electroplax, for the presence of a disulfide bond in the vicinity of the acetylcholine binding site (KARLIN, 1969). These findings have been extended to frog (MITTAG, TORMAY, 1970; LINDSTROM, SINGER, LENNOX, 1973; BEN HAIM, LANDAU, SILMAN, 1973), chick (RANG, RITTER, 1971), and rat muscle (BEN HAIM, LANDAU, SILMAN, 1973; ALBUQUERQUE et al., 1968). This disulfide is readily reduced by dithiothreitol and reoxidized by a number of oxidizing agents. Reduction results in a dramatic change in the specificity of the receptor toward mono- and bis-quaternary ammonium compounds. Reoxidation reverses all of these changes.

At least one of the sulfhydryls formed by reduction can be alkylated by quaternary ammonium maleimide derivatives three orders of magnitude faster than by uncharged maleimides (KARLIN, 1969). The reactions of the quaternary ammonium maleimides are retarded in the presence of reversibly binding receptor ligands. These quaternary ammonium maleimides are "affinity labels" of the reduced receptor. A correlation of rates of reaction and lengths between the reactive group and the quaternary ammonium group places at least one of the sulfhydryls formed by reduction about 1 nm from the negative subsite of the binding site. There is no covalent reaction at the binding site in the absence of prior reduction, and hence it is unlikely that there are preexisting free sulfhydryl groups in this site.

Another group of affinity alkylating and acylating agents reacts with the reduced receptor in the electroplax, causing an activation of the receptor and a considerable depolarization of the membrane that cannot be reversed by washing (SILMAN, KARLIN, 1969; BARTELS, WASSERMANN, ERLANGER, 1971). The covalently attached activating moieties are shorter than the maleimide moieties that fix the receptor in an inactive conformation. This suggests that in the active conformation, the distance from the reducible disulfide to the negative subsite of the binding site is less than in the inactive conformation (KARLIN, 1969).

A schematic model of the receptor based on these interpretations is shown in Figure 1 (KARLIN, 1973). The receptor is represented as a dimer, the minimal oligomer accounting for a Hill coefficient close to 2 (KARLIN, 1967; CHANGEUX, PODLESKI, 1968). Each protomer has a molecular weight of about 200,000 and is composed of two to four different polypeptide chains, one of which has a molecular weight of about 40,000 and bears all or part of the acetylcholine binding site (discussed further below). The receptor spans the membrane, providing a channel for ion permeation (discussed below).

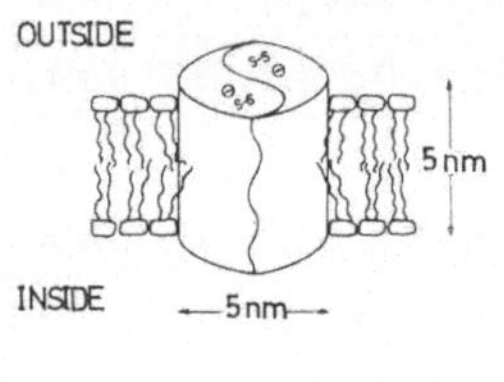

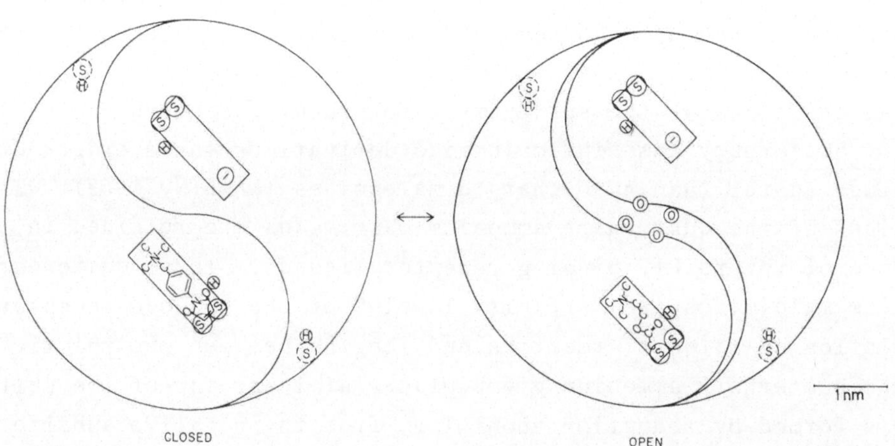

Fig. 1. A schematic model of the acetylcholine receptor. Top: The receptor spanning the membrane. Bottom: The extracellular aspect of a receptor dimer

The binding site for acetylcholine is represented as a slot in the outer surface of each protomer. Close to one end of the slot is the negative subsite, and overlapping the other end is the reducible disulfide group. The region in the vicinity of the disulfide is considered to be involved in hydrophobic interaction with ligands. Explicitly included is a hydrogen-bond donor site. It was surmised that an essential aspect of "the transition from inactive to active state involves a conformational change around the quaternary ammonium-negative subsite ion pair and a linked decrease in the distance between the negative subsite and the hydrophobic subsite, stabilized in potent activators by a bridge of the correct length positively interacting at both subsites" (KARLIN, 1969). The two conformations are represented (active = open), each with one site occupied. The site in the inactive conformation is shown occupied by MBTA acting as a competitive inhibitor but also being well positioned to affinity-alkylate one of the two SH groups, if the disulfide group were reduced. The site in the active conformation is shown occupied by acetylcholine, and the length of the site is decreased by 0.3 nm, corresponding roughly to the difference between the length of the covalently attached moiety of MBTA (nonactivating) and those of α-bromoacetylcholine or p-nitrophenyl ester of p-carboxyphenyltrimethylammonium (activating). An unreactive SH group has been placed some distance from the negative subsite. The channel, assumed to be a pathway of facilely exchangeable coordination sites for sodium and potassium ions, is shown lying on the axis between the subunits. The binding of acetylcholine, it is suggested, is translated into a relative shift of the subunits which generates the channel in the interface.

ASSAY BY AFFINITY ALKYLATION

The reaction of 4-(N-maleimido)benzyltri[^{3}H]-methyl ammonium iodide ([^{3}H]MBTA) with the reduced receptor is the basis for an assay procedure that has been applied to intact electroplax (KARLIN et al., 1971), to membrane fragments (KARLIN, COWBURN, REITER, 1973), and to solubilized and purified receptor (KARLIN, COWBURN, 1973). The specific reaction of [^{3}H]MBTA with the receptor is taken as that which is blocked by a variety of receptor protecting agents, including the affinity reoxidizing agents, dithiobischoline (BARTELS et al., 1970), the reversible ligands, hexamethonium and carbamylcholine, and the principle α-neurotoxin from *Naja naja siamensis*. The specificity is about 15% with intact cells and nearly 100% with purified receptor. In all cases, the protecting agents block the labeling of a unique polypeptide component of molecular weight of about 40,000 (REITER et al., 1972; KARLIN, COWBURN, 1973; WEILL, McNAMEE, KARLIN, 1974).

CHANNEL VERSUS CARRIER

In intact electroplax dissected from the organ of Sach of *Electrophorus*, it was determined that there are about 10 pmoles of specific [^{3}H]MBTA reactive sites per gram of cell or about 2 pmoles per milligram of protein (KARLIN et al., 1971). From conductance data of RUIZ-MANRESA and GRUNDFEST (1971) in electroplax and the density of sites, it can be calculated that during peak synaptic activation at least 10^6 sodium ions enter the cell per second per site. This would require an improbable turnover rate for a mobile carrier and, thus, a channel mechanism is most likely for receptor-controlled ion permeation. The receptor is probably oligomeric and spans the membrane. The binding of acetylcholine may cause a shift of subunits resulting in a channel of facilely exchangeable coordination sites for sodium and potassium ions spanning the phospholipid membrane (KARLIN, 1973). Channel models have been proposed for transport proteins, in general, by JARDETZKY (1966) and SINGER (1971), and recent evidence supports such a model in the case of Na^+, K^+-transport ATPase (KYTE, 1974).

SOLUBILIZATION AND PURIFICATION

Receptor from the electric tissue of *Electrophorus electricus* and that of *Torpedo californica* has been extracted and purified by nearly identical procedures (KARLIN, COWBURN, 1973; McNAMEE, WEILL, KARLIN, in press; WEILL, McNAMEE, KARLIN, 1974). The tissue is disrupted in 1 mM EDTA. A crude membrane fraction is obtained, washed with 1 M NaCl, and extracted with buffered 3% Triton X-100 at 4°C for 1 hr. The extract contains all of the original receptor and a small fraction of the original acetylcholinesterase activity. The receptor in this extract is totally adsorbed to an affinity gel consisting of p-carboxyphenyltrimethylammonium groups linked to agarose beads and selectively eluted by carbamylcholine solution (Table 1). An approximately 200-fold purification is commonly obtained by affinity chromatography in the case of *Electrophorus* receptor (Table 1). The overall purification is about 2,000-fold. The extract of *Torpedo* has a considerably greater specific activity than that of *Electrophorus* and the purification on

Table 1. Affinity column purification of ACHR from *Electrophorus electricus*

Column Treatment	Protein (mg)[a]	(%)	ACHE (pmoles)[b]	(%)	ACHR (pmoles)[c]	(%)
Triton extract	1,145	100.0	2,144	100.0	15,300	100.0
Not bound	1,136	99.2	1,036	48.3	0	0
Eluted with 50 mM NaCl	10.8	0.9	83	3.9	0	0
Eluted with 150 mM NaCl	1.16	0.1	468	21.8	1,958	12.8
Eluted with 50 mM CARB + 100 mM NaCl	1.22	0.1	41	1.9	3,813	25.0
Total recovery	1,148	100.3	1,628	75.9	5,771	37.8

[a] Lowry protein.

[b] Catalytic sites of acetylcholinesterase.

[c] Sites reacting with [^{3}H]MBTA.

the affinity gel is about 20- to 40-fold. The final specific activities are about the same for the two species.

The receptor can be further purified by centrifugation in a 5% to 20% sucrose density gradient containing 0.2% Triton X-100. The receptor moves as a symmetrical peak, with an apparent sedimentation coefficient of 9.5 S (McNAMEE, WEILL, KARLIN, in press). A 50% increase in specific activity is usually observed after centrifugation, yielding a specific activity of 4 to 5 nmole MBTA sites per milligram protein (see discussion below). The receptor is resolved from the remaining traces of acetylcholinesterase and less than 0.02% of the protein in the pooled receptor fractions is active esterase. The specific activity of receptor is constant within experimental error across the main part of the peak, suggesting near homogeneity.

SPECIFIC ACTIVITY OF PURIFIED RECEPTOR

The specific activity of receptor preparations depends on the assay used. It has been reported that about twice the number of α-neurotoxin molecules are bound to purified receptor both from *Torpedo*

(RAFTERY et al., 1974) and from *Electrophorus* (MEUNIER et al., 1974; CHANG, 1974) as are acetylcholine molecules. We have found twice the number of *Naja naja siamensis* α-neurotoxin molecules bound to receptor both from *Torpedo* (WEILL, McNAMEE, KARLIN, 1974) and from *Electrophorus* (McNAMEE, WEILL, KARLIN, in press) as [^{3}H]MBTA-reactive sites. In fact, nearly all the [^{3}H]MBTA reaction is blocked by preincubation of the receptor with toxin, but less than half of the toxin binding is blocked by prereaction with MBTA (McNAMEE, WEILL, KARLIN, in press). Specific activities based on toxin binding, therefore, may overestimate the number of acetylcholine binding sites. This is not the only possibility, however, and an unambiguous determination of the component(s) to which the α-neurotoxins bind is required. It would appear from present data that there is one acetylcholine binding site per about 200,000 daltons of receptor protein.

SUBUNIT COMPOSITION

The apparent molecular weight of one subunit of the receptor was first determined by polyacrylamide gel electrophoresis of a reduced, dodecyl sulfate-extract of *Electrophorus* electroplax affinity labeled with [^{3}H]MBTA. A component of about 40,000 daltons is uniquely and specifically labeled (REITER et al., 1972). Subsequently, a second

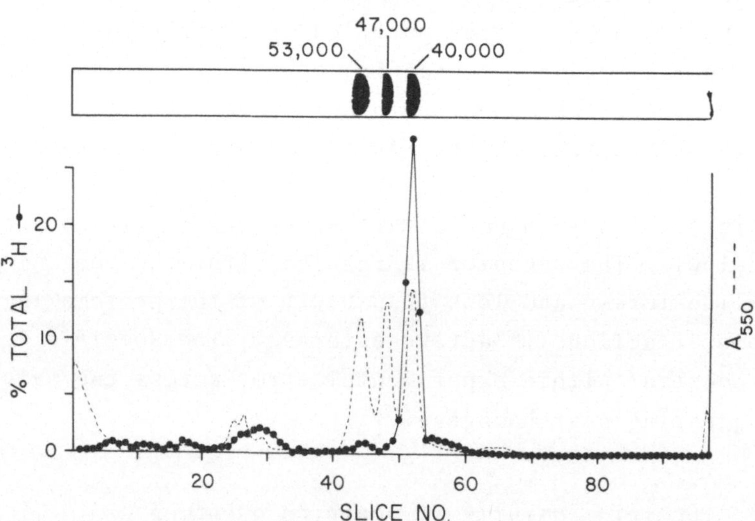

Fig. 2. Acrylamide gel electrophoresis in sodium dodecyl sulfate of purified receptor from *Electrophorus electricus* affinity labeled with [^{3}H]MBTA. Top: Photograph of gel stained with Coomassie brilliant blue. Bottom: ^{3}H activity (45,000 cpm total) in 1 mm slices of the same gel superimposed on the densitometer trace obtained before slicing

component of about 54,000 daltons was identified in purified receptor preparations from *Electrophorus* (LINDSTROM, PATRICK, 1974). Two components of about 40,000 and 50,000 daltons have also been reported by MEUNIER et al. (1974) and BIESECKER (1973). Invariably, we find three components of 40,000, 47,000, and 53,000 daltons (Figure 2) (KARLIN, COWBURN, 1973; McNAMEE, WEILL, KARLIN, in press). Further purification of our receptor preparation, as by sucrose density gradient centrifugation, does not alter the gel pattern. Receptor purified from *Torpedo californica* yields in our hands four components on dodecyl sulfate acrylamide gels (Figure 3) (WEILL, McNAMEE, KARLIN, 1974). These have apparent molecular weights of 39,000, 48,000, 58,000, and 64,000. A variety of compositions have been reported for receptor from species of *Torpedo*: 45,000, 37,000, and 32,000 (RAFTERY et al., 1974); 65,000, 50,000, and 40,000 (RAFTERY et al., in press); 46,000 (CARROLL, ELDEFRAWI, EDELSTEIN, 1973).

As mentioned before, the 40,000 dalton component of *Electrophorus* receptor is uniquely and specifically alkylated by [^{3}H]MBTA (KARLIN, COWBURN, 1973). When purified receptor is reduced and labeled with [^{3}H]MBTA, denatured in dodecyl sulfate and dithiothreitol, and run on dodecyl sulfate-acrylamide gels, all radioactivity is in the 40,000 dalton component (Figure 2). Remarkably, the 39,000 dalton component of *Torpedo* receptor is also uniquely and specifically alkylated by

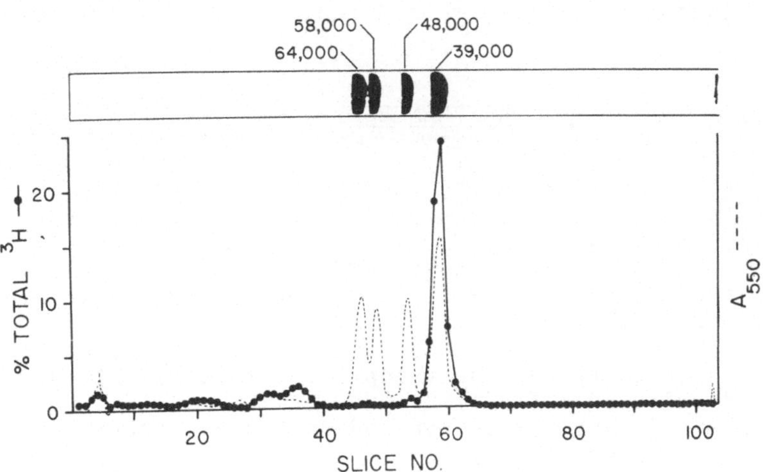

Fig. 3. Acrylamide gel electrophoresis in sodium dodecyl sulfate of purified receptor from *Torpedo californica* affinity labeled with [^{3}H]MBTA. Top: Photograph of gel stained with Coomassie brilliant blue. Bottom: ^{3}H activity (45,000 cpm total) in 1 mm slices of the same gel superimposed on the densitometer trace obtained before slicing

[^{3}H]MBTA (Figure 3). In both species, therefore, an approximately 40,000 dalton component bears all or part of the acetylcholine binding site. This is, so far, the only component for which a function has been demonstrated and which is proven to be a subunit of the receptor. The next higher molecular weight components in both species have similar masses and are likely to be subunits with similar functions. It would appear likely that the two types of subunits of about 40,000 and 48,000 daltons form a phylogenetically stable core structure of the receptor.

CONCLUSION

The acetylcholine receptors from two types of electric fish have been isolated in several laboratories. The isolated receptors retain the ability to bind acetylcholine and other rapidly reversible ligands, to bind α-neurotoxins, and to react with binding-site-directed alkylating agents. The binding site appears to be associated with a polypeptide of about 40,000 daltons.

The ability of isolated receptor to control membrane permeability is not well tested, although some promising steps in that direction have been reported (HAZELBAUER, CHANGEUX, 1974; RAFTERY et al., in press; McNAMEE, WEILL, KARLIN, in press). The mechanism of the control of ion permeation is one of the challenging problems in membrane biology. The investigation of the acetylcholine receptor will undoubtedly make an important contribution toward the solution of this problem.

REFERENCES

ALBUQUERQUE, E. X., SOKOLL, M. D., SONESSON, B., THESLEFF, S.: Studies on the nature of the cholinergic receptor. Eur. J. Pharmacol. 4, 40-46 (1968).

BARTELS, E., DEAL, W., KARLIN, A., MAUTNER, H. G.: Affinity oxidation of the reduced acetylcholine receptor. Biochim. Biophys. Acta 203, 568-571 (1970).

BARTELS, E., WASSERMANN, N. H., ERLANGER, B. F.: Photochromic activators of the acetylcholine receptor. Proc. Nat. Acad. Sci. U.S.A. 68, 1820-1823 (1971).

BEERS, W. H., REICH, E.: Structure and activity of acetylcholine. Nature 228, 917-922 (1970).

BEN HAIM, D., LANDAU, E. M., SILMAN, I.: The role of a reactive disulfide bond in the function of the acetylcholine receptor at the frog neuromuscular junction. J. Physiol. (Lond.) 234, 305-325 (1973).

BENNETT, M. V. L.: Electric organs. In: Fish Physiology (ed. W. S. HOAR, D. J. RANDALL) Vol. V, pp. 347-491. New York and London: Academic Press 1971.

BIESECKER, G.: Molecular properties of the cholinergic receptor purified from *Electrophorus electricus*. Biochem. Wash. 12, 4403-4409 (1973).
CARROLL, R. C., ELDEFRAWI, M. E., EDELSTEIN, S. J.: Studies on the structure of the acetylcholine receptor from *Torpedo marmorata*. Biochem. Biophys. Res. Commun. 55, 864-872 (1973).
CHANG, H. W.: Purification and characterization of acetylcholine receptor-I from *Electrophorus electricus*. Proc. Nat. Acad. Sci. U.S.A. 71, 2113-2117 (1974).
CHANGEUX, J. P., PODLESKI, T. R.: On the excitability and cooperativity of the electroplax membrane. Proc. Nat. Acad. Sci. U.S.A. 59, 944-950 (1968).
ELDEFRAWI, M. E., ELDEFRAWI, A. T.: Purification and molecular properties of the acetylcholine receptor from *Torpedo* electroplax. Arch. Biochem. Biophys. 159, 362-373 (1973).
HAZELBAUER, G. L., CHANGEUX, J. P.: Reconstitution of a chemically excitable membrane. Proc. Nat. Acad. Sci. U.S.A. 71, 1479-1483 (1974).
JARDETZKY, O.: Simple allosteric model for membrane pumps. Nature 211, 969-970 (1966).
KARLIN, A.: Chemical modification of the acetylcholine receptor. J. Gen. Physiol. 54, 245s-264s (1969).
KARLIN, A.: Minireview: The acetylcholine receptor: Progress report. Life Sciences 14, 1385-1415 (1974).
KARLIN, A.: Molecular interactions of the acetylcholine receptor. Fed. Proc. 32, 1847-1853 (1973).
KARLIN, A.: On the application of "a plausible model" of allosteric proteins to the receptor for acetylcholine. J. Theoret. Biol. 16, 306-320 (1967).
KARLIN, A., COWBURN, D. A.: The affinity-labeling of partially purified acetylcholine receptor from electric tissue of *Electrophorus*. Proc. Nat. Acad. Sci. U.S.A. 70, 3636-3640 (1973).
KARLIN, A., COWBURN, D. A., REITER, M. J.: Molecular properties of the acetylcholine receptor. In: "Drug Receptors," a Biological Council Symposium on Drug Action (ed. H. RANG) pp. 193-208. London: Macmillan 1973.
KARLIN, A., PRIVES, J., DEAL, W., WINNIK, M.: The affinity labeling of the acetylcholine receptor in the electroplax. J. Mol. Biol. 61, 175-188 (1971).
KYTE, J.: The reactions of sodium and potassium ion-activated adenosine triphosphatase with specific antibodies. J. Biol. Chem. 249, 3652-3660 (1974).
LEE, C. Y.: Chemistry and pharmacology of polypeptide toxins in snake venoms. Ann. Rev. Pharmacol. 12, 265-286 (1972).
LINDSTROM, J., PATRICK, J.: Purification of the acetylcholine receptor by affinity chromatography. In: Synaptic Transmission and Neuronal Interaction (ed. M. V. L. BENNETT) pp. 191-216. New York: Raven Press 1974.
LINDSTROM, J., SINGER, S. J., LENNOX, E. S.: The effects of reducing and alkylating agents on the acetylcholine receptor activity of frog sartorius muscle. J. Membrane Biol. 11, 217-226 (1973).
MAUTNER, H. G., BARTELS, E., WEBB, G. D.: Sulfur and selenium isologs related to acetylcholine and choline. IV. Activity in the electroplax preparation. Biochem. Pharmacol. 15, 187-193 (1966).
McNAMEE, M. G., WEILL, C. L., KARLIN, A.: Further characterization of purified acetylcholine receptor and its incorporation into phospholipid vesicles. In: Protein-Ligand Interactions (ed. H. SUND, G. BLAUER). Berlin: Verlag Walter de Gruyter, in press.
MEUNIER, J. C., SEALOCK, R., OLSEN, R., CHANGEUX, J. P.: Purification and properties of the cholinergic receptor protein from

Electrophorus electricus electric tissue. Eur. J. Biochem. 45, 371-394 (1974).

MITTAG, T. W., TORMAY, A.: Disulfide bonds in nicotinic receptors. Fed. Proc. 29, 547 abs. (1970).

NACHMANSOHN, D.: Chemical and molecular basis of nerve activity. New York: Academic Press 1959.

PODLESKI, T. R.: Molecular forces acting between ammonium ions and acetylcholine receptor. Biochem. Pharmacol. 18, 211-225 (1969).

RAFTERY, M. A., BODE, J., VANDLEN, R., MICHAELSON, D., DEUTSCH, J., MOODY, T., ROSS, M. J., STROUD, R. M.: Structural and functional studies of an acetylcholine receptor. In: Symposium on Protein-Ligand Interactions (ed. H. SUND, G. BLAUER). Berlin: Verlag Walter de Gruyter. in press.

RAFTERY, M. A., SCHMIDT, J., VANDLEN, R., MOODY, T.: Large-scale isolation and characterization of an acetylcholine receptor. In: Neurochemistry of Cholinergic Receptors (ed. E. DeROBERTIS, J. SCHACHT) pp. 5-18. New York: Raven Press 1974.

RANG, H. P., RITTER, J. M.: The effect of disulfide bond reduction on the properties of cholinergic receptors in chick muscle. Molec. Pharmacol. 7, 620-631 (1971).

REITER, M. J., COWBURN, D. A., PRIVES, J. M., KARLIN, A.: Affinity labeling of the acetylcholine receptor in the electroplax: Electrophoretic separation in sodium dodecyl sulfate. Proc. Nat. Acad. Sci. U.S.A. 69, 1168-1172 (1972).

RUIZ-MANRESA, F., GRUNDFEST, H.: Synaptic electrogenesis in eel electroplaques. J. Gen. Physiol. 57, 71-92 (1971).

SILMAN, H. I., KARLIN, A.: Acetylcholine receptor: Covalent attachment of depolarizing groups at the active site. Science 164, 1420-1421 (1969).

SINGER, S. J.: The molecular organization of biological membranes. In: Structure and Function of Biological Membranes (ed. L. I. ROTHFIELD) pp. 145-222. New York: Academic Press 1971.

WEILL, C. L., McNAMEE, M. G., KARLIN, A.: Affinity-labeling of purified acetylcholine receptor from *Torpedo californica*. Biochem. Biophys. Res. Commun. 61, 997-1003 (1974).

Effects of Colicin Ia on Membrane Functions of *Escherichia coli**

J. Konisky[1], M. J. R. Gilchrist[1], D. Nieva Gómez[2], and R. B. Gennis[2, 3]

Department of Microbiology[1], Chemistry[2], and Biochemistry[3], University of Illinois, Urbana, Illinois 61801

INTRODUCTION

The colicins are proteins produced by several members of the bacterial family *Enterobacteriaceae* that are antibiotically active against *Escherichia coli*. The sensitivity of *Escherichia coli* to colicins requires the presence of a specific colicin receptor on the cell surface. Different colicins adsorb to different receptors, as evidenced by the fact that bacterial mutants can be isolated that have lost the ability to adsorb certain colicins but not others. Among the colicins, three general modes of action have been described (for a recent review of colicin action, see LURIA, 1973):

- Specific inhibition of protein biosynthesis by colicins E3 (NOMURA, 1963) and D (TIMMIS, 1972).
- Specific inhibition of protein biosynthesis by colicin E2 (NOMURA, 1963).
- General effect on energy metabolism by colicins E1 and K (FIELDS, LURIA, 1969a, 1969b), Ia and Ib (LEVISOHN, KONISKY, NOMURA, 1968), A (JETTEN, VOGELS, 1973), and JF546 (FOULDS, 1971).

In this paper, three lines of evidence are presented that support the hypothesis that colicin Ia affects the normal functioning of the electron transport chain in *Escherichia coli*. This conclusion is derived from experiments demonstrating that (1) colicin Ia inhibits active transport while stimulating respiration, (2) colicin Ia causes a structural change in the membrane (as assayed by changes in the fluorescence properties of a reporter probe) that is mimicked by agents

*This work was supported by U.S. Public Health Service Research Grants AI 10106 (JK) and HL 16101 (RBG). M. J. R. Gilchrist holds a predoctoral fellowship from Procter and Gamble. D. Nieva Gómez is supported by the University of Illinois Research Board. The technical assistance of C. Green is greatly appreciated.

known to affect functions of the electron transport chain, and (3) a mutant with an uncharacterized alteration in electron transport is colicin Ia tolerant.

COLICIN Ia AND ITS BACTERIAL RECEPTOR

Colicin Ia has been purified to homogeneity and several of its structural features have been determined (KONISKY, RICHARDS, 1970; KONISKY, 1972). The Ia molecule is a simple protein that consists of a single polypeptide chain with a molecular weight of 78,000. Two structural features that deserve particular attention are (1) the rather high axial ratio of Ia and (2) the highly charged nature of the molecule. Based on the determined sedimentation and diffusion coefficients and the calculated partial specific volume, an f/f_o of 1.82 can be calculated for colicin Ia, suggesting that the molecule has a high axial ratio. Supportive evidence has been obtained through electron microscopic studies of purified colicin Ia (KONISKY, 1973), in which it was shown that Ia molecules negatively stained with phosphotungstate appear as rounded structures with a diameter of approximately 210 Angstroms. An anhydrous spherical molecule of colicin Ia would be expected to have a diameter of 57 Angstroms. Although the definitive shape of native colicin Ia cannot be considered as established at this time, the physical and microscopic information is most consistent with an elongated "plate-like" structure. Under the pH conditions (close to neutral) used in mode-of-action studies, colicin Ia is highly charged. This is apparent from both its high pI (about 10) and its amino acid composition. Of the approximately 712 amino acids in its structure, 69 are lysine and 49 are arginine. The ratio of polar to apolar amino acids is quite high as expected for an extended molecule.

We have recently used biologically active ^{125}Iodine-labeled colicin Ia to study several aspects of the interaction of this colicin with its bacterial receptor (KONISKY, COWELL, 1972; KONISKY, COWELL, GILCHRIST, 1973; KONISKY, LIU, 1974). The binding of ^{125}I-colicin Ia has been shown to be a saturable process having an average association constant of 10^{10} - 10^{11} M^{-1} at 37°C for the approximately 2,000 specific colicin receptors. In studies utilizing cell envelopes fractionated into outer and inner membranes, we were able to demonstrate the specific binding of colicin Ia to the outer membrane only. Outer membranes derived from a mutant strain known to be lacking functional colicin receptors did not bind colicin Ia. Thus, the specific Ia receptor is a component of the outermost *E. coli* membrane. We have recently been successful in solubilizing this receptor with Triton X-100, and a partial characterization of its properties has been described (KONISKY, LIU, 1974). One important finding was that the receptor is trypsin sensitive. Although these studies demonstrate that at least one class of colicin Ia receptors are protein components of the outer membrane, it is important to point out that these experiments do not rule out the existence of other classes of Ia receptors located, for example, in the cytoplasmic membrane.

An interesting facet of the mode of action of colicins is the single-hit curve obtained when one plots bacterial survivors versus the amount of colicin added to the culture. Such a curve for colicin Ia treated cells is shown in Figure 1, which gives results of experiments that actually determine the amount of colicin adsorbed to cells versus cell survival. The figure illustrates that up to an amount corresponding to a binding of approximately 40 molecules per cell, the killing obeys single-hit kinetics. There is no evidence that killing requires the cooperation of more than a single adsorbed colicin molecule; this situation would have led to a shoulder in the curve. From such data, one can calculate that for the killing of 95% of the cells, approximately 1 in every 14 adsorptions is lethal. In other words, the probability that an adsorbed colicin will trigger those events leading to cell death is 0.07.

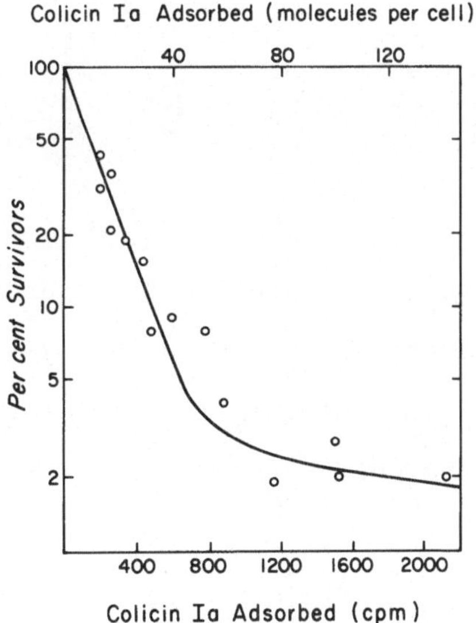

Fig. 1. Survival of strain W 3110 str-r as a function of adsorbed ^{125}I-colicin Ia. (Source: KONISKY, COWELL, 1972)

EFFECT OF Ia ON MEMBRANE FUNCTION

Earlier studies have shown that treatment of cells with colicin Ia leads to multiple effects (LEVISOHN, KONISKY, NOMURA, 1968). Treatment with Ia led to an inhibition of the incorporation of $^{32}P_i$ into DNA and RNA by more than 95%. Partial inhibition of $^{32}P_i$ incorporation into phospholipid and nucleotide fractions was observed. Incorporation of $^{32}P_i$ into the terminal phosphate of ATP was reduced to approximately half the control level. Protein synthesis, as measured by incorporation of (^{14}C) leucine, was abolished. In two cases, a stimulatory effect was observed: first, a slight enhancement in the rate of respiration was seen; second, the level of ^{32}P containing nonnucleotide organic phosphate compounds was increased in colicin Ia treated cells to approximately twice the level found in control cells. These results led to the suggestion that the mode of action of colicin Ia and Ib involved a primary interference with energy metabolism which subsequently led to the general effects described above.

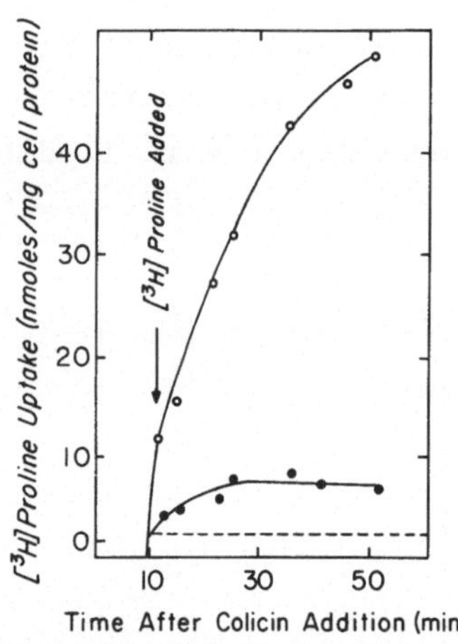

Fig. 2

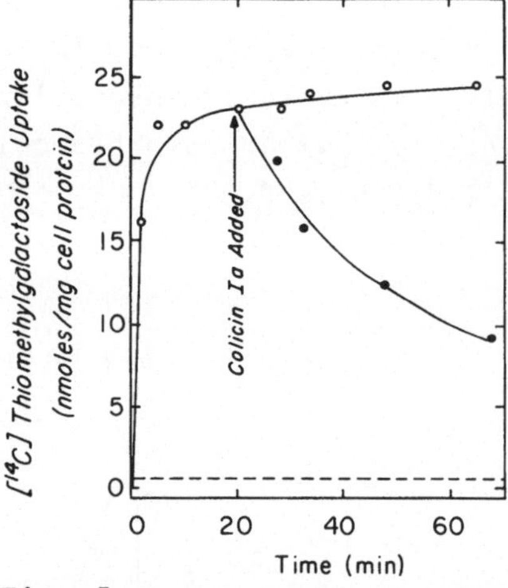

Fig. 3

Fig. 2. Effect of colicin Ia on proline transport. Cells were grown with 0.15% glucose as carbon source. Ten minutes after addition of either colicin Ia (final concentration, 0.097 µg/ml) or buffer, [^{3}H]proline (5.3 µC/µM, 1.44 x 10^{-4} M) was added and transport determined at 37°C. Assay of the cultures 10 minutes after colicin Ia addition showed 0.0012% survivors. The dotted line shows the level of isotope predicted if no concentrative effect were occurring. o = control; • = colicin Ia treated

Fig. 3. Effect of colicin Ia on accumulation of [^{14}C]-TMG. Cells were grown in Tris-S medium containing 0.5% glycerol and 1 x 10^{-4} M isopropyl thio B-D-galactoside. The experiment is as described in Figure 2 except that [^{14}C]TMG (16 µC/µM, 10^{-4} M) transport was assayed at 37°C. The final concentration of colicin Ia was 0.12^4 µg/ml. o = control; • = colicin Ia treated

The effect of colicin Ia on energy-dependent processes extending to bacterial active transport is illustrated in Figure 2. It is evident that treatment of cells with colicin Ia for 10 minutes leads to an inhibition in their subsequent ability to actively transport proline. Figure 3 shows that Ia treatment leads to the almost immediate efflux of substrates accumulated by active transport. It is clear that cells preloaded with thio-methyl-B-D-galactoside (TMG) rapidly lose accumulated substrate upon colicin addition. Essentially the same results are found when proline is used as a substrate. In addition to its effect on proline and TMG, colicin Ia treatment induces the leakage of intracellular potassium ion. This is shown in Figure 4, which

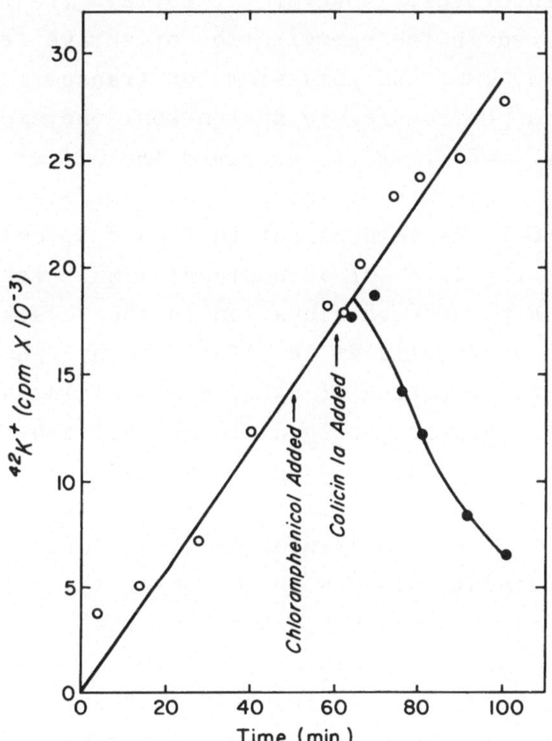

Fig. 4. Effect of colicin Ia on uptake of potassium ion. Cells grown on Tris-S medium with 0.15% glucose as carbon source were washed and resuspended in Tris-S from which KCl was omitted. To this cell suspension was added $^{42}K^+$, to a final concentration of 1.44 x 10^6 cpm/ml. The final suspension consisted of Tris-S containing 1.6 x 10^{-3} M potassium ion and 0.15% glucose. The cell concentration was adjusted to 102 Klett units (No. 42 filter). After 50 minutes, chloramphenicol was added to a final concentration of 100 µg/ml. Ten minutes later, colicin was added to a final concentration of 0.089 µg/ml. Assay after a 10-minute exposure to colicin showed 0.029% survivors. For sampling, 0.1 ml aliquots were removed, filtered, and washed with 0.1 M Tris-HCl buffer, pH 7.4. The experiment was carried out at 37°C. o = control; • = colicin treated

demonstrates the effect of colicin Ia on $^{42}K^{+}$ loading by colicin-sensitive cells. The effects of colicin Ia on active transport are not due to a direct effect on the activity of substrate carriers. For example, the rate of in vivo ortho-nitrophenyl-β-D-galactopyranoside (ONPG) hydrolysis in colicin Ia treated cells is found to be 92% of the rate found in control cells. It is more likely that colicin-treated cells are unable to "energize" active transport systems.

The alteration in the function of the various transport systems described could result from a specific effect of colicin Ia on active transport or a generalized leakiness that colicin Ia induces in the cell membrane. If the cell does, in fact, become leaky, then substrates of the phosphotransferase system might be expected to exhibit altered transport properties. However, if colicin treatment affects some component involved in the energization of active transport and common to the proline, TMG, and potassium ion transport systems, then substrates of the phosphotransferase system would be expected to be transported normally. We therefore examined the effect of colicin Ia on accumulation of a substrate of the phosphotransferase system α-methylglucoside (α-MG). As is apparent in Figure 5, colicin Ia treatment fails to reduce the level of accumulated α-MG. Instead, colicin Ia addition induces a pronounced elevation in the levels of accumulated substrate. When the extracellular level of α-methylglucoside is adjusted to 2.2 mM by the addition of (^{12}C) α-methylglucoside, the (^{14}C) α-methylglucoside that has been accumulated by colicin-treated cells is released. Thus, the additional isotope accumulated after colicin treatment is present inside the cells in a form that is susceptible to exchange. The fact that α-methylglucoside accumulation is not diminished by colicin treatment, along with the observation that the accumulated isotope is exchangeable, indicates that colicin Ia does not induce a generalized cell leakiness.

It is likely that the colicin-induced accumulation seen in Figure 5 is mediated by the phosphoenolpyruvate phosphotransferase system (PTS), since (1) the accumulated sugar is phosphorylated normally, (2) colicin-induced accumulation is reduced in mutants defective in the PTS system, and (3) the enhanced uptake is inhibited by NaF. Although it is possible that the increased levels of α-MG are due to an enhanced accumulation of phosphoenolpyruvate in colicin-treated cells, recent experiments have shown that the level of PEP found in cells is not limiting with respect to α-MG accumulation (KLEIN, BOYER, 1972). Alternatively, it is possible that colicin Ia treatment leads to preferential inhibition of α-MG efflux.

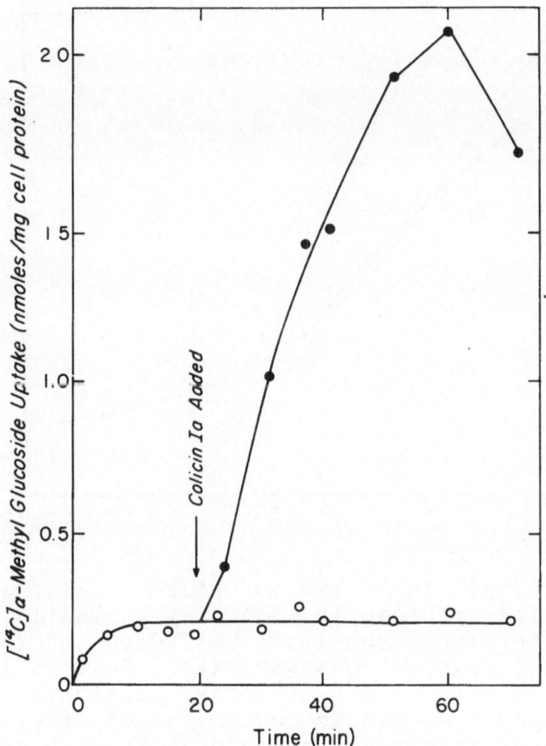

Fig. 5. Effect of colicin Ia on α-methylglucoside accumulation. Cells were grown as described with 0.5% glycerol as a carbon source. The experiment is as described in Figure 2 except that $[^{14}C]$α-methylglycoside (25 μCi/mM, 2 x 10^{-5} M) was allowed to accumulate for 20 minutes before colicin Ia addition (final concentration 0.1 μg/ml). Survivors were 0.001%. o = control; • = colicin Ia treated

It was previously reported that colicin Ia stimulated respiration by 5% to 10% (LEVISOHN, KONISKY, NOMURA, 1968). Since electron transport is known to play an important role in active transport (for a review, see HAROLD, 1972), it is of interest to reexamine this subject in more detail. As shown in Figure 6, the rate of oxygen uptake in the presence of glucose is increased 50% by the addition of colicin Ia. In contrast, a reduction in the rate of oxygen uptake is observed when succinate substrate is used. In the succinate medium, the rate of oxygen uptake is reduced to nearly zero within 15 minutes after colicin addition. The subsequent finding that colicin Ia treated cells are unable to accumulate succinate has resolved this apparent discrepancy.

The pattern of inhibition observed in colicin Ia treated cells is reminiscent in several aspects of the action of uncouplers, such as 2,4 dinitrophenol, azide, and carbonylcyanide m-chlorophenylhydrazone (CCCP). For example, these uncouplers have been shown to lead to an

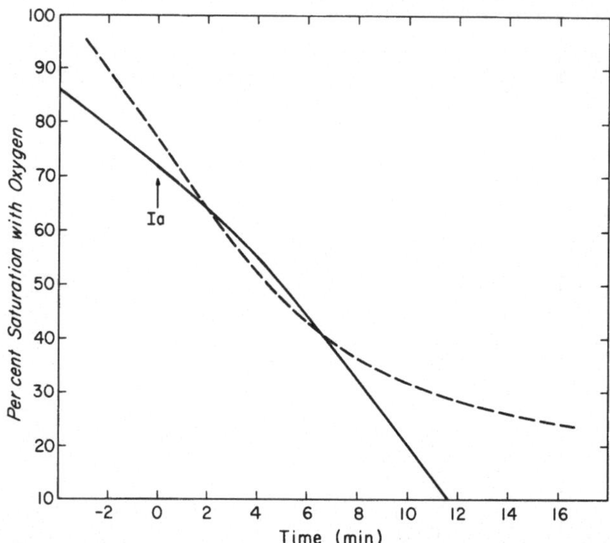

Fig. 6. Effect of colicin Ia on oxygen uptake. Cells were grown in Tris-S media with either glucose (0.15%) or glycerol (0.5%) as carbon source. For each experiment, substrate was added and the rate of respiration allowed to reach a constant rate. Colicin Ia or buffer was next added and the rate of oxygen uptake monitored using a Clark oxygen electrode system. A colicin concentration of 0.11 µg/ml was utilized. The survivor rate was less than 1% in each case. ——— = glucose; ------ = succinate

inhibition of macromolecular synthesis, active transport of amino acids and TMG, and oxidative phosphorylation (for a review of the action of uncouplers on *E. coli*, see HAROLD, 1972). Also, 2,4 dinitrophenol and azide treatment have led to enhanced levels of α-MG accumulation (HOFFEE, ENGLESBURG, 1962).

Since uncouplers are thought to function by fostering the diffusion of protons across the cell membrane, we examined the effect of colicin Ia treatment on proton permeability. Figure 7 shows the results of an experiment in which washed cells, suspended in weak Tris buffer containing 50 mM KCl, were pretreated with either buffer (Figure 7A, C) or colicin Ia (Figure 7B, D). Next, the suspensions were given an (HCl) acid load sufficient to lower the pH to 6.2, and subsequent changes in pH were monitored by a pH electrode. As shown most clearly in Figures 7C and 7D, over a period of 5 minutes both suspensions have a slow rise in pH as protons enter the cells. The rate of pH rise is identical in both suspensions. Similar results are obtained when the pH is followed for as long as 15 minutes, by which time the pH approached the level seen prior to acid loading. Identical results are obtained in suspensions containing 1 mM KCl. The addition of colicin

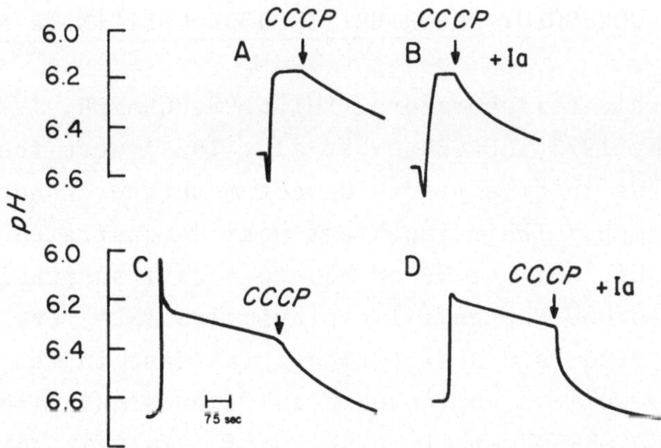

Fig. 7. Effect of colicin Ia and CCCP on proton permeability. Cells grown in synthetic medium containing 0.4% glycerol as carbon source were harvested at 2 x 10^8/ml and washed twice with 0.1 M Tris buffer, pH 8.0, and twice with 50 mM KCl. The cells were next resuspended in 50 mM KCl at a concentration of 3 x 10^9/ml. In Figures 7 A and C, 200 µl of 50 mM KCl were added. After 20 minutes at 23°C, 20 µl of 0.12 M Tris buffer, pH 8.0 were added, followed immediately by 10 µl of 0.03 M HCl to bring the pH down to about 6.2. At the indicated time, CCCP (final concentration, 2 µM) was added. The pH changes were monitored continuously. Figures 7 B and D were treated exactly as A and C, except that 9.5 µg of colicin Ia were added 20 minutes before addition of the Tris buffer. When assayed 20 minutes after Ia addition, the level of survivors was 0.14%

Ia to cells pretreated with ethylenediamine tetraacetate (EDTA) (LEIVE, 1968), which is known to increase membrane permeability to several substances, does not influence the results.

Although colicin Ia treatment has no effect on proton permeability per se, it does influence the pH changes associated with the subsequent addition of CCCP. Although CCCP addition leads to an increased rate of pH rise in both suspensions, the rate is greater in the colicin Ia treated culture. The colicin Ia enhanced rate of pH change due to CCCP was not observed when either colicin-resistant (no adsorption) or colicin-tolerant (adsorption, but no killing) cells replaced sensitive cells. The most reasonable explanation of these results is that colicin Ia-induced leakage of cell K^+ facilitates the CCCP-induced movement of protons into the cells. A similar interpretation has been used to explain the finding that CCCP-induced proton permeability is enhanced by pretreatment with valinomycin, a macrocyclic antibiotic that fosters K^+ permeability (PAVLASOVA, HAROLD, 1969). The results and interpretation of the experiment described in Figure 7 are similar to those described previously for colicin E1 (FEINGOLD, 1970) and colicin A (JETTEN, VOGELS, 1973).

USE OF A FLUORESCENT REPORTER TO PROBE COLICIN Ia ACTION

Cramer and his collaborators (PHILLIPS, CRAMER, 1973; CRAMER, PHILLIPS, KEENAN, 1973; HELGERSON et al., 1974) were the first to use fluorescent probes in attempts to detect membrane changes associated with colicin action. Their important work demonstrated that the addition of colicin E1, but not E2 or E3, to a cell suspension containing the fluorescent probe N-phenyl-1-napthylamine (NPN) evoked an increase in fluorescent yield, and a significant increase in the polarization of fluorescence, as well as an increase in fluorescent lifetime. Based on these results, HELGERSON et al. (1974) proposed that colicin E1 treatment results in an increase in the microviscosity of the *E. coli* cell envelope. These authors, furthermore, proposed that this increase in microviscosity might be due to the physical insertion of the colicin E1 molecule into the cell membrane or envelope.

Since the approach used by CRAMER and his colleagues serves as an important adjunct to function studies, we have carried out similar experiments with colicin Ia. As shown in Figure 8, addition of low concentrations of colicin Ia to a washed cell suspension containing NPN leads to a rapid increase in fluorescence intensity. There is no increase in fluorescence when a colicin I resistant strain (no colicin adsorption) is employed. Figure 8 also shows that CCCP, under conditions where it can be shown to be active as an uncoupler, induces a fluorescence increase in whole cells of *E. coli*. A similar increase in NPN fluorescence was observed when cells were treated with sodium azide (2 mM), potassium cyanide (10 mM), amytal (8 mM), or when nitrogen was bubbled through the cuvette. Addition of either chloramphenicol (100 μg/ml) or potassium arsenate (20 mM) had no effect, indicating that the fluorescence changes are induced specifically by agents that affect the activities of the electron transport chain.

Figure 9 shows that the increase in fluorescence intensity is accompanied by a blue shift in the emission spectrum. Such shifts can result from either a change in the probe environment, in which the local solvent polarity decreases, or an increase in microviscosity (RADDA, 1971). With all the agents that caused an increase in fluorescence intensity, we have observed a similar shift in emission spectra.

When the polarization of NPN fluorescence was measured at various times after the addition of colicin Ia to sensitive cells, results such as shown in Figure 10 (left) were observed. This demonstrates that Ia treatment leads to an almost immediate increase in polarization, which reaches a maximum level in 8 to 10 minutes. CCCP induces a similar

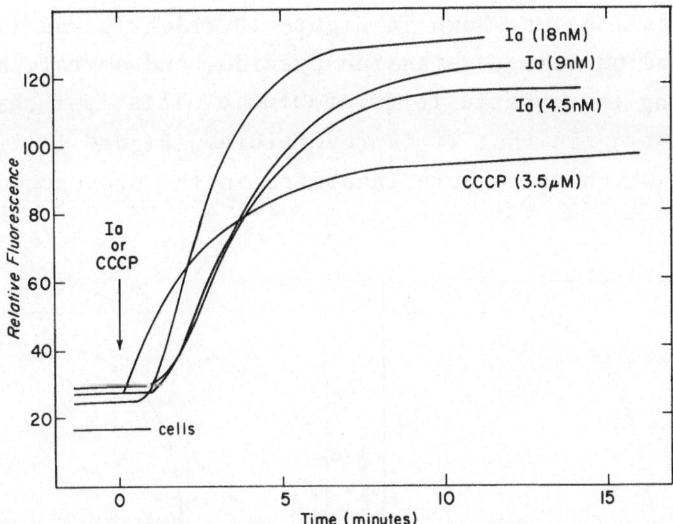

Fig. 8. Colicin Ia and CCCP-induced fluorescence changes. Cells were grown in M9 medium (MILLER, 1972) with glucose 0.13% as carbon source. The cells were harvested at a Klett of 80 to 100 (No. 42 filter), washed once with M9 (minus glucose), and resuspended in an original volume of M9 (minus glucose). The cells were kept on ice before use. Fluorescence intensity measurements were performed at 37°C on a Perkin-Elmer spectrofluorometer. The excitation wavelength was 340 nm with an excitation slit of 2 to 4 nm, while the emission wavelength was 420 nm with an emission slit of 40 nm. For measurement of emitted light, an interference filter was used that cut off light of a wavelength less than 390 nm. NPN was used at 3.0 μM

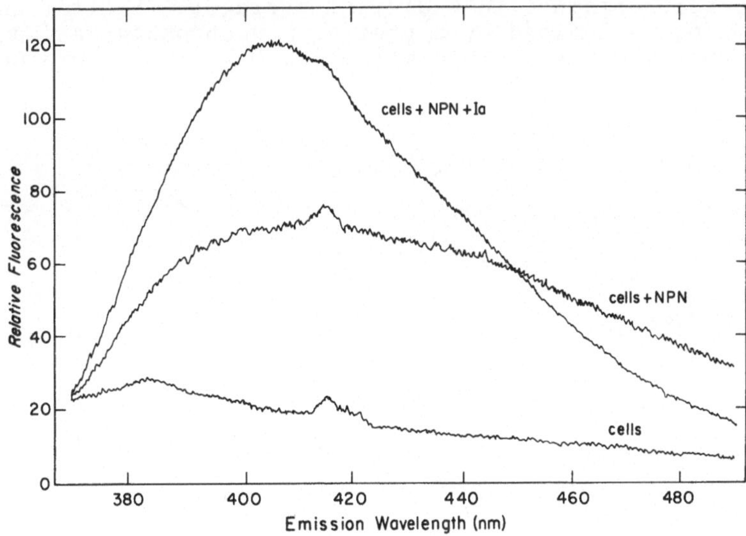

Fig. 9. NPN fluorescence emission spectra from cell suspensions with and without colicin Ia. Excitation wavelength was 340 nm. Emission was defined by means of a grating monochromator and an interference filter with a cutoff wavelength of 350 nm. Spectra were not corrected for the spectral response of the system. The scale of the sample (NPN + Ia) has been reduced threefold

change in polarization, as shown in Figure 10 (right), and like effects were seen with sodium azide, potassium cyanide, and amytal, as well as by simply allowing the cuvette to go anaerobic. This last case is particularly interesting in that it is reversible. Figure 11 illustrates an experiment in which cells were incubated in the presence of D-

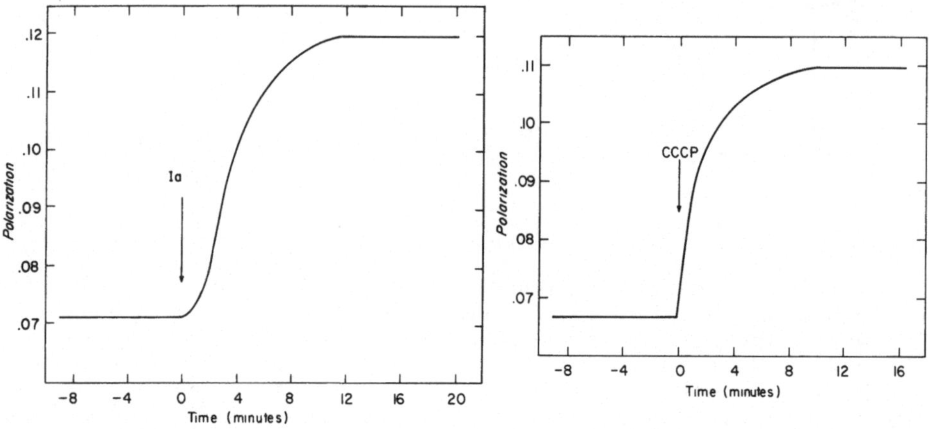

Fig. 10. Colicin Ia and CCCP-induced changes in fluorescence polarization. Cells were prepared as described in Figure 8. Fluorescence polarization was measured at 90°C. Parallel and perpendicularly polarized light intensities were measured simultaneously by means of two photomultiplier tubes. The ratio of these intensities, as well as a correction number for differences in photomultiplier gains and other factors, were obtained in a digital averager. Excitation wavelength was 340 nm, as defined by a grating monochromator and a Corning 7-51 filter. Emission was defined with a $NaNO_2$ filter and a Corning 3-73 filter

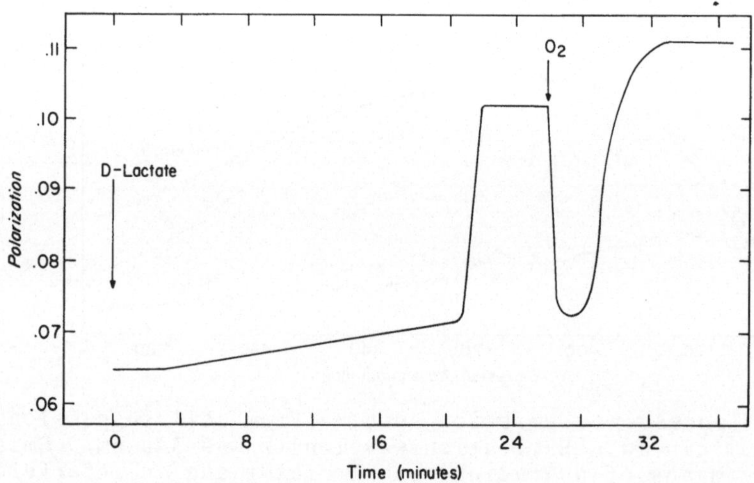

Fig. 11. Effect of anoxia on NPN fluorescence polarization. The experiment was performed as described in Figure 10

lactate (30 mM) and the polarization of the probe was then monitored. A slight increase in polarization was observed over the first 20 minutes, followed by an abrupt rise at approximately 21 minutes which reached a maximum in another minute or so. We assume that this change is due to anoxia, since subsequent addition of an oxygen pulse led to an immediate decrease in polarization to the base line level. After 1 to 2 minutes, a further increase in polarization was observed. The reversible polarization changes observed in this experiment correlate to shifts in the emission spectrum of the probe. Thus, whereas the increase in polarization seen at 21 minutes is accompanied by a blue shift (max. 407 nm; see Figure 9 for comparison), the oxygen-induced decrease in polarization at 25 minutes is accompanied by a red shift (max. 445 nm).

The anoxia-induced changes in fluorescence are dependent on the presence of an oxidizable substrate (lactate, glycerol, glucose, succinate, etc.). This is in contrast to the changes seen with colicin Ia, or the other agents examined above, in which the same quantitative changes in fluorescence occur whether oxidizable substrate is present or not. In contrast to the situation seen with anoxia-induced polarization, oxygen addition to colicin Ia, CCCP, azide, cyanide, or amytal-treated cells does not reverse the observed changes in fluorescence polarization.

The increase in polarization observed in these experiments could be the result of several factors: (1) an increase in dye binding, (2) a decrease in emission lifetime, (3) an increase in microviscosity, or (4) a combination of the above. We ruled out a decrease in lifetimes as an explanation, since the fluorescence lifetimes actually increase under conditions that increase polarization. Preliminary evidence indicates that upon treatment of the cells with such agents as CCCP, cyanide, and colicin Ia, there is a substantial increase in the fraction of dye bound. It is not yet clear whether the increase in dye binding will account for all or only part of the observed polarization increase. It is possible that part of the increase in polarization is due to an increase in membrane microviscosity induced by colicin Ia and the other agents named. Our results do not rule out the possibility that the changes observed may be caused by probe migration to a region of higher hydrophobicity or microviscosity. That the implied change in membrane conformation is due to the insertion of colicin molecules into the cell envelope is, we contend, but one of several plausible interpretations. Because of our findings that colicin Ia-induced fluorescence changes are mimicked by a wide variety of agents that disrupt functions of the electron transport chain, we suggest, rather, that changes noted reflect an effect of colicin Ia on electron chain function.

A COLICIN Ia TOLERANT MUTANT

The selection of *Escherichia coli* strains insensitive to the action of a particular colicin led to the isolation of two mutant classes--resistant and tolerant. Colicin-resistant mutants are defective in their capacity to adsorb colicin. Colicin-tolerant mutants retain a capacity to adsorb colicin, even though the mutant can survive colicin treatment.

The isolation and characterization of colicin-tolerant mutants is a potentially powerful tool in furthering our understanding of colicin action. The identification of structural and/or metabolic alterations in such mutants could serve to identify functions necessary for colicin action, and, thus, shed light on the mechanism of action itself. Recently, we described the isolation of a mutant (Tol I) which is uniquely tolerant to colicin Ia and the closely related colicin Ib (CARDELLI, KONISKY, 1974). Although Tol I is refractory to colicin Ia killing, it retains fully active colicin Ia receptors and some of our findings, which follow, suggest that the activity of the electron transport chain is reduced in Tol I.

The action of colicin Ia is in many ways similar to the action of azide. Furthermore, our preliminary genetic data showed that the Tol I gene mapped close to the azide resitant locus (azi). This prompted us to examine the sensitivity of Tol I to azide. As shown in Table 1, Tol I is, indeed, more resistant to azide than its parent strain X36. For example, whereas 12.5 mM sodium azide completely inhibited growth of X36, Tol I grew to a level corresponding to 43% of that attained by the same strain incubated in the absence of azide. Mutational analysis has shown that azide and phenethylalcohol (PEA) resistance are closely related (YURA, WADA, 1968). Therefore, it was not altogether surprising that Tol I was also found to be less sensitive to PEA than its parent (see Table 1). Although the one Tol I mutant that we have isolated exhibits azide resistance, the reciprocal relationship does not hold. For example, we have isolated a mutant of strain X36 that exhibits the same degree of azide resistance as Tol I but is not colicin I tolerant. Furthermore, we have recently carried out transductional analysis that shows that Tol I can be cotransduced with serB (90 minutes) and is, thus, genetically distinct from azi which maps at

Table 1. Sensitivity of strains X36 and Tol I to various agents

Exp.	Addition	Growth (percent of no addition) X36	Tol I
I	None	100	100
	Azide 2.5 mM	24	38
	Azide 12.5 mM	0	43
	Azide 37.5 mM	0	22
II	None	100	100
	PEA 0.1%	94	99
	PEA 0.2%	49	77
	PEA 0.3%	29	68
III	None	100	100
	Neomycin 6.25 μg/ml	9	91
	Neomycin 12.50 μg/ml	7	89
	Neomycin 62.50 μg/ml	2	24
IV	None	100	100
	DNP 62.5 μM	76	81
	DNP 125 μM	62	62
	DNP 250 μM	35	30
	DNP 500 μM	0	0
V	None	100	100
	Tetracycline 1.25 μg/ml	38	54
	Tetracycline 2.50 μg/ml	24	30
	Tetracycline 6.25 μg/ml	17	24
	Tetracycline 12.50 μg/ml	0	0

2 minutes on the *E. coli* genetic map. Table 1 also shows that Tol I is resistant to the antibiotic neomycin C. This finding was surprising since neomycin C is thought to be a specific inhibitor of protein synthesis. Tol I was found to exhibit parental-like sensitivity to 2,4 dinitrophenol and tetracycline.

Since azide and colicin Ia seem to interfere with functions of the electron transport chain, we next examined the ability of Tol I to oxidize a variety of substrates. Table 2 shows that although the mutant can oxidize glucose and glycerol at near normal levels, it is clearly deficient in lactate-, succinate-, and malate-driven respiration. The reduced levels of respiration are not due to the inability of Tol I to take up these substrates.

Since active transport in *E. coli* is coupled to a functioning electron transport chain, it was of interest to determine the capability of Tol I to carry out active transport. Since the antibiotic activity of neomycin C is thought to require its transport into the cell, it seemed a reasonable possibility that Tol I was able to survive neomycin because of some defect in active transport. As can be seen in

Table 2. Respiration in whole cells of strains X36 and Tol I[a]

Strain	Grown	Substrate	O_2 Consumption (ng-atoms/min/mg cell protein)
X36	Broth	Glucose (11 mM)	1010
Tol I	Broth	Glucose (11 mM)	865
X36	Glycerol	Glycerol (44 mM)	618
Tol I	Glycerol	Glycerol (44 mM)	542
X36	Broth + lactate	D-Lactate (22.5 mM)	1184
Tol I	Broth + lactate	D-Lactate (22.5 mM)	554
X36	Glycerol + succinate	Succinate (16.9 mM)	269
Tol I	Glycerol + succinate	Succinate (16.9 mM)	59
X36	Broth + malate	Malate (8.6 mM)	142
Tol I	Broth + malate	Malate (8.6 mM)	60

[a]The rate of oxygen consumption was determined from data of the type shown in Figure 6.

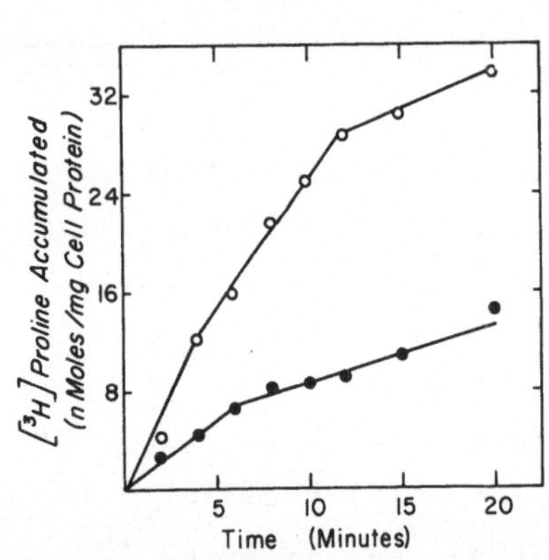

Fig. 12

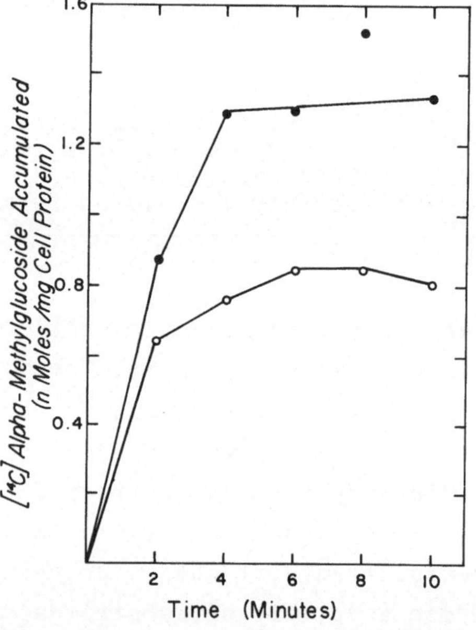

Fig. 13

Fig. 12. Proline transport in colicin Ia sensitive (X36) and tolerant (Tol I) strains. The strains were grown to 2 x 10^8/ml in synthetic medium containing 0.4% glycerol as carbon source. The cells were next washed twice in medium minus glycerol and resuspended to a Klett of 100 (filter 42) in medium minus glycerol but containing chloramphenicol (100 μg/ml). Next, [^{3}H]proline (5.3 μCi/μM, 1.44 x 10^{-4} M) was added and transport was determined at 23°C. X36 = o——o; Tol I = ●——●

Fig. 13. α-MG Transport in colicin Ia sensitive (X36) and tolerant (Tol I) strains. The procedure is as described in Figure 12 with the exception that ^{14}C-α-MG (sp. act. 52.2 mC/mM, 2.44 x 10^{-5} M) was added instead of proline. X36 = o——o; Tol I = ●——●

Figure 12, glycerol-grown Tol I is defective in its ability to transport proline. A similar finding was observed when we examined active transport of leucine, alanine, glycine, and TMG. It is known that high glucose represses the activity of the electron transport chain (HEMPFLING, 1970; BROMAN, DOBROGOSZ, WHITE, 1974). Furthermore, it is known that in the absence of electron transport, active transport is driven by ATP generated by substrate level phosphorylation (see HAROLD, 1972). Therefore, if the lower levels of active transport seen in Tol I are due to inefficiency of the electron transport chain, transport should be normal in glucose-grown cells. This is, indeed, what was found (data not shown).

As discussed above, transport of α-MG is mediated by the PTS system by group translocation. It should, therefore, be transported normally in Tol I. As can be seen in Figure 13, accumulation of α-MG is actually increased in Tol I.

We also examined the ability of membrane vesicles derived from Tol I and X36 to carry out substrate oxidation and active transport of proline. The results correlate very well with our in vivo studies. Table 3 shows that D-lactate, succinate, and α-glycerolphosphate oxidase activities are reduced in Tol I vesicles. Active transport of proline is reduced in the mutant when either D-lactate or succinate is used as a substrate. Since proline transport is also reduced when it is driven by the artificial electron donor system, ascorbate plus phenazine methosulfate, it is unlikely that Tol I is defective solely at the level of specific dehydrogenases. Since proline transport in

Table 3. Respiration and proline transport in X36 and Tol I vesicles[a]

Substrate	O_2 Consumption (ng-atoms/min/mg vesicle protein) X36	Tol I	Proline uptake (nmol/min/mg vesicle protein) X36	Tol I
None	<1	<1	0.016	0.017
D-Lactate (20 mM)	97	41	0.52	0.21
Succinate (20 mM)	210	41	0.09	0.054
Ascorbate (20 mM) plus phenazine methosulphate (0.1 mM)			0.20	0.076
α-Glycerolphosphate (20 mM)	86	38		

[a]Vesicles were prepared and transport was measured according to KABACK (1971). Oxygen consumption was determined as in Figure 6.

vesicles is known to be coupled to the activity of the electron transport chain, these studies further strengthen our hypothesis that the mutant is defective in some aspect of electron transport.

Previously, we demonstrated that the fluorescence properties of NPN can be correlated to functions of the electron transport chain. Furthermore, we have provided evidence that the activity of the electron transport chain is reduced in the Tol I mutant. It was, therefore, of interest to examine whether the changes in fluorescence observed in treated wild-type cells were mimicked by Tol I cells alone The results, as shown in Figure 14, demonstrate that this is, indeed, the case. Thus, the addition of NPN to strain X36 leads to an initial rise in fluorescence that soon stabilizes. In the case of Tol I, however, the fluorescence increases over a period of 5 minutes to a level comparable to that found in CCCP-treated strain X36.

It is important to note that the pleiotropic phenotype of the Tol I strain is due to a single mutation. This is supported by several lines of evidence. Tol I is a spontaneous mutant selected for growth in the presence of colicin I. All seven independent revertants that were selected for growth on acetate (Tol I cannot utilize acetate as a sole carbon source) simultaneously reverted to full sensitivity to colicin Ia and neomycin and regained normal capacity to transport amino acids.

Several reasons for Tol I insensitivity to colicin Ia can be postulated. First, a well functioning electron transport chain to-

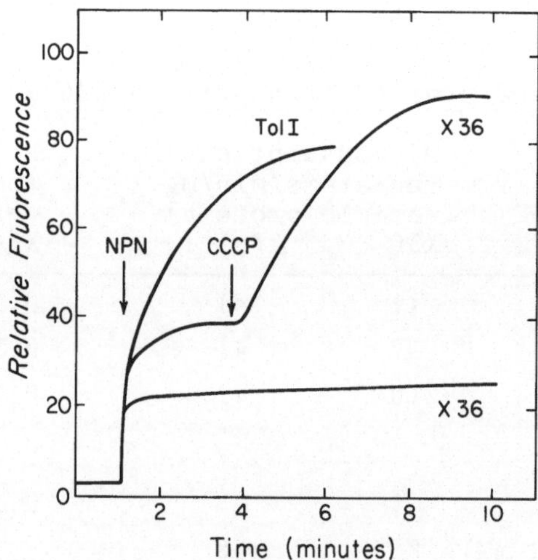

Fig. 14. NPN fluorescence in the presence of strain X36 (Ia sensitive) and Tol I (Ia tolerant). The experiment is as described in Figure 8

gether with proper energy coupling may be necessary for colicin Ia to exert its action. For example, the intriguing possibility exists that colicin Ia must be transported across or, at least, into the cytoplasmic membrane by a process energized by electron transport. Alternatively, it is possible that a target of colicin action is some component of the electron transport-energy coupling system that by mutation has become simultaneously refractory to colicin Ia action and less efficient in energy transduction. One cannot help but note that in many respects the Tol I mutant resembles its colicin-treated parental strain, X36. For example, both are defective in active transport, accumulate excess α-MG, and have similar interaction with NPN. This leads to the possibility that the mutant may have an alteration in a primary target of colicin Ia. However, such a suggestion must be tempered by the finding that, unlike the mutant, Ia-treated cells have a high rate of respiration.

MODE OF ACTION

The results obtained in these studies provide evidence that the many changes seen in colicin Ia-treated cells reflect the colicin's ability to disrupt energy metabolism. Thus, under growth conditions in which active transport is most certainly driven by dehydrogenase-linked electron transport (see KLEIN, BOYER, 1972), colicin Ia treatment has been shown to both inhibit active transport and stimulate respiration. Furthermore, we have recently carried out experiments that demonstrate that colicin Ia inhibits active transport of proline in an uncA mutant. This finding is particularly relevant since it is known that active transport in such mutants is solely respiration driven (GIBSON, COX, 1974). It would seem, therefore, that the colicin is able to uncouple electron transport from the generation of the energized intermediate (a proton gradient or altered membrane constituents; see HAROLD, 1974; LOMBARDI et al., 1974) thought to drive active transport. Alternatively, colicin Ia might interfere with coupling of the energized intermediate to active transport.

At present, we do not understand the mechanism whereby colicin Ia, CCCP, azide, cyanide, amytal, or anoxia lead to changes in the fluorescence parameters of NPN. In keeping with current dogma, it might be suggested that these agents have the common property of either blocking the generation of or dissipating an "energized membrane state." Furthermore, it might be supposed that the collapsed membrane state has a higher microviscosity or other altered physical properties. Whatever the explanation, the correlation is clear that agents active against the electron transport chain per se, as well as uncouplers, lead to fluorescent changes that are indistinguishable from those obtained with colicin Ia treatment. These results strengthen our notion that colicin Ia treatment interferes with energy coupling.

THE TRANSMISSION SYSTEM

Any explanation for the mode of action of colicin I must explain how the biological activity of these colicins is transmitted from the receptor to the sensitive target--presumably the cell membrane. Although we know little of the molecular events taking place between the initial interaction of colicin Ia and its cell wall receptor and the observed changes in cell functions, we can nevertheless suggest several possibilities (see Figure 15).

The cell wall (outer membrane plus murein) of *E. coli* is approximately 10 to 12 nm in width and, in general, is separated from the cytoplasmic membrane (inner membrane) by a space (periplasmic space) of

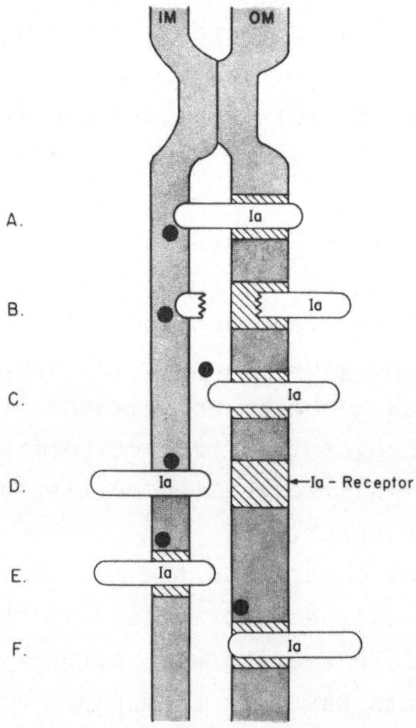

Fig. 15. Schematic presentation of transmission models. (IM, inner membrane; OM, outer membrane.) The darkened circle represents a hypothetical molecule activated in colicin Ia treated cells

4.0 to 4.5 nm. As discussed above, physical studies as well as electron microscopic examination of colicin Ia have suggested that the molecule is oblate in shape with the approximate diameter of 20 nm. Thus, it is conceivable that colicin I molecules can span the cell envelope and directly cause membrane alterations while remaining adsorbed to their receptors in the cell wall (see Figure 15A). It should be noted that where physical characterization has been possible, all of those colicins which are thought to affect membrane-associated functions are highly elongated molecules with dimensions sufficient to span the cell envelope (KONISKY, 1973).

The action of diphtheria toxin on cells is a two-step mechanism in which the adsorption of toxin molecules to a surface receptor is followed by cleavage of the toxin molecule releasing an active fragment into the cell interior. A similar mechanism may be operative in the case of the I colicins. Enzymatic cleavage could result in the release from the receptor of an active fragment, which would then be released into the periplasmic space where it could interact with the outer layers of the cytoplasmic membrane (Figure 15B).

It is also possible that the adsorption of I colicins to the specific cell-wall receptors is an intermediate step in the transport of the colicin molecules to some membrane target (Figure 15D). Perhaps the colicin I receptor complex in toto is dissociated from the cell wall and transported to the membrane (Figure 15E). Alternatively, the colicin I molecule may be transported in a series of steps via intermediate transport molecules.

Another possibility (Figure 15F) is that the Ia molecule does not leave its cell-wall receptor but is able to elicit a change in the outer membrane which is subsequently communicated to the inner membrane. For example, it is known that there are approximately 200 envelope sites where the inner and outer membranes are in close apposition. These so-called adhesion sites might be the vehicle whereby a colicin-induced alteration in the outer membrane might be transmitted to the ultimate target in the cytoplasmic membrane.

Contrary to the examples discussed here are possible mechanisms by which the colicin molecule itself does not interact with the final membrane target. For example, the interaction of I colicin with its receptor may trigger an activation of membrane-modifying enzymes, such as phospholipases in the cell wall; or lipases, proteases, ATPases, etc., in the periplasmic space (Figure 15C); or the membrane itself (Figures 15A,B,D,E, and F).

REFERENCES

BROMAN, R. L., DOBROGOSZ, W. J., WHITE, D. C.: Stimulation of cytochrome synthesis in *Escherichia coli* by cyclic AMP. Arch. Biochem. Biophys. 162, 595-601 (1974).

CARDELLI, J., KONISKY, J.: Isolation and characterization of an *Escherichia coli* mutant tolerant to colicins Ia and Ib. J. Bacteriol. 119, 379-385 (1974).

CRAMER, W. A., PHILLIPS, S. K., KEENAN, T. W.: On the role of membrane phase in the transmission mechanism of colicin E1. Biochemistry 12, 1177-1181 (1973).

FEINGOLD, D. S.: The mechanism of colicin E1 action. J. Membrane Biol. 3, 272-276 (1970).

FIELDS, K. L., LURIA, S. E.: Effects of colicins E1 and K on transport systems. J. Bacteriol. 97, 57-63 (1969).

FIELDS, K. L., LURIA, S. E.: Effects of colicins E1 and K on cellular metabolism. J. Bacteriol. 97, 64-77 (1969).

FOULDS, J.: The mode of action of a bacteriocin from *Serratia marcescens*, J. Bacteriol. 107, 833-843 (1971).

GIBSON, F., COX, G. B.: The use of *Escherichia coli* K 12 in studying electron transport and oxidative phosphorylation. Essays Biochem. 9, 1-29 (1974).

GILL, D. M., PAPPENHEIMER, A. M. JR.,: Structure-activity relationships in diphtheria toxin. J. Biol. Chem. 246, 1492-1495 (1971).

HAROLD, F. M.: Conservation and transformation of energy by bacterial membranes. Bact. Revs. 36, 172-230 (1972).

HAROLD, F. M.: Chemiosmotic interpretation of active transport in bacteria. In: The Mechanism of Energy Transduction in Biological Systems (ed. D. E. GREEN). Ann. N.Y. Acad. Sci. 227, 297-311 (1974).

HELGERSON, S. L., CRAMER, W. A., HARRIS, J. M., LYTLE, F. E.: An increase in micro-viscosity of the *E. coli* cell envelope caused by colicin E1. Biochemistry 13, 3057-3061 (1974).

HEMPFLING, W. P.: Repression of oxidative phosphorylation in *Escherichia coli* B by growth in glucose and other carbohydrates. Biochem. Biophys. Res. Commun. 41, 9-15 (1970).

HOFFEE, P., ENGELSBURG, E.: Effect of metabolic activity on the glucose permease of bacterial cells. Proc. Nat. Acad. Sci., U.S.A. 48, 1759-1765 (1962).

JETTEN, A. M., VOGELS, G. D.: Effects of colicin A and staphylococcin 1580 on amino acid uptake into membrane vesicles of *Escherichia coli* and *Staphylococcus aureus*. Biochem. Biophys. Acta 311, 483-495 (1973).

KABACK, H.R.: Bacterial membranes. In: Methods in Enzymology (ed. W.B. JAKOBY) Vol.XXII, pp.99-120. New York: Academic Press 1971.

KLEIN, W. L., BOYER, P. D.: Energization of active transport by *Escherichia coli*. J. Biol. Chem. 247, 7257-7265 (1972).

KONISKY, J.: Characterization of colicin Ia and Ib chemical studies of protein structure. J. Biol. Chem. 247, 3750-3755 (1972).

KONISKY, J.: Structure of colicins. In: Function and Structure of Colicins (ed. L. HAGER) pp. 41-58. New York: Academic Press 1973.

KONISKY, J., COWELL, B. S.: Interaction of colicin Ia with bacterial cells. Direct measurement of Ia-receptor interaction. J. Biol. Chem. 274, 6524-6529 (1972).

KONISKY, J., COWELL, B. S., GILCHRIST, M. J.: Colicin Ia and Ib binding to *Escherichia coli* envelopes and partially purified cell walls. J. Supramol. Struct. 1, 208-219 (1973).

KONISKY, J., LIU, C-T.: Solubilization and partial characterization of the colicin I receptor of *Escherichia coli*. J. Biol. Chem. 249, 835-840 (1974).

KONISKY, J., RICHARDS, F. M.: Characterization of colicin Ia and Ib. Purification and some chemical properties. J. Biol. Chem. 245, 2972-2978 (1970).

LEIVE, L.: Studies on the permeability changes produced on coliform bacteria by ethylenediaminetetraacetate. J. Biol. Chem. 243, 2373-2380 (1968).

LEVISOHN, R., KONISKY, J., NOMURA, M.: Interaction of colicins with bacterial cells. IV. Immunity breakdown studied with colicin Ia and Ib. J. Bacteriol. 96, 811-821 (1968).

LOMBARDI, F. J., REEVES, J. P., SHORT, S. A., KABACK, H. R.: Evaluation of the chemiosmotic interpretation of active transport in bacterial membrane vesicles. In: The Mechanism of Energy Transduction in Biological Systems (ed. D. E. GREEN). Ann. N.Y. Acad. Sci. 227, 312-327 (1974).

LURIA, S. E.: Colicins. In: Bacterial Membranes and Walls (ed. L. LEIVE) Vol. I, pp. 293-320. New York: Marcel Dekken, Inc. 1973.

MILLER, J. H.: Experiments in molecular genetics. Cold Spring, N.Y.: Cold Spring Harbor Laboratory 1972.

NOMURA, M.: Mode of action of colicins. Cold Spring Harbor Symposia on Quantitative Biology, 28, 315-324 (1963).

PAVLASOVA, E., HAROLD, F. M.: Energy coupling in the transport of B-galactosides by *Escherichia coli*: Effect of proton conductors. J. Bacteriol. 98, 198-204 (1969).

PHILLIPS, S. K., CRAMER, W. A.: Properties of the fluorescence probe response associated with the transmission mechanism of colicin E1. Biochemistry 12, 1170-1176 (1973).

RADDA, G. K.: Enzyme and membrane conformation in biochemical control. Biochem. J. 122, 385-396 (1971).

TIMMIS, K., HEDGES, A. J.: The killing of sensitive cells by colicin D. Biochem. Biophys. Acta 262, 200-207 (1972).

YURA, T., WADA, C.: Phenethyl alcohol resistance in *Escherichia coli* I. Resistance of strain C600 and its relation to azide resistance. Genetics 59, 177-190 (1968).

Energy Coupling

Energy Coupling and Solute Transfer

A. A. Eddy
Department of Biochemistry, University of Manchester, Institute of Science and Technology, Manchester M60 1QD, England

It is a commonplace among biologists that certain living cells concentrate specific nutrients that merely penetrate or are excluded from other types of cells. The scale of these effects is illustrated by the following examples. Whereas glucose penetrates the human erythrocyte by facilitated diffusion, without becoming concentrated, portions of the gut can accumulate glucose in amounts making the ratio of the cellular-to-extracellular sugar concentrations at least a factor of 20. Microorganisms provide other familiar examples of uphill, "active," transport. Thus, the yeast *Saccharomyces cerevisiae* probably absorbs galactose without concentrating it, whereas certain strains of *Escherichia coli* concentrate it up to 10^5 fold. Since a ratio of 10^8 or 10^9 might be produced if one equivalent of ATP was hydrolyzed per mole of solute absorbed, an interesting question arises as to the thermodynamic efficiency of natural transport systems, especially those known to produce relatively small concentration gradients.

The carbohydrate is probably absorbed as such in each of the above examples, even when it is subsequently metabolized. This is in contrast with the bacterial phosphotransferase systems where the carbohydrate is phosphorylated as it enters the cells (SIMONI, ROSEMAN, 1973).

Two hypotheses dominate current discussions about the mechanism of active transport processes that do not involve chemical conversion of the substrate. First, ejection of absorption of the solute might be directly linked to an exergonic chemical reaction, the change in free energy being partly conserved as an osmotic gradient of the solute itself.

Clearly, such a system might be studied by classical biochemical techniques. The prototype is the erythrocyte sodium pump, which is known to comprise an ATPase system synergistically activated by Na^+ and K^+. The activation of other mammalian ATPase systems by specific amino acids has been considered in the same context (LULY, VERNA, 1974).

Another outstanding proposal concerns the demonstration that specific modes of electron transport drive the accumulation of various solutes in bacterial membrane vesicles. It has been suggested that the alternate oxidation and reduction of a "carrier" substance might power the mechanism (KABACK, 1972).

Second, a great deal of current work points in the direction of an alternative interpretation, however, of the energy coupling mechanism during the absorption of specific carbohydrates, amino acids, and other simple solutes (MITCHELL, 1970; SCHULTZ, CURRAN, 1970; HAROLD, 1972). This interpretation, the co-substrate hypothesis, is based on an extension of the principles underlying penetration by facilitated diffusion. It was first applied to Na^+-dependent transport in mammalian cells, and subsequently to H^+-dependent systems of mitochondria, bacteria, and fungi.

The mechanism is based on two components. The first component is a carrier moving the primary solute, the sugar or amino acid, across the relevant membrane, together with its partners, one or more co-substrate ions. The latter are principally either Na^+ or H^+ in known examples. It is the spontaneous flow of the co-substrate ions across the cell membrane, down their own gradient, this being appropriately defined in terms of ionic activities and the prevailing electrical potentials acting across the membrane, that drives the accumulation of the primary solute. The second component of the mechanism is situated in a different part of the membrane. Here an amount of co-substrate ion equivalent to that absorbed with the solute is expelled across the membrane. This takes place through the ion pumps that all cells and various cell organelles possess. There is evidence that these pumps are activated by ATP in certain instances, and by a mechanism that probably depends directly on electron transport in other instances. According to this model, the ion pumps themselves are enzymes, working in conjunction with a selective "carrier" that facilitates the simultaneous passage through the membrane of both the solute to be concentrated and its attendant co-substrate ions. Other postulates are as follows.

(1) The membrane as a whole is relatively impermeable to the co-substrate ions and, at least in certain instances, allows electrical coupling to occur between the respective movements of charge through the "carrier" and the ion pump.

(2) The solute is absorbed with a specific number of equivalents of the co-substrate ions. In general, the power input would increase with that number.

(3) The absorption and possible concentration of the solute can occur, in principle, independently of the ion pumps and, hence, of the availability of ATP or whatever other substrate the ion pumps utilize. A specific prediction is that when the co-substrate ions flow spontaneously into the cells, so that charge neutralization can occur, the primary solute is concentrated in the cellular phase even when this is depleted of energy metabolites. The magnitude of the solute gradient formed would be a predictable function of the gradients of the co-substrate ions and any other ionic species participating.

(4) Both solute absorption and ion pumping would be influenced in a characteristic manner by specific ionophores.

The theoretical framework provided by these hypotheses has stimulated a great deal of experimental work. Much new information has been obtained about the factors governing solute accumulation in various systems. The problem has not been solved, however, and some of the new controversies are outlined in other chapters of this book.

REFERENCES

HAROLD, F. M.: Conservation and transformation of energy by bacterial membranes. Bacteriological Rev. 36, 172-230 (1972).

KABACK, R.: Transport across isolated bacterial cytoplasmic membranes. Biochim. Biophys. Acta 265, 367-416 (1972).

LULY, P., VERNA, R.: Stimulation of (Na^+-K^+)-ATPase of rat liver plasma membrane by amino acids. Biochim. Biophys. Acta 367, 109-113 (1974).

MITCHELL, P.: Membranes of cells and organelles: Morphology, transport and metabolism. In: Organization and Control in Prokaryotic and Eukaryotic Cells. XXth Symp. Soc. Gen. Microbiol., 121-166 (1970).

SCHULTZ, S. G., CURRAN, P. F.: Coupled transport of sodium and organic solutes. Physiol. Rev. 50, 637-718 (1970).

SIMONI, R. D., ROSEMAN, S.: Sugar transport. VII. Lactose transport in *Staphylococcus aureus*. J. Biol. Chem. 248, 966-974 (1973).

Studies of Respiratory Control in Submitochondrial Particles and Reconstituted Systems*

Peter C. Hinkle, Yen-sheng L. Tu, and Jung Ja Kim
Section of Biochemistry, Molecular and Cell Biology, Cornell University, Ithaca, New York 14850

INTRODUCTION

The stimulation of respiration by uncoupling agents, called respiratory control by analogy with the stimulation of respiration by ADP plus phosphate, has been very useful in studies of the mechanism of energy coupling in oxidative phosphorylation. LOOMIS and LIPMANN (1948) originally reported that the rate of respiration of isolated mitochondria is increased by addition of 2,4-dinitrophenol. This led to the concept that such "uncoupling" agents break down a high-energy intermediate and thus relieve the back pressure of the coupling reactions on the oxidation reactions of the respiratory chain (SLATER, 1953). LEE and ERNSTER (1966) extended these observations by demonstrating that sonically fragmented vesicles of mitochondria, submitochondrial particles, also showed a stimulation of respiration by uncoupling agents, and that the controlled respiration of such preparation corresponded to the ability to couple respiration to various energy-linked functions.

MITCHELL (1961, 1963) suggested in his chemiosmotic hypothesis of oxidative phosphorylation that the coupling intermediate that was broken down by uncoupling agents was an electrochemical gradient of protons across the inner mitochondrial membrane. Numerous studies have shown a correlation between the effectiveness of uncoupling agents as proton conductors and in stimulating controlled respiration (BIELAWSKI et al., 1966; HOPFER et al., 1968; SKULACHEV, 1971; BAKKER et al., 1973). In terms of the chemiosmotic hypothesis, the rate of respiration is controlled by the size of the electrochemical proton gradient across the coupling membrane, which in turn is controlled by the bal-

*This study was supported by Public Health Research Grant HL 14483 and Career Development Award GM 22427.

ance of proton translocation by the respiratory chain and the backflow of protons by "leaks." This hypothesis predicts, therefore, that as the respiratory chain is inhibited by an inhibitor or by low substrate concentration, the degree of respiratory control should decrease, since the rate of proton translocation would be less while the leaks would be unchanged, assuming that there are several respiratory chains per vesicle. However, there are conflicting reports. VALLIN (1968) reported that inhibition of respiration in submitochondrial particles with rotenone lowered the stimulation by uncoupling agents, whereas LEE, ERNSTER, and CHANCE (1969) found that inhibition by cyanide, rotenone, or antimycin did not lower respiratory control, but limiting NADH concentrations or inhibition of succinate oxidation by malonate did. More recently, HUNTER (1974) reported that cyanide increased the apparent respiratory control in submitochondrial particles, but his effect occurred only if the cyanide was added after NADH and the rates of respiration were not linear since the cyanide was not given time to equilibrate with cytochrome oxidase.

We have studied uncoupler-stimulated oxidation reactions in reconstituted complexes of the respiratory chain and have correlated the stimulation of the proton-translocating reactions with the proton conducting activity of uncoupling agents. To attempt to resolve the conflicting observations described above, we have re-examined uncoupler-stimulated respiration in submitochondrial particles after partially inhibiting the oxidation chains with various inhibitors.

RESULTS

A titration of the effect of cyanide on rates of NADH oxidation by submitochondrial particle preparation from bovine heart mitochondria, ETP_H (BEYER, 1967), before and after addition of the uncoupler carbonyl cyanide m-chlorophenylhydrazone (CCCP), is shown in Figure 1. Cyanide was added 3 minutes before NADH, and oxygen uptake was measured with a Clark oxygen electrode. The rates were linear before and after addition of CCCP. Oligomycin was added in these experiments to obtain maximum stimulation of respiration by uncouplers (LEE, ERNSTER, 1966). Very similar results were found in the absence of oligomycin, however, since the submitochondrial particle preparation is not deficient in F_1 ATPase. The respiratory control ratios, i.e., the rate in the presence of CCCP divided by the rate in the absence of CCCP, are shown in Figure 2 as a function of the oxidation rate in the presence of CCCP. The respiratory control ratio (RCR) increased linearly with oxidation

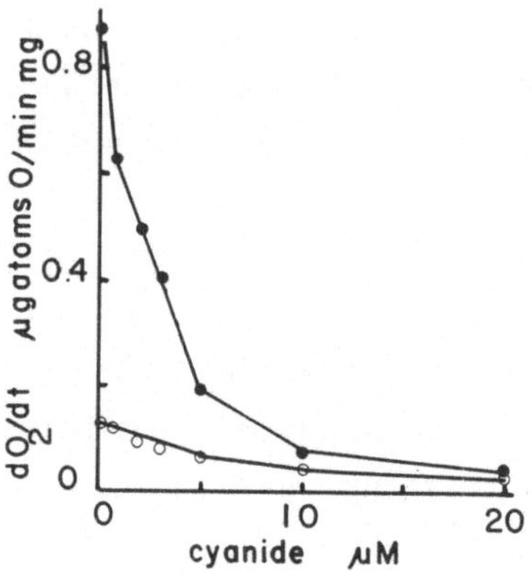

Fig. 1

Fig. 2

Fig. 1. Inhibition of respiration by cyanide. The reaction mixture contained 250 mM sucrose, 10 mM K^+ morpholinopropane sulfonate, pH 7.0, 1.2 mg ETP_H, 1.2 μg oligomycin, and 0.6 mM NADH in a final volume of 1.2 ml. The ETP_H were preincubated with cyanide at the levels shown for 3 minutes, after which respiration was initiated by addition of NADH. After about 1 minute, 4 μM CCCP were added. The rates of respiration with (closed circles) and without (open circles) CCCP are shown

Fig. 2. Effect of respiration rate on respiratory control. The Respiratory Control Ratios (RCR) calculated from the data in Figure 1 are shown versus respiration rate in the presence of CCCP (V_{CCCP})

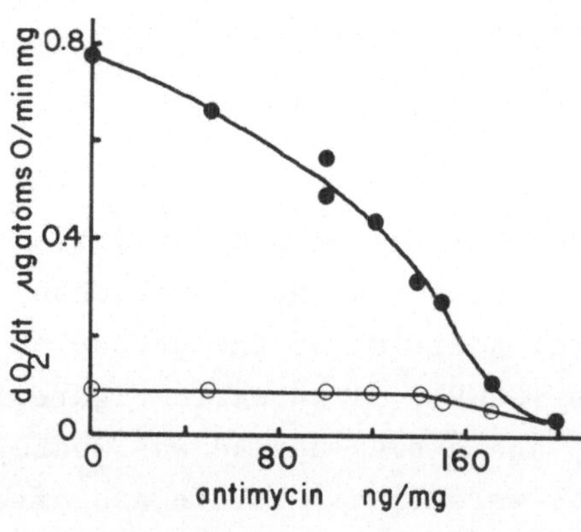

Fig. 3

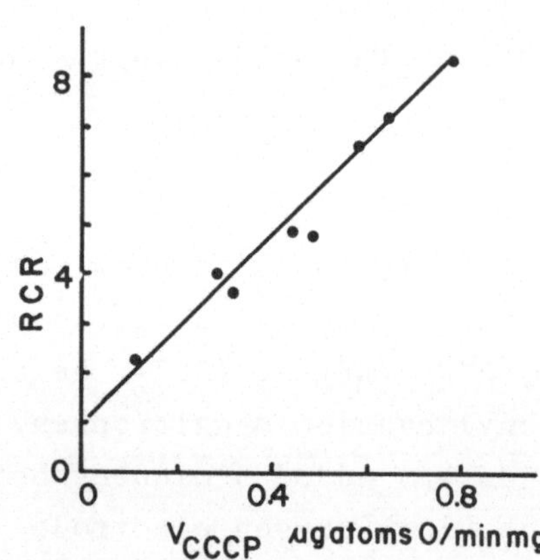

Fig. 4

Fig. 3. Inhibition of respiration by antimycin A. The reaction mixture was as described in Figure 1, except that antimycin A was used as the inhibitor in place of cyanide

Fig. 4. Effect of respiration rate on respiratory control. The data from Figure 3 are recalculated as respiratory control ratio versus rate of respiration in the presence of CCCP

rate from close to 1.0 at the highest cyanide concentration to 6.8 at the maximum oxidation rate without cyanide. The slope of the line is 0.00658 (ng atoms O/min mg)$^{-1}$.

A titration of NADH oxidation with antimycin A carried out in the same way is shown in Figure 3. The degree of inhibition is sigmoid with antimycin A concentration (SLATER, 1963). A plot of the respiratory control ratio at different degrees of inhibition (Figure 4) shows that the control ratio increased linearly with respiratory rate as in the case of cyanide inhibition. The intercept at zero respiratory rate is 1.0 and the slope is 0.00959 (ng atoms O/min mg)$^{-1}$.

A titration of NADH oxidation with rotenone is shown in Figure 5. The plot of the respiratory control ratio versus respiration rate (Figure 6) gives a straight line as with the other inhibitors, except that the intercept at zero respiration rate is about 1.4 instead of 1.0. In another experiment with a different preparation of ETP_H this intercept was 1.2 and a value slightly above 1 may not be significant. The slope in Figure 6 is 0.00862 (ng atoms O/min mg)$^{-1}$.

A possible explanation for the results of LEE, ERNSTER, and CHANCE (1969), who found no effect of inhibitors on respiratory control in submitochondrial particles, is that the halftime for inhibition by rotenone, antimycin, or cyanide at the low concentrations used in these studies is about 1 minute. A prior incubation with the inhibitor is therefore essential in order to obtain linear rates of respiration.

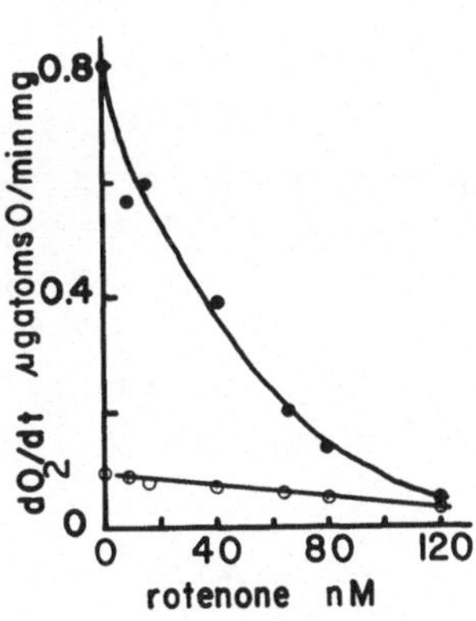

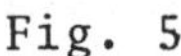

Fig. 5

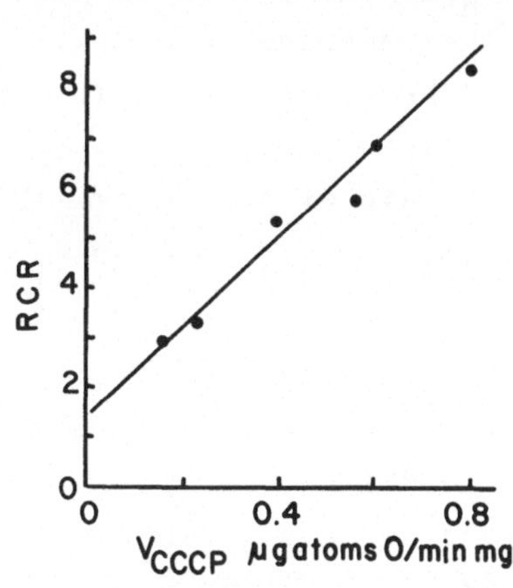

Fig. 6

Fig. 5. Inhibition of respiration by rotenone. The reaction mixture is as described in Figure 1, except that rotenone was used as the inhibitor in place of cyanide

Fig. 6. Effect of respiration rate on respiratory control. The data from Figure 5 are recalculated as respiratory control ratio versus rate of respiration in the presence of CCCP

The above authors calculated respiration rates from the time between addition of substrate and anaerobiosis. If the inhibitors were not adequately equilibrated, the longer time needed to reach anaerobiosis without uncouplers would have allowed more complete inhibition which would have appeared to increase the respiratory control.

Our results show that when the inhibitors cyanide, antimycin A, or rotenone are used to partially block respiration in well coupled submitochondrial particles, respiratory control is lost as it is when respiration is inhibited by limiting substrate concentration. Furthermore, the control ratio is apparently linearly dependent on the respiration rate. The following analysis of the factors controlling respiration in submitochondrial particles discusses some interesting aspects of these data.

Consider the steady state of respiration in submitochondrial particles, where respiration drives proton uptake and the protons leak out of the vesicles at an equal rate. If no permeant acids are present, the decay of a pH gradient in mitochondria and submitochondrial particles is very closely first order, indicating that the proton flux through the leak pathway is proportional to the pH gradient. As described by MITCHELL and MOYLE (1967a), we can write the leak rate as

$$J_{H^+}^{leak} = C_m \, \Delta\bar{\mu}_{H^+} \tag{1}$$

where J_H^+ is the efflux of protons plus the influx of hydroxide ions, c_m is the membrane conductance to protons plus hydroxide ions, and $\Delta\bar{\mu}_{H^+}$ is the electrochemical potential difference of protons across the membrane.

The rate of proton translocation by the respiratory chain is

$$J_{H^+}^{pump} = V \, H^+/0 \tag{2}$$

where V is the rate of oxidation and $H^+/0$ is the proton coupling ratio, which is 6 in the case of NADH oxidation (MITCHELL, MOYLE, 1967b; HINKLE, HORSTMAN, 1971).

To calculate at what point this pump and leak system will reach a steady state, it is necessary to know the rate of respiration as a function of the electrochemical proton gradient. This function is not known, but there are some indications that it may be linear. PADAN and ROTTENBERG (1973) measured the $\Delta\bar{\mu}_{H^+}$ and rate of respiration in rat liver mitochondria at various levels of 2,4-dinitrophenol and observed a linear decrease in $\Delta\bar{\mu}_{H^+}$ as respiration increased. The highest value of $\Delta\bar{\mu}_{H^+}$ they measured, however, was 150 mv, which is considerably lower than the estimates of MITCHELL and MOYLE (1969), SKULACHEV (1971), and our own unpublished observations, which show a value of about 250 mv. Another indication that the respiratory rate is a linear function of $\Delta\bar{\mu}_{H^+}$ comes from studies of the effect of ΔG_{ATP} on the respiratory rate of submitochondrial particles (THAYER, HINKLE, 1974). We found that the rate of NADH oxidation decreased linearly with increasing ΔG_{ATP} and that the respiratory rate in the presence of CCCP corresponded to the point where $\Delta\bar{\mu}_{H^+}$ is zero.

One experiment that indicates that respiration is maximally activated at values of $\Delta\bar{\mu}_{H^+}$ above zero is the one in which ADP + P_i stimulate respiration maximally in rat liver mitochondria but cannot lower $\Delta\bar{\mu}_{H^+}$ very much because ATP synthesis requires energy, as was discussed by PADAN and ROTTENBERG (1973). MITCHELL and MOYLE (1969) measured the change of $\Delta\bar{\mu}_{H^+}$ on addition of ADP to respiring mitochondria and found that it was only 30 mv lower than during controlled respiration. In submitochondrial particles, however, maximum rates of respiration are obtained only with uncouplers. In any case, as a working hypothesis Eq. (3) can be used to describe the rate of respiration,

$$V = V_{CCCP}\ (1 - b\ \Delta\bar{\mu}_{H^+}) \qquad (3)$$

where V is the oxidation rate in the presence of the proton gradient $\Delta\bar{\mu}_{H^+}$, V_{CCCP} is the uncoupled oxidation rate when $\Delta\bar{\mu}_{H^+} = 0$, and b is a coefficient that can be calculated from our data as 0.003 mv^{-1} (THAYER, HINKLE, 1974) and from PADAN and ROTTENBERG (1973) as 0.0063 mv^{-1}. To use Eq. (3) to analyze the respiratory control experiments, the proton flux from the leak is equated with the flux from the respiratory chain since the system is in a steady state:

$$J_{H^+}^{leak} = J_{H^+}^{pump} = C_m\ \Delta\bar{\mu}_{H^+} = V(H^+/0) \qquad (4)$$

Rearranging Eq. (4),

$$\Delta\bar{\mu}_{H^+} = V\ \frac{H^+/0}{C_m} \qquad (5)$$

Substituting Eq. (5) into Eq. (3),

$$V = V_{CCCP}\ (1 - \frac{Vb\ H^+/0}{C_m} \qquad (6)$$

and

$$RCR = \frac{V_{CCCP}}{V} = 1 + \frac{b\ H^+/0}{C_m}\ V_{CCCP} \qquad (7)$$

Thus the assumption that the respiration rate is proportional to $(1 - b\ \Delta\bar{\mu}_{H^+})$ leads to the conclusion that the respiratory control ratio caused by uncoupling agents is proportional to the rate of respiration in the presence of the uncoupler, and that the slope of the line relating RCR to V_{CCCP} is

$$\frac{b\ H^+/0}{C_m}.$$

The constant b was previously evaluated as 0.003 mv^{-1}, $H^+/0$ is 6.0 for NADH respiration (MITCHELL, MOYLE, 1967b; HINKLE, HORSTMAN, 1971), and C_m is in the range 0.5 - 1.5 nmoles/mg mv min (MITCHELL, MOYLE, 1967a; HINKLE, HORSTMAN, 1971; HINKLE, 1974). Thus, the value of

$$\frac{b\ H^+/0}{C_m}$$

should be between 0.0036 and 0.0011 (ng atoms 0/min mg)$^{-1}$. This compares well with the values of 0.00658 (Figure 2), 0.00959 (Figure 4), and 0.00862 (Figure 6).

Studies of energy coupling in reconstituted systems have now progressed to include the mitochondrial ATPase complex and the three coupling regions of the respiratory chain. As outlined in Table 1, each of these systems shows an uncoupler-stimulated reaction analogous to respiratory control. This effect can be shown to be caused by the dissipation of a proton gradient across the liposome membrane by several lines of evidence. Table 2 shows the effect of ionophores on the rate of respiration by cytochrome oxidase vesicles. An uncoupler or

Table 1. Uncoupler-stimulated reactions in reconstituted systems

System	CCCP-stimulated reaction	Reference
ATPase vesicles	$ATP + H_2O \rightarrow ADP + P_i$	KAGAWA, RACKER (1971)
Cytochrome oxidase vesicles	$2\ Cyt\ c_{red} + \frac{1}{2}\ O_2 \rightarrow 2\ Cyt\ c_{ox} + H_2O$	HINKLE, KIM, RACKER (1972)
NADH-CoQ reductase vesicles	$NADH + M^+ + CoQ_1 \rightarrow NAD + CoQ_1H_2$	RAGAN, HINKLE (1974)
CoQ-Cyt c reductase vesicles	$CoQ_2 + 2\ Cyt\ c_{ox} \rightarrow CoQ_2 + 2\ Cyt\ c_{red}$	HINKLE, LEUNG (1974)

Table 2. Respiratory control in cytochrome oxidase vesicles[a]

Additions	Oxygen uptake (ng atoms/min)	Respiratory control ratios
None	53	-
4 μM CCCP	124	2.3
100 μM 1799	205	3.9
0.5 μg valinomycin	59	1.1
0.4 μg nigericin	59	1.1
Valinomycin + nigericin	196	3.8

[a]Respiration was measured with a Clark oxygen electrode, with cytochrome oxidase vesicles (HINKLE, KIM, RACKER, 1972) (20 μg protein) in 50 mM KP_i, pH 7.0, 20 mM Na ascorbate, and 1.2 mg cytochrome c in a final volume of 1.2 ml.

the combination of valinomycin plus nigericin in a potassium medium gives maximum release of respiration. The synergetic effect of valinomycin, a potassium ionophore and nigericin, and electrically neutral potassium for hydrogen exchange carrier is similar to studies with submitochondrial particles (CHANCE, MONTAL, 1972) and strongly suggests that the $\Delta\bar{\mu}_{H^+}$ must be dissipated for release of respiration.

Titrations of cytochrome oxidase vesicles with polylysine, sodium sulfide, and cytochrome c during measurement of uncoupler-stimulated respiration is shown in Figure 7. Cyanide was not included because nonlinear rates were obtained on adding uncoupler even after extensive preincubation of the vesicles with the inhibitor. The slope of the line in Figure 7 is 0.011 (ng atom 0/mg lipid min)$^{-1}$. To apply Eq. 7 to cytochrome oxidase vesicles values of b can be taken as 0.003 mv^{-1}, the same as in submitochondrial particles since there are no estimates of this parameter in cytochrome oxidase vesicles. The value of $H^+/0$ is 2 (HINKLE, KIM, RACKER, 1972), and C_m can be estimated (MITCHELL, MOYLE, 1967a) as 0.008 ng ions H^+/mv min mg lipid from previous determinations of the internal buffering power as 13.6 ng ions H^+/mg lipid pH unit (HINKLE, 1973) and a typical halftime for decay of a pH gradient of 2 min. The value of

$$\frac{b\ H^+/0}{C_m}$$

is then estimated as 0.0075 (ng atoms 0/mg lipid min)$^{-1}$, which compares well with the value of 0.011 from Figure 7. The underestimation may result from heterogeneity in the cytochrome oxidase preparation, which could contain lipid vesicles that do not have cytochrome oxidase since

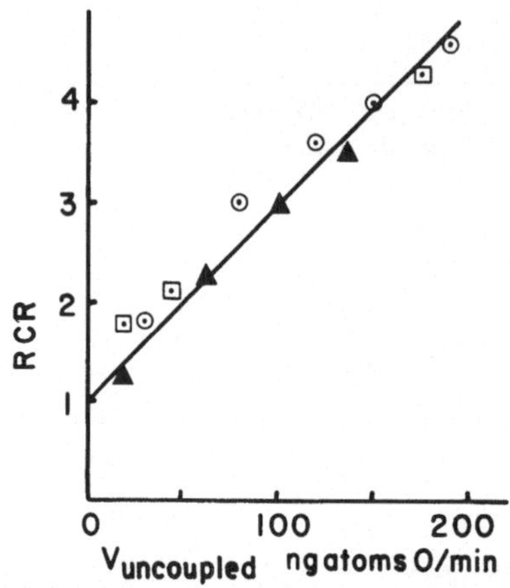

Fig. 7. Effect of respiration rate on respiratory control in reconstituted cytochrome oxidase vesicles. Respiration was measured as described in Figure 1 in a medium containing 40 mM potassium phosphate, pH 7.0, 17 mM sodium ascorbate, and cytochrome oxidase vesicles (HINKLE et al., 1972) (20 µg protein or 0.66 mg phospholipid) and 0.8 mg cytochrome c. Uncoupling was achieved with 0.5 µg valinomycin plus 0.4 µg nigericin. Respiration was partially inhibited by varying the cytochrome c concentration from 800 to 50 µg (circles), adding sodium sulfide from 10 to 40 µg (squares) or adding polylysine from 50 to 150 µg (triangles)

an excess of lipid was used in the reconstitution. This would mean that the internal buffering power of the active vesicles would be less than the total and that the proton conductance C_m of the active vesicles is less than was calculated.

It should be possible to reconstitute cytochrome oxidase vesicles with only one oxidase complex per liposome. Such vesicles would not show a loss of respiratory control if an irreversible or only slowly reversible inhibitor is used to partially inhibit respiration, because the proton leaks in each vesicle would be associated with only one oxidase and the ratio of leaks to pumps would not change. Reconstitutions were tried at very low levels of cytochrome oxidase, but the results were similar to those shown in Figure 7. It is possible that (1) the reconstitution with a small amount of enzyme creates a small population of vesicles with several oxidase complexes per vesicle, or (2) that the inhibitors used are too rapidly reversible to demonstrate this phenomenon.

CONCLUSIONS

Our results do not confirm previous reports that respiratory control is unchanged by partial inhibition of respiration. Instead, we found that as respiration was inhibited, the extent of stimulation by uncouplers decreased. Quantitatively, the effect could be described as the balance between proton translocation by the respiratory chain and the leakage of protons back across the membrane. This analysis includes the assumption that the rate of respiration is a linear function of the electrochemical potential difference of protons across the membrane over a range from 0 to 5 kcal/g ion. This assumption, which is supported by other observations, is the central postulate of the field of nonequilibrium thermodynamics (KATCHALSKY, CURRAN, 1967), and further studies along these lines may demonstrate a more general application of this discipline to oxidative phosphorylation. It may be, however, that the applicability of these equations is limited to a few preparations, such as submitochondrial particles and some reconstituted systems, and that other systems such as whole liver mitochondria have different rate-limiting reactions and require a different analysis.

REFERENCES

BAKKER, E. P., VANDENHEUVEL, E. J., WIECHMANN, A. H. C. A., VANDAM, K.: Biochim. Biophys. Acta 292, 78-87 (1973).

BEYER, R. E.: Methods in Enzymology 1o, 186 (1967).

BIELAWSKI, J., THOMPSON, T. E., LEHNINGER, A. L.: Biochem. Biophys. Res. Commun. 24, 948 (1966).

CHANCE, B., MONTAL, M.: In: Current Topics in Membranes and Transport (ed. F. BRONNER, A. KLEINZELLER) Vol. II, p. 99. New York: Academic Press 1972.

HINKLE, P. C.: Proc. New York Acad. Sci. 227, 159-165 (1973).

HINKLE, P. C.: Unpublished observations (1974).

HINKLE, P. C., HORSTMAN, L. L.: J. Biol. Chem. 246, 6024-6028 (1971).

HINKLE, P. C., KIM, J. J., RACKER, E.: J. Biol. Chem. 247, 1338 (1972).

HINKLE, P. C., LEUNG, K. H.: In: Membrane Proteins in Transport and Phosphorylation (ed. G. F. AZZONE, M. E. KLINGENBERG, E. QUAGIARIELLO, N. SILIPRANDI), pp. 73-78. Amsterdam: North-Holland Publishing Co. 1974.

HOPFER, U., LEHNINGER, A. L., THOMPSON, T. E.: Proc. Nat. Acad. Sci. U.S.A. 59, 484 (1968).

HUNTER, D. R.: Biochem. Biophys. Res. Commun. 57, 1063-1068 (1974).

KAGAWA, Y., RACKER, E.: J. Biol. Chem. 246, 5477 (1971).

KATCHALSKY, A., CURRAN, P. F.: Nonequilibrium Thermodynamics in Biophysics. Cambridge: Harvard University Press 1967.

LEE, C. P., ERNSTER, L.: BBA Library 7, 218 (1966).

LEE, C. P., ERNSTER, L., CHANCE, B.: Eur. J. Biochem. 8, 153-163 (1969).

LOOMIS, W. F., LIPMANN, F.: J. Biol. Chem. 173, 807 (1948).

MITCHELL, P.: Nature 191, 144 (1961).

MITCHELL, P.: In: Cell Interface Reactions (ed. H. D. BROWN). New York: Scholar's Library 1963.

MITCHELL, P.: Biol. Rev. 41, 445 (1966).

MITCHELL, P., MOYLE, J.: Biochem. J. 104, 588 (1967a).

MITCHELL, P., MOYLE, J.: Biochem. J. 105, 1147 (1967b).

MITCHELL, P., MOYLE, J.: Eur. J. Biochem. 7, 471 (1969).
PADAN, E., ROTTENBERG, H.: Eur. J. Biochem. 431-437 (1973).
RAGAN, C. I., HINKLE, P. C.: Submitted (1974).
SKULACHEV, V.: Current Topics in Bioenergetics 4, 127 (1971).
SLATER, E. C.: Nature 172, 975 (1953).
SLATER, E. C.: Biochim. Biophys. Acta 301, 129-154 (1963).
THAYER, W. S., HINKLE, P. C.: Submitted (1974).
VALLIN, I.: Biochim. Biophys. Acta 162, 477-486 (1968).

Energy Coupling in the Plasma Membrane of *Neurospora*: ATP-Dependent Proton Transport and Proton-Dependent Sugar Cotransport

Carolyn W. Slayman and Clifford L. Slayman
Departments of Human Genetics and Physiology, Yale School of Medicine, New Haven, Connecticut 06510

INTRODUCTION

In the plasma membranes of eukaryotic cells, there are two distinct mechanisms by which energy can be supplied to "active" transport systems. The first involves the reaction of a high-energy compound (such as ATP) directly with the transport system; the second, which has been called "cotransport" by animal physiologists, makes use of the downhill movement of one substance along its electrochemical gradient, with the physical coupling between the two fluxes thought to reside in a ternary complex between the carrier and the two substances. The best-known examples of these two mechanisms in animal cells are the ATP-driven transport of Na^+ and K^+ (reviewed by SKOU, 1971) and the Na^+-driven transport of amino acids and sugars (reviewed by CHRISTENSEN, 1970; SCHULTZ, CURRAN, 1970). In these instances, and in general, cotransport mechanisms of the second type depend ultimately upon primary, covalent bond-driven mechanisms of the first type.

Although in animal cells the common link between the two kinds of systems is the sodium ion, there is increasing evidence in other kinds of cells that it may be the proton. Attention was first called to ATP-dependent H^+ transport and H^+-dependent substrate cotransport in mitochondria, chloroplasts, and bacterial membranes by Peter MITCHELL in 1961; and although the idea remains controversial, there is a considerable body of experimental data that can most easily be interpreted in MITCHELL's terms. The arguments and counterarguments arising from work on bacteria are summarized elsewhere in this volume (see HAROLD, this volume; KABACK et al., this volume). The present paper will concentrate on a somewhat different approach to the same questions in the simple eukaryotic organism, the fungus *Neurospora crassa*.

ATP-DEPENDENT H^+ TRANSPORT

The chief experimental advantage of *Neurospora* is that--among microorganisms that have been well studied genetically and biochemically--it is unique in having cells large enough to be punctured with micropipette electrodes. One can therefore make direct measurements of membrane potential and resistance--properties that are important in determining the movement of ions across the membrane.

The earliest measurements showed that (1) the resting membrane potential of *Neurospora* is -180 to -250 mV, too large to be a conventional ionic diffusion potential (SLAYMAN, 1965a); and (2) it is extremely sensitive to metabolic inhibitors (SLAYMAN, 1965b; SLAYMAN, LONG, LU, 1973). Both of these characteristics are illustrated in Figure 1, which is a simultaneous recording of membrane potential from two cells bathed in standard dimethylglutarate buffer at pH 5.8. The resting potentials in these two cells were -235 and -222 mV; at the upward arrow, 3 mM KCN was added and the potential fell very rapidly by about 165 mV; at the downward arrow, CN was washed out, and the potential recovered gradually, with an overshoot, to the control value. The cycle was repeated once in this particular experiment and can be repeated indefinitely.

One might suspect, from the fact that *Neurospora* is an obligate aerobe with a conventional cytochrome chain (WEISS et al., 1970;

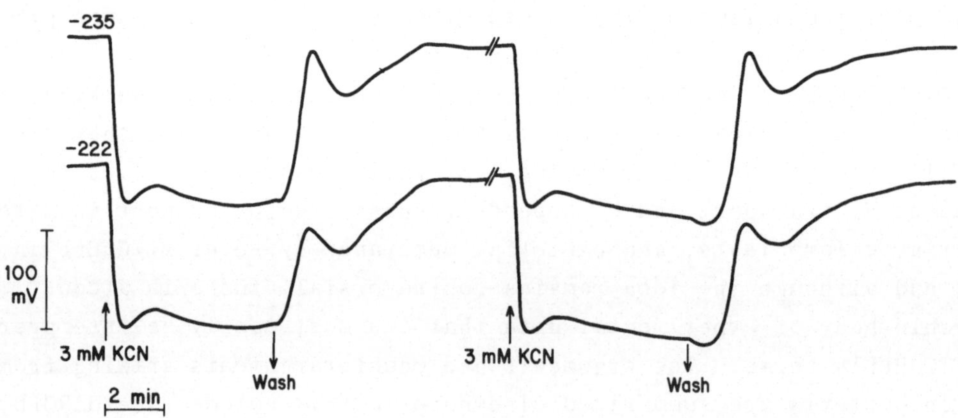

Fig. 1. Effect of cyanide on the membrane potential of *Neurospora*. Simultaneous records from two separate hyphae. Preparation bathed in standard DMG buffer [20 mM dimethylglutaric acid, 25 mM K(OH), 1 mM $CaCl_2$, 1% glucose, pH 5.8]; 3 mM KCN (neutralized with HCl) added at the up arrows, and washed out at the down arrows. Seven minutes omitted at the break in the records. Depolarization (cell interior becoming less negative) is indicated by a downward deflection. Ambient temperature approximately 24°C

LAMBOWITZ et al., 1972; LAMBOWITZ, SMITH, SLAYMAN, 1972a,b), that CN acts at the level of cytochrome oxidase, blocking ATP synthesis coupled to the cytochrome chain and causing cellular ATP to fall. Figure 2 shows that this is the case and indicates, furthermore, that there is a quantitative correspondence between the rate at which ATP falls and the rate of change of the membrane potential. In one set of experiments, CN (at saturating concentrations, 1 to 25 mM) was injected into a well mixed suspension of cells, and aliquots of cells were harvested at rapid intervals between 0 and 60 sec and assayed for ATP by the luciferase method. The results are plotted as the solid points in Figure 2A. The decay of cellular ATP can be fitted well by a simple exponential curve with a time constant of 5.7 sec (SLAYMAN, LONG, LU,

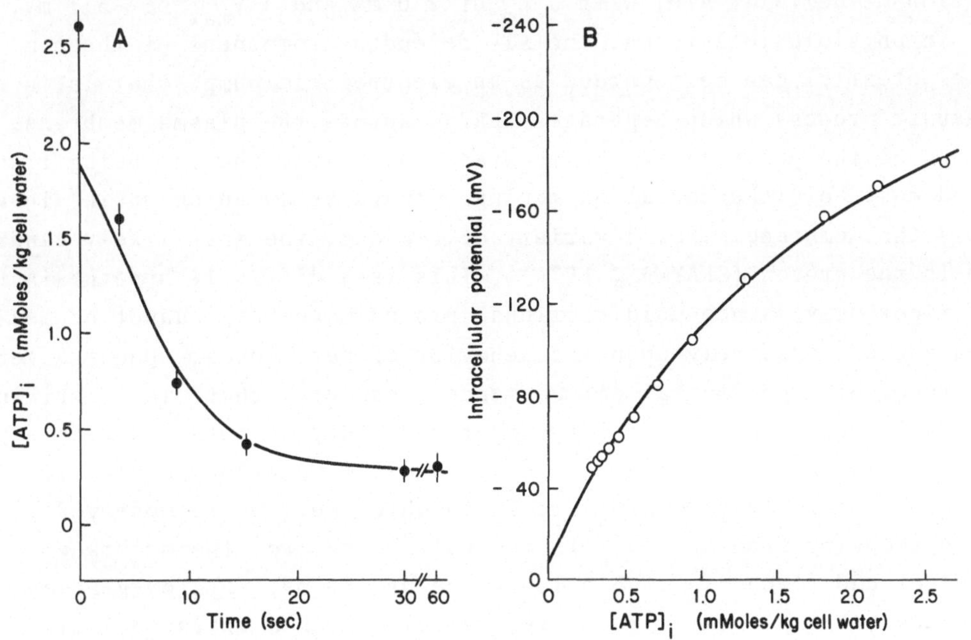

Fig. 2. Relationship between membrane potential and cellular ATP level at the onset of cyanide inhibition. (A) Plot of both ATP (●) and voltage (——) against time. Each ATP value is the average for 12 to 21 determinations at CN concentrations of 1 to 25 mM; vertical bars, ± 1 S.E.M. Curve of voltage is the time average of 23 trials on 16 different hyphae. ATP data obtained at 25°C, voltage data at 24°C. (B) Plot of voltages directly against ATP concentrations. Values for ATP taken from an exponential fit to the data in part A, adjusted to 24°C; time points for interpolation (sec) were 0, 1.25, 2.5, 2.5-sec intervals to 22.5, 30. Curve fitted according to the equation

$$V_m = V_o + \frac{V_{pm} \cdot [ATP]_i}{K_{1/2} + [ATP]_i}$$

with $V_o = 7.0 \pm 6.6$ mV, $V_{pm} = 311 \pm 24$ mV, and $K_{1/2} = 2.0 \pm 0.4$ mM. Data from SLAYMAN, LONG, and LU (1973)

1973). In a parallel set of experiments, individual hyphae were impaled with microelectrodes, CN was introduced into the chamber (at a high concentration, 25 mM, to minimize mixing delay), and the membrane potential was recorded as a function of time. The average response in 23 trials is given as the solid curve in Figure 2A, where, after a short lag, the membrane potential can be seen to fall with the same time-course as ATP. The results demonstrate that the membrane potential in *Neurospora* is closely dependent upon ATP. More quantitatively, when the values of membrane potential from Figure 2A are replotted as a function of cellular ATP concentration (Figure 2B), one sees that the potential has two distinct components: (1) a small, ATP-independent component--probably a diffusion potential--amounting to -7 mV under these conditions, and (2) a much larger component which saturates as a function of cellular ATP, with a K_m of 2.0 mM and a V_{max} of -311 mV.

In physiological terms, the ATP-dependent component of the membrane potential can be regarded as an electrogenic pump--that is, a transport process which separates charge across the plasma membrane. Because of the polarity of the membrane potential, the ion being transported must be either a cation moving outward or an anion moving inward across the membrane. For a variety of reasons, the most likely candidate is the proton (SLAYMAN, 1970). This is a difficult hypothesis to test rigorously, since unidirectional proton movements cannot be measured and one must rely upon the behavior of net fluxes. Under a very wide range of experimental circumstances, however, there is a striking correlation between the net outward proton flux, as measured with a pH electrode in the extracellular medium, and the membrane potential. One instance is illustrated in Figure 3, in which cells were observed during recovery from anoxia. In the voltage record, the membrane potential had reached a stable low value of -40 mV; when oxygen was reintroduced, the potential recovered rapidly, with a characteristic overshoot during the first minute, and finally stabilized at -175 mV. Similarly, the net outward proton flux, measured in a parallel experiment with a cell suspension, had reached a very low value (about 0.4 mmoles/liter cell water•min) in the anoxic cells. When oxygen was added, the proton flux increased rapidly, again with an overshoot in the first minute, and then leveled off at about 8 mmoles/liter cell water•min (a normal value for aerated cells). Such similarity of time courses cannot, of course, prove conclusively that the proton flux and the membrane potential are causally related; but with all other reasonable ion fluxes ruled out on independent grounds (SLAYMAN, 1970), each new example of similar behavior strengthens the case that the substrate for the electrogenic pump is the proton.

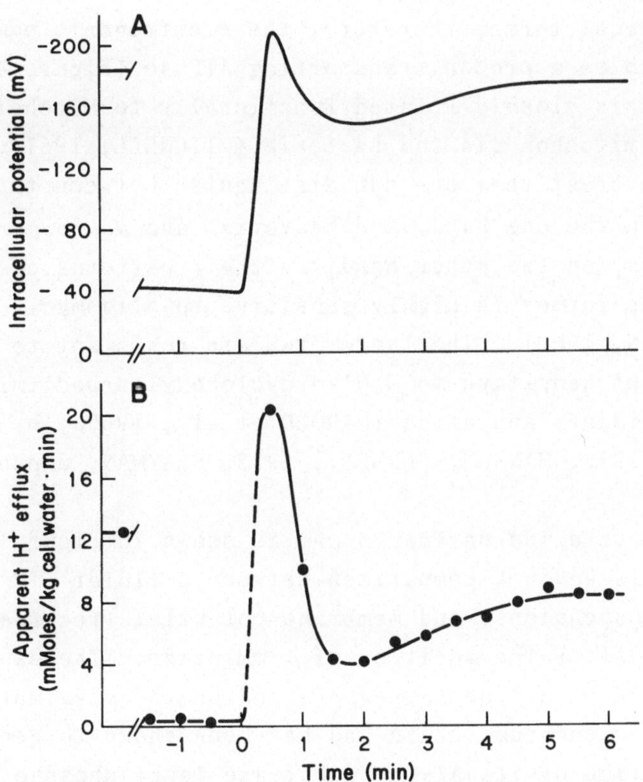

Fig. 3. Comparison of membrane potential (A) and rate of net H^+ efflux (B) following a period of anoxia. In the voltage measurement, cells were maintained in a standard salt solution (10 mM KCl, 1 mM $CaCl_2$, 2% sucrose) saturated with carbon monoxide (1 atm) for 20 min before O_2-saturated solution was washed through (zero time). In the H^+-flux experiments, cells were incubated with a diluted DMG buffer solution [4 mM dimethylglutaric acid, 5 mM K(OH), 20 mM KCl, 1 mM $CaCl_2$, 1% glucose, pH 5.8] in a closed chamber for 10 to 15 min and had consumed all of the oxygen about 3 min before a pulse of O_2-saturated solution was injected. Plot B is the average for single trials on five separate preparations. Analysis was carried out as for Figure 9. A, ambient temperature; B, 25°C

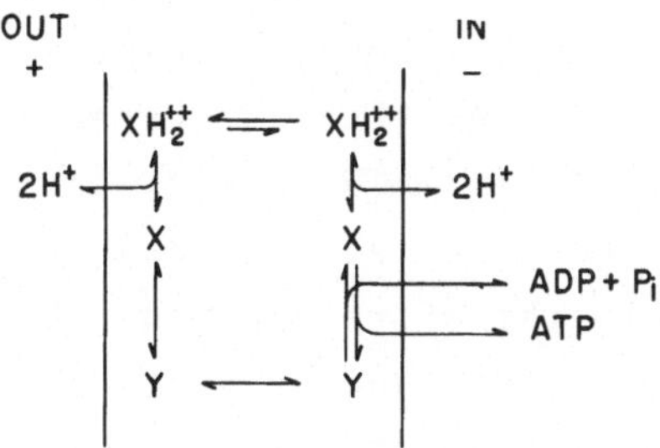

Fig. 4. Cyclic carrier model for an electrogenic proton pump that uses energy from ATP hydrolysis to drive H^+ outward, with a stoichiometry of 2 H^+:1 ATP (the stoichiometry reported for mitochondria and consistent with pump voltages in *Neurospora*). Unequal arrows indicate that the normal function of the pump is to drive XH_2^{++} outward, polarizing the membrane negatively at the inside surface. (Taken from SLAYMAN, 1974.)

In biochemical terms, therefore, the electrogenic pump of *Neurospora* appears to be a proton-transporting ATPase (Figure 4); and seen in this way, it is closely related functionally to Mitchell's view of the ATPases of mitochondria and bacteria (MITCHELL, 1961). It is therefore of interest that one can distinguish between the mitochondrial ATPase, on the one hand, and bacterial and *Neurospora* plasma-membrane ATPases, on the other hand, by their patterns of response to inhibitors. The former is highly sensitive to oligomycin and rutamycin (LARDY, FERGUSON, 1969). The latter two are resistant to oligomycin and rutamycin but sensitive to N,N'-dicyclohexylcarbodiimide (DCCD), Dio 9, chlorhexidine, and azide (HAROLD et al., 1969a, b; EVANS, 1970; ROISIN, KEPES, 1973; HANSON, KENNEDY, 1973; SLAYMAN, unpublished results).

The effect of azide on *Neurospora* is shown in Figure 5. This pair of experiments is again a comparison between cellular ATP levels (measured in cell suspensions) and membrane potential (recorded from individual hyphae), after the addition of 1 mM azide. The experiments were performed both on wild-type *Neurospora* and on the *poky* mutant, which has a defective cytochrome chain and has been shown to generate a substantial portion of its ATP by substrate-level phosphorylation linked to a CN-resistant oxidase (SLAYMAN et al., 1974). The aim in studying *poky* was to minimize, as much as possible, the lowering of cellular ATP from the inevitable direct action of azide on cytochrome oxidase. Even in *poky*, azide did cause a moderately rapid (τ = 8 sec), but partial (50%), reduction in cellular ATP; and this reduction would be expected to cause a parallel fall in membrane potential (to about 75% of its control value, as shown in the upper curve, calculated from the previously established dependence of membrane potential on cellular ATP). But in fact, the membrane potential was affected much more drastically, falling with an initial time constant of 2 sec and reaching a minimum of 6% to 7% of its control value. A similar phenomenon was seen in wild-type *Neurospora* although here, as expected, azide had a much larger effect on cellular ATP levels, and consequently there was a smaller difference between the calculated and measured values of the membrane potential. It is therefore apparent that azide has a rapid, direct action on the plasma membrane of *Neurospora*, in addition to a slower action on ATP synthesis in the mitochondria. Separate experiments, not included in Figure 5, have shown that azide does not bring about any immediate change in membrane resistance and thus is not simply short-circuiting the membrane. We suspect that azide acts as a direct inhibitor of the membrane ATPase system, a finding compatible

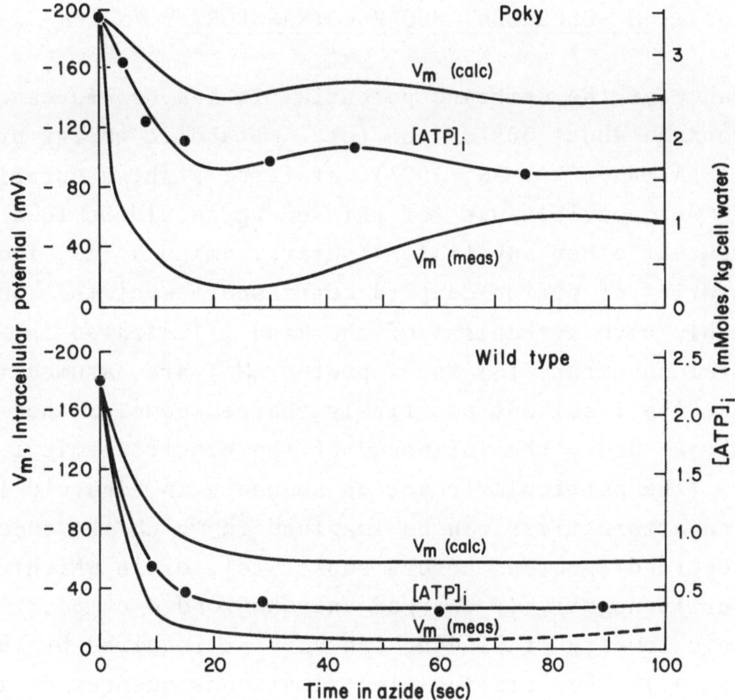

Fig. 5. Effect of sodium azide on membrane potential and ATP in *poky* and wild-type *Neurospora*. Data obtained from cells bathed in standard DMG buffer (see legend to Figure 1); 1 mM sodium azide (neutralized with HCl) added at zero time. Calculated voltage curves were obtained from the equation in the legend for Figure 2, with V_0 assumed to be -7 mV, $K_{1/2}$ assumed to be 2.0 mM (see SLAYMAN, LONG, and LU, 1973), and with the values of V_{pm} adjusted to compensate for the normal variation of control values of membrane potential and cellular ATP from one series of experiments to another (in this case, V_{pm} for wild-type = -304 mV and V_{pm} for *poky* = -295 mV)

with the reports of ROISIN and KEPES (1973) and HANSON and KENNEDY (1973) that azide inhibits the solubilized membrane ATPase from *Escherichia coli*. As mentioned previously, DCCD (10^{-4}M) has a similar effect on the membrane potential of *Neurospora*, although it acts much more slowly than azide, with a time constant of approximately 100 sec. (This may well reflect a problem of limited access to the membrane. Intact cells of Gram-negative bacteria are not sensitive to DCCD unless their outer lipopolysaccharide layer is first disrupted by treatment with tris-EDTA [BRAGG, HOU, 1974]. Less is known about factors controlling the penetration of small molecules through the *Neurospora* cell wall, but no pretreatment yet tried, including tris-EDTA, has accelerated the action of DCCD on the membrane potential.) Among other inhibitors tested, oligomycin, rutamycin, and aurovertin (all at 10 μg/ml) had no significant effect on the membrane potential.

H^+-DEPENDENT SUGAR COTRANSPORT

Maintenance of the membrane potential in *Neurospora* can be calculated to consume about 30% of the total metabolic energy production of the cell (SLAYMAN, SLAYMAN, 1973), at first sight a surprisingly high figure. One possible use for this energy would be to drive the uphill movement of other substrates (sugars, amino acids, inorganic ions) via a series of proton-coupled cotransport systems. Such systems would presumably have mechanisms of the kind illustrated in Figure 6, where a neutral substrate (S) and a proton (H^+) are assumed to bind to a carrier, and the resultant positively charged complex then crosses the cell membrane under the influence of the electrochemical gradient for protons. (The particular reaction sequence in Figure 6 is somewhat arbitrary, and alternatives can be imagined in which the order of binding is reversed [protons before substrate], or in which the unloaded carrier is negatively charged and the loaded complex is neutral. Also, the stoichiometry of binding can vary as required by the charge of the substrate.) The detailed electrical consequences of a cotransport system of this kind will depend upon the magnitudes of its EMF and resistance in relation to the other circuit parameters of the membrane, but since all of the possible reaction sequences involve the net flow of charge across the membrane, the immediate result of turning on a proton-dependent cotransport system should be a depolarization of the membrane.

The particular transport system in which we chose to look for such a depolarization was glucose transport system II, first described by SCARBOROUGH (1970), SCHNEIDER and WILEY (1971), and NEVILLE, SUSKIND, and ROSEMAN (1971) and shown to be derepressed during carbon starvation. This system transports a family of structurally related hexoses,

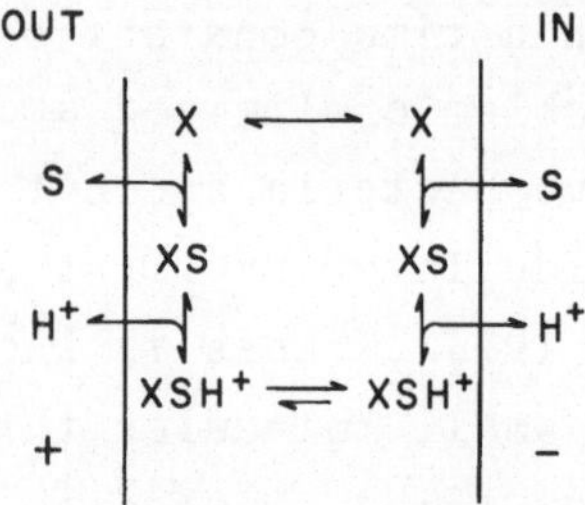

Fig. 6. One possible cyclic carrier model for cotransport of an uncharged substrate (S) along with protons. Unequal arrows indicate the net counterclockwise driving of the cycle by the pre-existing membrane potential. (Taken from SLAYMAN, 1974.)

including glucose, 2-deoxyglucose, and 3-0-methylglucose. 3-O-methylglucose, which is not phosphorylated (SCHNEIDER, WILEY, 1971), is concentrated to steady-state ratios (intracellular concentration: extracellular concentration) as high as 5,900 (LOWENDORF, SLAYMAN, SLAYMAN, in press), a value easily compatible with the electrochemical gradient for protons calculated to exist across the plasma membrane. (The resting membrane potential in carbon-starved cells averages -225 mV. The pH gradient in starved cells is not known, but in normal cells the intracellular pH has been estimated by dye distribution to be 6.8 in standard medium at pH 5.8 [SLAYMAN, unpublished results]. Therefore, the proton electrochemical gradient is 225 + 59 log 10 = 284 mV, equivalent to a concentration gradient for a nonelectrolyte of 65,000.)

Figure 7 shows that the addition of glucose to carbon-starved cells (derepressed for transport system II) does cause a pronounced depolarization of the membrane, consistent with the cotransport hypothesis, while the addition of glucose to normal cells (which lack sys-

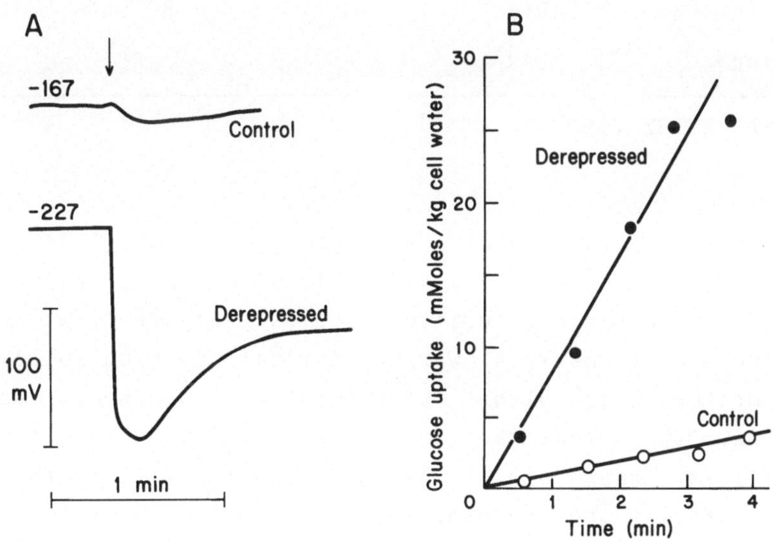

Fig. 7. Effect of glucose transport via system II on the membrane potential in *Neurospora*. (A) Voltage records from normal (upper) and derepressed (lower) cells. Arrow indicates addition of 1 mM glucose; depolarization indicated by a downward deflection. The time constant for the steep fall in derepressed cells is 0.7 sec. Ambient temperature approximately 24°C. (B) Uptake of 1 mM ^{14}C-glucose by normal (lower) and derepressed (upper) cells. Influxes (slopes) were 0.97 and 8.3 mmoles/liter cell water·min for normal and derepressed cells, respectively. All cells were maintained in 0.3x Vogel's medium + 2.3 mM $CaCl_2$. Derepression was carried out by preincubating the cells for 2 to 3 hr in Vogel's medium without added sugar. Temperature, 25°C. (Data partly from SLAYMAN and SLAYMAN, 1974.)

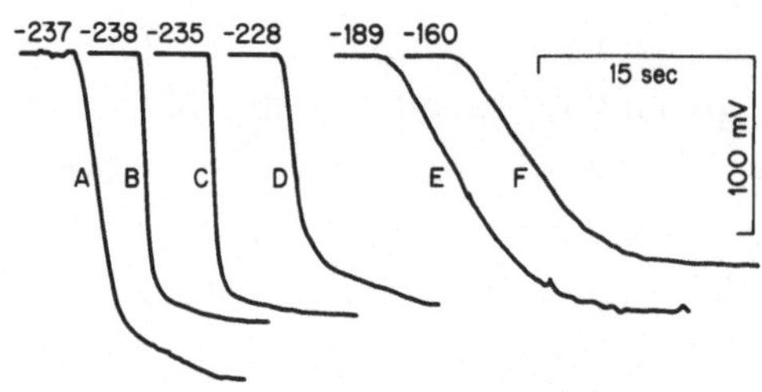

Fig. 8. Speed of glucose-depolarization compared with that of cyanide-depolarization. (A-D) Initial voltage change produced in derepressed cells by addition of 1 mM glucose. Cells bathed in 0.3x Vogel's medium + 2.3 mM $CaCl_2$. Records A and D obtained from different hyphae, B and C from successive trials on a single hypha. (E-F) Initial voltage change produced in normal (repressed) cells by 5 mM KCN. Cells bathed in 10 mM potassium phosphate buffer (pH 5.8); 1 mM $CaCl_2$, 1% glucose. Records from two different hyphae

Table 1. Dependence of transport and depolarization on the extracellular sugar concentration

	Glucose ($K_{1/2}$ [μM])	3-0-Methylglucose ($K_{1/2}$ [μM])
Transport	30-50[a]	89-90[a]
Depolarization	42	80

[a]SCHNEIDER and WILEY (1971).

tem II) has only a very small effect (SLAYMAN, SLAYMAN, 1974). The depolarization seen in carbon-starved cells is extremely rapid, with a time constant in the range of 0.5 to 1 sec. In Figure 8, it is compared with the significantly slower depolarization which follows the addition of metabolic inhibitors (time constant = 6.0 sec). Control experiments have shown that the glucose depolarization is not related to any significant change of cellular ATP, and that it has the same substrate specificity and the same dependence upon substrate concentration as does system II assayed directly (Table 1; see also SLAYMAN, SLAYMAN, 1974).

The results described up to this point demonstrate that glucose uptake is mediated by a charge-carrying system. Identification of the charge carried comes from an independent set of experiments, in which a pH electrode was used to monitor changes in extracellular pH during sugar uptake. Sample traces are shown in Figure 9 for three different sugars, glucose, 2-deoxyglucose, and 3-0-methylglucose. The latter two sugars gave the clearest results. In both cases, addition of the sugar

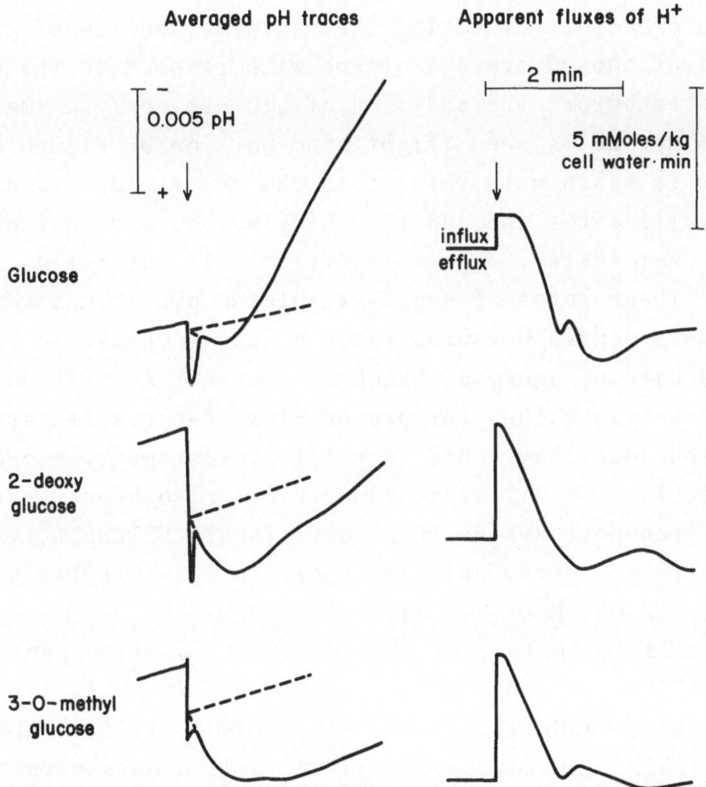

Fig. 9. Effect of glucose, 2-deoxyglucose, and 3-0-methylglucose on proton fluxes in *Neurospora*. Cells derepressed as described for Figure 7. *Left-hand graphs*: averaged pH traces for 2 to 6 separate trials with each sugar. Sugars added at the arrow; the initial steep downward deflection and rebound in all three traces are artifacts associated with injection and mixing of sugar solutions. Upward movement of the traces represents net H^+ release by the cells; downward movement, net H^+ uptake. Control rates of H^+ release: 0.48, 0.59, and 0.52 mmoles/liter cell water•min, respectively, for glucose, 2-deoxyglucose, and 3-0-methylglucose. *Right-hand graphs*: apparent fluxes of H^+ caused by sugar addition; calculated by first taking the difference between each experimental curve and the extrapolated baseline (dashed line) in the pH traces, and then differentiating with respect to time. Temperature, 25°C. Data for 3-0-methylglucose from SLAYMAN and SLAYMAN (1974)

was followed immediately by a shift of the pH trace toward more alkaline values, consistent with the notion that protons were being taken up by the cells. Quantitatively, as shown by the derivatives of these curves (right-hand portion of Figure 9), the apparent rates of proton uptake reached 4.0 and 4.3 mmoles/liter cell water•min for 2-deoxyglucose and 3-0-methylglucose. The glucose trace was more complicated. Initially there was a slight indication of a shift toward more alkaline extracellular pH values, but the initial shift was soon obscured by a massive *efflux* of acid, reflecting the rapid metabolism of glucose. If the derivative of the pH trace is taken with respect to the baseline rate that existed before the addition of glucose, only a small apparent rate of proton uptake is seen (right-hand portion of Figure 9). But if the derivative is taken with respect to the new steady-state baseline rate that prevails after the addition of glucose, apparent proton uptake is somewhat faster, 3.8 mmoles/liter cell water•min (not shown in Figure 9). These rates of sugar-stimulated proton uptake are compared with independently measured rates of sugar uptake in Table 2. Except for the case of glucose, which is aberrant even if one accepts the higher of the two values for proton flux, the results are compatible with the idea that there is a 1:1 stoichiometry between protons and sugar molecules. A 1:1 stoichiometry has also been reported for β-galactoside transport system in *E. coli* (WEST, MITCHELL, 1973), and a 2:1 stoichiometry for amino acid transport in yeast (EDDY, NOWACKI, 1971; SEASTON, INKSON, EDDY, 1973).

A characteristic feature of the pH curves in all organisms studied so far (WEST, 1970; WEST, MITCHELL, 1972; EDDY, NOWACKI, 1971; SEASTON, INKSON, EDDY, 1973; KASHKET, WILSON, 1973; KOMOR, 1973; SLAYMAN, SLAYMAN, 1974) is their time-course: apparent proton uptake reaches a peak rate almost immediately and then, over a period of a minute or so, tapers off and returns to the baseline rate (Figure 9). Sugar uptake, on the other hand, is linear with time for several minutes (Figure 7). It seems reasonable to believe that the underlying stoichiometry of the cotransport system does not change, but instead that an initial acidification of the cell interior leads to an increased proton efflux,

Table 2. Stoichiometry of the cotransport system[a]

	Initial rate of sugar uptake (mMoles/liter cell water•min)	Initial rate of proton uptake (mMoles/liter cell water•min)	
		Col. 1	Col. 2
Glucose, 1 mM	8.8	1.1	3.8
2-Deoxyglucose, 1 mM	4.6	4.0	4.6
3-0-Methylglucose, 2 mM	3.8	4.3	4.3

[a]The rates of proton uptake in column 1 were calculated as described for Figure 9, with reference to the baseline existing before the addition of sugar. The rates in column 2 were calculated by the same method, but with reference to the steady-state baseline after the addition of sugar.

presumably in metabolizing cells via an acceleration of the electrogenic H^+ pump. In this regard it is interesting that the initial depolarization observed during cotransport in *Neurospora* is not sustained, but that the membrane potential returns part way toward its original value with essentially the same time-course as seen in the pH trace (compare Figures 7, 9). Such repolarization lends some support to the idea that the electrogenic pump accelerates after the first minute or so of cotransport.

DISCUSSION

The ultimate aim of this work is a complete description of the coupling between the electrogenic pump and the cotransport system, and will require measurements of membrane potential (and resistance or current-voltage relationship), proton fluxes, and sugar fluxes as affected by changes in the various driving forces that influence the system (availability of ATP; sizes of the voltage gradient and the pH and sugar gradients). In this sense, the work on *Neurospora*--where the electrical properties of the membrane can be measured directly--complements work on bacteria, where the potential can be estimated by fluorescent probes but only after making a series of assumptions to calibrate the measurement (LARIS, HOFFMAN, 1974; LARIS, PERSHADSINGH, 1974; KASHKET, WILSON, 1974).

We are of course interested to know whether other transport systems in *Neurospora* may also be coupled to the membrane potential. An indirect argument can be made that this is the case: at least four other transport systems (for neutral amino acids [WILEY, MATCHETT, 1966], K^+ [SLAYMAN, SLAYMAN, 1968], and two systems for inorganic phosphate [LOWENDORF, SLAYMAN, SLAYMAN, in press]) are very sensitive to the uncouplers CCCP and dinitrophenol (LOWENDORF, SLAYMAN, SLAYMAN, unpublished results); the measurements were made in *poky*, where cellular ATP levels are kept high by substrate-level phosphorylation (SLAYMAN et al., 1974). One obvious interpretation of these results is that the uncouplers act at the plasma membrane to short-circuit the membrane potential. Consistent with this idea, both dinitrophenol and CCCP cause a rapid fall in membrane potential and membrane resistance (SLAYMAN, 1965b; GRADMANN, SLAYMAN, unpublished results). (The situation is complicated, however, by a subsequent rise in resistance and by a slow recovery of the potential, so that the mode of action of uncouplers is complex and needs to be studied further.)

In summary, our current working hypothesis is that the plasma membrane of *Neurospora* contains a primary electrogenic H^+ transport system and, coupled to it, a series of H^+-dependent cotransport systems that are driven principally by the membrane potential. The cotransport systems can be either constitutive (K^+, PO_4 I) or derepressible (glucose, PO_4 II), can contain shockable binding proteins (neutral amino acid system; WILEY, 1970) or can be resistant to osmotic shock (K^+; SLAYMAN, unpublished results), and can transport either cations, anions, or neutral molecules. The same mechanism of coupling has been put forth for bacteria, yeast, algae, and higher plants (reviewed in SLAYMAN, 1974); the common factor linking these various organisms is that their cells are exposed to a variable environment so that, unlike animal cells, they can neither generate nor depend on a constant sodium gradient.

REFERENCES

BRAGG, P. D., HOU, C.: Reconstruction of energy-dependent transhydrogenase in ATPase-negative mutants of *Escherichia coli*. Biochem. Biophys. Res. Commun. 50, 729-736 (1974).

CHRISTENSEN, H. N.: Linked ion and amino acid transport. In: Membranes and Ion Transport (ed. E. E. BITTAR) Vol. I, pp. 265-394. New York: Wiley-Interscience 1970.

EDDY, A. A., NOWACKI, J. A.: Stoichiometrical proton and potassium ion movements accompanying the absorption of amino acids by the yeast *Saccharomyces carlsbergensis*. Biochem. J. 122, 701-711 (1971).

EVANS, D. J., Jr.: Membrane Mg^{2+}-(Ca^{2+})-activated adenosine triphosphatase of *Escherichia coli*: Characterization in the membrane-bound and solubilized states. J. Bacteriol. 104, 1203-1212 (1970).

HANSON, R. L., KENNEDY, E. P.: Energy-transducing adenosine triphosphatase from *Escherichia coli*: Purification, properties, and inhibition by antibody. J. Bacteriol. 114, 772-781 (1973).

HAROLD, F. M.: The varieties of bacterial transport systems; this volume.

HAROLD, F. M., BAARDA, J., BARON, C., ABRAMS, A.: Inhibition of membrane-bound adenosine triphosphatase and of cation transport in *Streptococcus faecalis* by N,N'-dicyclohexylcarbodiimide. J. Biol. Chem. 244, 2261-2268 (1969a).

HAROLD, F. M., BAARDA, J., BARON, C., ABRAMS, A.: Dio 9 and chlorhexidine: Inhibitors of membrane-bound ATPase and of cation transport in *Streptococcus faecalis*. Biochim. Biophys. Acta 183, 129-136 (1969b).

VON JAGOW, G., WEISS, H., KLINGENBERG, M.: Comparison of the respiratory chain of *Neurospora crassa* wild type and the mi-mutants mi-1 and mi-3. Eur. J. Biochem. 33, 140-157 (1973).

KABACK, H. R.: Transport in isolated bacterial membrane vesicles; this volume.

KASHKET, E. R., WILSON, T. H.: Proton-coupled accumulation of galactoside in *Streptococcus lactis* 7962. Proc. Nat. Acad. Sci. 70, 2866-2869 (1973).

KASHKET, E. R., WILSON, T. H.: Protonmotive force in fermenting *Streptococcus lactis* 7962 in relation to sugar accumulation. Biochem. Biophys. Res. Commun. 59, 879-886 (1974).

KOMOR, E.: Proton-coupled hexose transport in *Chlorella vulgaris*. FEBS Letters 38, 16-18 (1973).

LAMBOWITZ, A. M., SLAYMAN, C. W., SLAYMAN, C. L., BONNER, W. D., Jr.: The electron transport components of wild-type and *poky* strains of *Neurospora crassa*. J. Biol. Chem. 247, 1536-1545 (1972).

LAMBOWITZ, A. M., SMITH, E. W., SLAYMAN, C. W.: Electron transport in *Neurospora* mitochondria. Studies on wild type and *poky*. J. Biol. Chem. 247, 4850-4858 (1972a).

LAMBOWITZ, A. M., SMITH, E. W., SLAYMAN, C. W.: Oxidative phosphorylation in *Neurospora* mitochondria. Studies on wild type, *poky*, and chloramphenicol-induced wild type. J. Biol. Chem. 247, 4859-4865 (1972b).

LARDY, H. A., FERGUSON, S. M.: Oxidative phosphorylation in mitochondria. Ann. Rev. Biochem. 38, 991-1034 (1969).

LARIS, P. C., HOFFMAN, J. F.: Determination of membrane potentials in human and amphiuma red blood cells by means of a fluorescent probe. J. Physiol. 239, 519-552 (1974).

LARIS, P. C., PERSHADSINGH, H. A.: Estimations of membrane potentials in *Streptococcus faecalis* by means of a fluorescent probe. Biochem. Biophys. Res. Commun. 57, 620-626 (1974).

LOWENDORF, H. A., SLAYMAN, C. L., SLAYMAN, C. W.: Phosphate transport in *Neurospora*. Kinetic characterization of a constitutive, low-affinity transport system. Biochim. Biophys. Acta (in press).

MITCHELL, P.: Coupling of phosphorylation to electron and hydrogen transfer by a chemiosmotic type of mechanism. Nature 191, 144-148 (1961).

MITCHELL, P.: Molecule, group, and electron translocation through natural membranes. Biochem. Soc. Symp. 22, 142-168 (1963).

NEVILLE, M. M., SUSKIND, S. R., ROSEMAN, S.: A derepressible active transport system for glucose in *Neurospora crassa*. J. Biol. Chem. 246, 1294-1301 (1971).

ROISIN, M. P., KEPES, A.: The membrane ATPase of *Escherichia coli*. II. Release into solution, allotopic properties and reconstitution of membrane-bound ATPase. Biochim. Biophys. Acta 305, 249-259 (1973).

SCARBOROUGH, G. A.: Sugar transport in *Neurospora crassa*. A second glucose transport system. J. Biol. Chem. 245, 3985-3987 (1970).

SCHNEIDER, R. P., WILEY, W. R.: Kinetic characteristics of two glucose transport systems in *Neurospora crassa*. J. Bacteriol. 106, 479-486 (1971).

SCHULTZ, S. G., CURRAN, P. F.: Coupled transport of sodium and organic solutes. Physiol. Rev. 50, 559-718 (1970).

SEASTON, A., INKSON, C., EDDY, A. A.: The absorption of protons with specific amino acids and carbohydrates by yeast. Biochem. J. 134, 1031-1043 (1973).

SINGH, A. P., BRAGG, P. D.: Effect of dicyclohexylcarbodiimide on growth and membrane-mediated processes in wild type and heptose-deficient mutants of *Escherichia coli* K12. J. Bacteriol. 119, 129-137 (1974).

SKOU, J. C.: Sequence of steps in the (Na + K)-activated enzyme system in relation to sodium and potassium transport. In: Current Topics in Bioenergetics (ed. D. R. SANADI) Vol. IV, pp. 127-190. New York: Academic Press 1971.

SLAYMAN, C. L.: Electrical properties of *Neurospora crassa*. Effects of external cations on the intracellular potential. J. Gen. Physiol. 49, 69-92 (1965a).

SLAYMAN, C. L.: Electrical properties of *Neurospora crassa*. Respiration and the intracellular potential. J. Gen. Physiol. 49, 93-116 (1965b).
SLAYMAN, C. L.: Movement of ions and electrogenesis in microorganisms. Amer. Zool. 10, 377-392 (1970).
SLAYMAN, C. L.: Proton pumping and generalized energetics of transport: A review. In: Membrane Transport in Plants (ed. J. DAINTY, U. ZIMMERMANN). Berlin-Heidelberg-New York: Springer-Verlag 1974.
SLAYMAN, C. L., LONG, W. S., LU, C. Y.-H.: The relationship between ATP and an electrogenic pump in the plasma membrane of *Neurospora crassa*. J. Membrane Biol. 14, 305-338 (1973).
SLAYMAN, C. W., REES, D. C., ORCHARD, P. P., SLAYMAN, C. L.: Generation of ATP in cytochrome-deficient mutants of *Neurospora crassa*. J. Biol. Chem. (in press).
SLAYMAN, C. L., SLAYMAN, C. W.: Net uptake of potassium in *Neurospora*. Exchange for sodium and hydrogen ions. J. Gen. Physiol. 52, 424-443 (1968).
SLAYMAN, C. L., SLAYMAN, C. W.: H^+-dependent cotransport and the electrogenic pump in the plasma membrane of *Neurospora*. Abstr. Ann. Meeting Am. Soc. Microbiol., p. 172 (1973).
SLAYMAN, C. L., SLAYMAN, C. W.: Depolarization of the plasma membrane of *Neurospora* during active transport of glucose: Evidence for a proton-dependent cotransport system. Proc. Nat. Acad. Sci. 71, 1935-1939 (1974).
WEISS, H., VON JAGOW, G., KLINGENBERG, M., BUCHER, T.: Characterization of *Neurospora crassa* mitochondria prepared with a grind-mill. Eur. J. Biochem. 14, 75-82 (1970).
WEST, I. C.: Lactose transport coupled to proton movements in *Escherichia coli*. Biochem. Biophys. Res. Commun. 41, 655-661 (1970).
WEST, I. C., MITCHELL, P.: Proton-coupled β-galactoside translocation in nonmetabolizing *Escherichia coli*. J. Bioenergetics 3, 445-462 (1972).
WEST, I. C., MITCHELL, P.: Stoichiometry of lactose-proton symport across the plasma membrane of *Escherichia coli*. Biochem. J. 132, 587-592 (1973).
WILEY, W. R.: Tryptophan transport in *Neurospora crassa*: A tryptophan-binding protein released by cold osmotic shock. J. Bacteriol. 103, 656-662 (1970).
WILEY, W. R., MATCHETT, W. H.: Tryptophan transport in *Neurospora crassa*. I. Specificity and kinetics. J. Bacteriol. 92, 1698-1705 (1966).

Isolated Bacterial Cytoplasmic Membrane Vesicles: A Model System for the Study of Active Transport

H. R. Kaback, G. Rudnick, S. Schuldiner, and S. A. Short
The Roche Institute of Molecular Biology, Nutley, New Jersey 07110

INTRODUCTION

Isolated bacterial cytoplasmic membrane vesicles have provided a useful model system for studies of active transport (KABACK, 1970a, b; KABACK 1972; HONG, KABACK, 1973; KABACK, 1973; KABACK, in press). Vesicles are devoid of the cytoplasmic constituents of the intact cell, and their metabolic activities are restricted to those provided by the enzymes of the membrane itself. This constitutes a considerable advantage over intact cells in the study of certain transport mechanisms, since active transport by membrane vesicles is practically nil in the absence of the appropriate exogenous energy source. Thus, the energy source for transport of a particular substrate can be determined by studying which substances stimulate accumulation. Moreover, metabolic conversion of the transport substrate and the energy source is minimal, allowing clear definition of the reactions involved.

Membrane vesicles isolated from a number of organisms catalyze the active transport of many metabolites at rates that are comparable to those of the parent whole cells in many cases (KABACK, 1970a, b; KABACK, 1972; HONG, KABACK, 1973; KABACK, 1973; KABACK, in press; SHORT, WHITE, KABACK, 1972; LOMBARDI, KABACK, 1972). In *E. coli* and *S. typhimurium* vesicles, most active transport systems are coupled primarily to the oxidation of D-lactate or reduced phenazine methosulfate (PMS), (or pyocyanine) via a membrane-bound cytochrome chain with oxygen as the terminal electron acceptor, and neither the generation nor utilization of ATP or other high-energy phosphate intermediates is apparently involved in the mechanism (KABACK, 1972; HONG, KABACK, 1973; KABACK, 1973; KABACK, in press). The energy-coupling site for transport in *E. coli* vesicles is localized in a segment of the respiratory chain between D-lactate dehydrogenase (D-LDH) and cytochrome b_1 (KABACK, 1972; HONG, KABACK, 1973; KABACK, 1973; KABACK, in press;

BARNES, KABACK, 1971), but the transport carriers are not electron transfer intermediates (HONG, KABACK, 1972).

This discussion is concerned with recent studies pertaining to respiration-linked active transport in membrane vesicles isolated from *E. coli*. Methods for the preparation of bacterial membrane vesicles, their morphology and other properties, and experimental observations related to the role of the P-enolpyruvate-P-transferase system in the vectorial phosphorylation of certain sugars will not be discussed, and the reader is referred to other publications (KABACK, 1970a, b; KABACK, 1972; HONG, KABACK, 1973; KABACK, 1973; KABACK, in press).

TRANSPORT ACTIVITY OF INDIVIDUAL MEMBRANE VESICLES

Although the transport activity of isolated membrane vesicles is comparable to that of the parent whole cells in many cases (SHORT, WHITE, KABACK, 1972; LOMBARDI, KABACK, 1972), quantitative comparisons between vesicles and whole cells are difficult to interpret. Despite evidence to the contrary (KABACK, 1970a, b; KABACK, 1972; HONG, KABACK, 1973; KABACK, 1973; KABACK, in press; SHORT, WHITE, KABACK, 1972; LOMBARDI, KABACK, 1972; KONINGS et al., 1973; ALTENDORF, STAEHELEN, 1974), several workers have argued that a significant number of membrane vesicles become inverted during preparation (HAROLD, 1972; MITCHELL, 1973; WEINER, 1974; FUTAI, 1974a; VAN THIENEN, POSTMA, 1973). Such inverted vesicles would not be expected to catalyze active transport, but they might oxidize certain electron donors, such as NADH, to which the membrane is presumably impermeable. This possibility is not trivial because the high degree of specificity of D-lactate as an electron donor for active transport has important implications for the mechanism of energy-coupling in this system (KABACK, 1972; HONG, KABACK, 1973; KABACK, 1973; KABACK, in press). The experiments described in the next few paragraphs have allowed a direct approach to this problem.

2-Hydroxy-3-butynoic acid (WALSH et al., 1972) irreversibly inactivates D- and L-lactate dehydrogenases (L-LDH) and D-lactate-dependent active transport in membrane vesicles isolated from *E. coli* (WALSH, ABELES, KABACK, 1972; WALSH, KABACK, 1974). The compound is a substrate for the membrane-bound, flavin-linked D-LDH, which undergoes 15 to 30 turnovers prior to inactivation. Inactivation is due to covalent attachment of a reactive intermediate to FAD at the active site of the enzyme. The proposed reaction sequence is shown in Figure 1 (Reaction II). Both the hydroxy function and the alkyne linkage in 2-hydroxy-3-butynoate are critical for inactivation. Thus, 3-butynoate has no effect on the enzyme; and vinylglycolate (2-hydroxy-3-butenoic acid) serves as a substrate for D-LDH and is an effective electron donor for transport.

Inactivation of D- and L-LDH and D-lactate-dependent transport by 2-hydroxy-3-butynoate is highly specific. Other membrane-bound dehydrogenases are not inhibited, transport in the presence of ascorbate-PMS is not affected, and α-glycerol-P-dependent transport in *Staphylococcus aureus* vesicles is not inactivated by the acetylenic hydroxy acid. Moreover, inactivation of D-lactate-dependent transport is blocked by D-lactate but not by succinate and NADH.

In view of this high degree of specificity, it was surprising when subsequent experiments demonstrated that hydroxy butynoate inactivates vectorial phosphorylation catalyzed by the P-enolpyruvate-P-transferase system, and that vinylglycolate (2-hydroxy-3-butenoate), a substrate for D-LDH, is 50 to 100 times more potent. The key to the puzzle came with the realization that Reaction I (in Figure 1), although inconsequential for D-LDH, leads to the formation of a highly reactive electrophile. Shortly thereafter, it was demonstrated that vinylglycolate inactivates Enzyme I of the P-transferase system and, by this means, blocks vectorial phosphorylation in whole cells and membrane vesicles of *E. coli* (WALSH, KABACK, 1974; WALSH, KABACK, 1973). The relative lack of potency of hydroxy-butynoate is due to inactivation of D- and

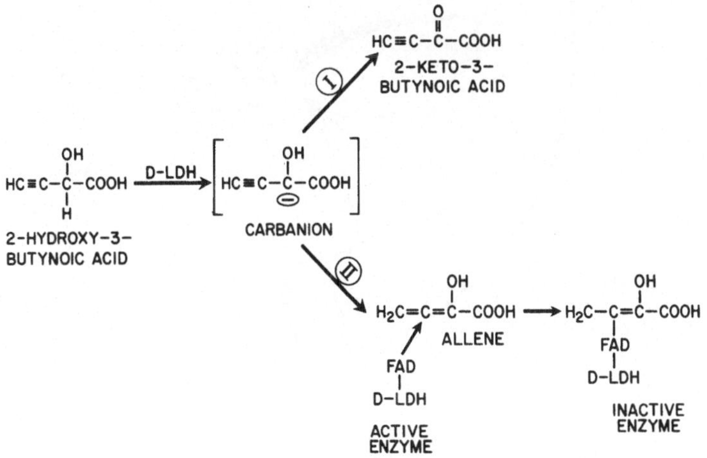

Fig. 1. Inactivation of D-lactate dehydrogenase (D-LDH) by 2-hydroxy-3-butynoic acid. FAD = flavin adenine dinucleotide

L-LDH's by this compound. Generation of 2-keto-3-butynoate (Figure 1) is limited therefore by inactivation of the enzymes that catalyze its formation. Vinylglycolate, on the other hand, is a noninactivating substrate, and the putative electrophile (2-keto-3-butenoic acid) is generated at a rapid rate and for extended periods of time. Synthesis of isotopically labeled vinylglycolate has allowed a detailed study of the biochemical properties of this compound (SHAW, submitted for publication; KABACK et al., 1974; SHORT et al., in press).

Prior to inactivation of the P-transferase system, vinylglycolate is transported by the lactate transport system. Subsequently, it is oxidized by membrane-bound D- and L-LDH's to yield a reactive electrophile (presumably 2-keto-3-butenoate) which then reacts with Enzyme I and many other sulfhydryl-containing proteins on the membrane (Figure 2).

There is considerable evidence supporting these conclusions (WALSH, KABACK, 1974; WALSH, KABACK, 1973; SHAW et al., submitted for publication; KABACK et al., 1974; SHORT et al., in press); however,

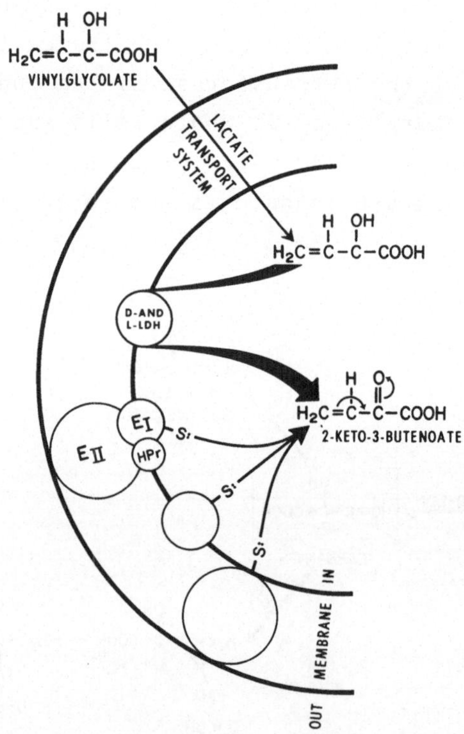

Fig. 2. Proposed sequence of events in uptake and covalent binding of vinylglycolate (2-hydroxy-3-butenoate) by *E. coli* membrane vesicles. D- and L-LDH, D- and L-lactate dehydrogenases; E II, Enzyme II of the P-enolpyruvate-P-transferase system; E I, Enzyme I of the P-transferase system; HPr, histidine-containing protein

only two points are critical for this discussion: (1) vinylglycolate transport is the limiting step for labeling the membrane proteins; and (2) almost all of the vinylglycolate taken up is covalently bound to the vesicles. In experimental terms, the rate of covalent binding of vinylglycolate is stimulated at least 10-fold by ascorbate-PMS; and stimulation is completely abolished by uncoupling agents or phospholipase treatment, neither of which affect vinylglycolate oxidation.

With this background, vinylglycolate can be utilized to estimate the transport activity of individual membrane vesicles. Using extremely high specific activity [^{3}H]vinylglycolate, vesicles have been labeled for an appropriate time in the presence of ascorbate-PMS and examined by radioautography in the electron microscope (KABACK et al., 1974; SHORT et al., in press). Each vesicle that takes up vinylglycolate is overlaid with exposed silver grains. Examination of the preparations reveals that 85% to 95% of the vesicles are labeled. It should be emphasized that this is a minimal estimation. Virtually all of the large vesicles are labeled, while the size of the smaller vesicles is such that their proximity to individual silver grains in the emulsion may be limiting. Moreover, essentially identical radioautographic results are obtained with [^{3}H]acetic anhydride, a reagent that reacts nonspecifically with the vesicles. These studies provide strong evidence that most, if not all, of the vesicles in the preparation catalyze active transport. This type of study is not possible with the usual transport substrates because they are not covalently bound by the vesicles, and are readily lost with even the most gentle manipulations.

ANAEROBIC TRANSPORT

One aspect of the respiration-dependent transport systems that has begun to be studied is their relationship to anaerobic transport. Obligate anaerobes or facultative anaerobes growing under anaerobic conditions transport nutrients; moreover, δ-amino-levulinic acid- or heme-requiring mutants of *E. coli* do not manifest transport defects (DEVOR et al., 1974). Although ATP is not apparently involved in active transport under aerobic conditions (KABACK, 1972; HONG, KABACK, 1973; KABACK, 1973; KABACK, in press), evidence has been presented that suggests that intact cells can utilize glycolytically generated ATP to drive transport under anaerobic conditions (KLEIN, BOYER, 1972; SCHAIRER, HADDOCK, 1972; BERGER, 1973; BUTLIN, 1973; PARNES, BOOS, 1973; OR, KANNER, GUTNICK, 1973; ABRAMS, SMITH, 1971). It is also possible, however, that anaerobic cells might use the same general type of transport mechanism as that used aerobically, with the exception that an alternative electron acceptor is used rather than oxygen.

Recent experiments (KONINGS, KABACK, 1973) demonstrate that anaerobic lactose transport in whole cells and membrane vesicles from *E. coli* may be coupled to the oxidation of α-glycerol-P or D-lactate with fumarate as an electron acceptor. Alternatively, anaerobic lactose transport may be coupled to the oxidation of formate utilizing nitrate as electron acceptor. Both anaerobic electron transfer systems are induced by growth of the organism under appropriate conditions. Components of both systems are loosely bound to the membrane, necessitating the use of a modified procedure for vesicle preparation in order to demonstrate anaerobic transport in the vesicles. Inclusion of ATP or an ATP-generating system during lysis and the subsequent steps in the preparation of vesicles does not stimulate transport. The results support the conclusion that at least one type of anaerobic transport is coupled primarily to electron flow.

ENERGY-DEPENDENT BINDING OF β-GALACTOSIDES TO THE LAC CARRIER PROTEIN

Fluorescent compounds that exhibit polarity-dependent fluorescence properties have been used to investigate the structure of biological membranes (RADDA, VANDERKOOI, 1972). Two such compounds, 1-anilino-8-naphthalene sulfonate (REEVES, LOMBARDI, KABACK, 1972) and dansyl phosphatidyl ethanolamine (SHECHTER, GULIK-KRZYWICKI, KABACK, 1972) have been used to study structural changes associated with transport in membrane vesicles of *E. coli*. However, the nonspecificity of these probes limits the type of information that can be obtained.

Fluorescent β-galactosides [1-(N-dansyl)amino-β-D-galactopyranoside (DG_0), 2-(N-dansyl)aminoethyl-β-D-thiogalactopyranoside (DG_2), 2-(N-dansyl)aminoethyl-β-D-galactopyranoside (oxy-DG_2), and 6-(N-dansyl)aminohexyl-β-D-thiogalactopyranoside (DG_6)] competitively inhibit lactose transport by membrane vesicles from *Escherichia coli* ML 308-225, but they are not actively transported (REEVES et al., 1973; SCHULDINER et al., in press; SCHULDINER et al., submitted for publication). An increase in the fluorescence of these dansylgalactosides is observed upon addition of D-lactate, imposition of a membrane diffusion potential (positive outside), or dilution-induced, carrier-mediated lactose efflux. The increase is not observed with 2-(N-dansyl)aminoethyl-β-D-thioglucopyranoside nor with membrane vesicles lacking the β-galactoside transport system. Moreover, the D-lactate-induced fluorescence increase is blocked and/or rapidly reversed by addition of β-galactosides, sulfhydryl reagents, certain inhibitors of D-lactate oxidation, or uncoupling agents. The fluorescence increase exhibits an emission maximum at 500 nm and excitation maxima at 345 nm and at 292 nm. The latter excitation maximum is absent unless D-lactate is added, indicating that the bound dansylgalactoside molecules are excited by energy transfer from the membrane proteins. Titration of vesicles with dansylgalactosides in the presence of D-lactate demonstrates that the *lac* carrier protein constitutes 3% to 6% of the total membrane protein, and that the affinity of the carrier for substrate is directly related to the length of the alkyl chain between the galactosidic and the dansyl moieties of the dansylgalactosides. In addition, there is excellent agreement between the affinity constants of the various dansylgalactosides as determined by fluorimetric titration and their apparent K_i's for lactose transport (K_D's and/or apparent K_i's are approximately 550 μM, 30 μM, 40 μM, and 5 μM for DG_0, DG_2, oxy-DG_2, and DG_6, respectively).

Polarization of fluorescence measurements with 2-(N-dansyl)aminoethyl-β-D-thiogalactopyranoside (DG_2) and 6-(N-dansyl)aminohexyl-β-D-thiogalactopyranoside (DG_6) demonstrate a dramatic increase in polarization on addition of D-lactate, which is reversed by anoxia or addition of lactose. These findings provide strong evidence for the contention that the fluorescence changes observed on "energization" of the membrane are due to binding of the dansylgalactosides *per se*, rather than binding followed by transfer into the hydrophobic interior of the membrane.

Fluorescence lifetime studies corroborate the results discussed above (SCHULDINER et al., manuscript in preparation). In the absence of D-lactate, DG_2 fluorescence exhibits a half-life of 3 nsec. On addition of D-lactate to membrane vesicles containing the *lac* transport system, a percentage of the DG_2 molecules exhibits a half-life of 18 nsec.

The results are consistent with the suggestion that the *lac* carrier protein is inaccessible to the external medium unless energy is provided, and that energy is coupled to one of the initial steps in transport. It is proposed furthermore that at least one aspect of "energization" may be the generation of a membrane potential (positive outside), resulting in increased accessibility of a negatively charged *lac* carrier protein to the external solvent.

The ultimate strength of the conclusions discussed above rests heavily on observations that indicate that although the dansylgalactosides bind specifically to the *lac* carrier protein, they are not actively transported. Recent experimental evidence (SCHULDINER et al., submitted for publication) provides more definitive evidence that these compounds are not transported to any extent whatsoever by either whole cells or isolated membrane vesicles: (1) Although DG_2 and oxy-DG_2 are competitive inhibitors of lactose transport in intact cells of *E. coli* and induce the synthesis of β-galactosidase in vitro, the compounds do not induce β-galactosidase in vivo. (2) *p*-Chloromercuribenzenesulfonate does not cause efflux of lactose from the intravesicular pool, but causes rapid reversal of D-lactate-induced dansylgalactoside fluorescence. This difference between the effects of *p*-chloromercuribenzenesulfonate on β-galactoside transport and on the fluorescence behavior of the dansylgalactosides indicates that the fluorescence changes do not reflect events that occur subsequent to transport of the fluorescent galactosides into the vesicles. (3) Dansylgalactosides inhibit dilution-induced, carrier-mediated efflux of lactose from the intravesicular pool. The results indicate that the dansylgalactosides inhibit lactose efflux by preventing the unloaded carrier from returning to the inner surface of the membrane.

Studies with photoaffinity-labeled β-galactosides, 2-nitro-4-azidophenyl-β-D-thiogalactopyranoside (APG) (RUDNICK, WEIL, KABACK, in press) and 2-(2-nitro-4-azidophenyl)aminoethyl-β-D-thiogalactopyranoside (APG_2) (RUDNICK, WEIL, KABACK, in preparation) provide independent support for the conclusions derived from the dansylgalactoside experiments. The rationale behind the use of this class of compounds (FLEET, PORTER, KNOWLES, 1969; KIEFER et al., 1973; RUOHO et al., 1973) is that irradiation with visible light causes photolysis of the azido group to form a highly reactive nitrene that then reacts covalently with the macromolecule to which the azido-containing ligand is bound.

APG is a competitive inhibitor of lactose transport in membrane vesicles isolated from *E. coli* ML 308-225, exhibiting an apparent K_i of 75 μM. The initial rate and steady-state level of [^{3}H]APG accumulation are markedly stimulated by the addition of D-lactate to vesicles containing the *lac* transport system, and kinetic studies reveal an apparent K_m of 75 μM. Membrane vesicles devoid of the *lac* transport system do not take up significant amounts of APG in the presence or absence of D-lactate. When exposed to visible light in the presence of D-lactate, APG irreversibly inactivates the *lac* transport system. Strikingly, photolytic inactivation is not observed in the absence of D-lactate.

Kinetic studies of the inactivation process yield a K_D of 77 μM. Since lactose protects against photolytic inactivation and APG does not inactivate amino acid transport, it is apparent that these effects are specific for the *lac* transport system. Analogous studies carried out with APG_2 demonstrate that this compound behaves similarly but has a higher affinity for the *lac* carrier protein (i.e., the K_i for competitive inhibition of lactose transport is approximately 30 μM, and the K_D for photolytic inactivation in the presence of D-lactate is approximately 30 μM). However, as opposed to APG, APG_2 is not actively transported by the vesicles.

In addition to the specificity of the inactivation process, the marked dependence of APG and APG_2 photoinactivation on D-lactate is noteworthy. As discussed above, studies with dansylgalactosides indicate that the *lac* carrier protein is not accessible from the outside of the vesicle membrane in the absence of energy-coupling, and that generation of a membrane potential (positive outside) causes the carrier to "move" to the outside of the membrane where it is able to bind ligand. It seems likely that the effect of D-lactate on azidophenylgalactoside photoinactivation is the result of a similar phenomenon.

RECONSTITUTION OF D-LACTATE-DEPENDENT ACTIVE TRANSPORT IN D-LACTATE DEHYDROGENASE MUTANTS

The membrane-bound D-LDH of *E. coli* has been solubilized and purified to homogeneity (KOHN, KABACK, 1973; FUTAI, 1973). The enzyme has a molecular weight of 75,000 ± 7%, contains approximately 1 mole of flavin adenine dinucleotide per mole of enzyme, and exhibits low activity toward L-lactate. Oxidized diphosphopyridine nucleotide (NAD) has no effect on the catalytic conversion of D-lactate to pyruvate.

While this work was in progress, REEVES et al. (1973) demonstrated that guanidine•HCl extracts from wild-type membrane vesicles containing D-LDH activity are able to reconstitute D-lactate-dependent oxygen consumption and active transport in membrane vesicles from *E. coli* and *S. typhimurium* mutants defective in D-LDH (dld^-). These studies have been confirmed and extended by SHORT et al. (1974) using the homogeneous preparation of D-LDH described above, and FUTAI (1974b) has independently confirmed many of the observations.

Reconstitution of D-lactate oxidation and D-lactate-dependent active transport is accomplished by diluting the enzyme preparation (dissolved in 0.1% Triton X-100 and 0.6 *M* guanidine•HCl) 25-fold or more into a suspension of *dld*$^-$ membrane vesicles. Presumably, dilution of the detergent and the guanidine·HCl decreases the solubility of the enzyme, causing it to bind to the vesicles. The reconstituted vesicles are centrifuged, resuspended in a small volume of buffer, and assayed for transport and dehydrogenase activities. Optimal reconstitution takes place in the presence of 0.6 *M* guanidine·HCl, at 25°C to 37°C, and at pH 6.6. The vesicles can be washed several times after reconstitution without loss of activity.

Reconstituted *dld*$^-$ membranes carry out D-lactate oxidation and catalyze the transport of a number of substrates when supplied with D-lactate. D-Lactate is not oxidized and will not support transport of any of these substrates in unreconstituted *dld*$^-$ vesicles. Binding of enzyme to wild-type membranes produces an increase in D-lactate oxidation but has little or no effect on the ability of the membranes to catalyze active transport. Reconstitution of *dld*$^-$ membranes with increasing amounts of D-LDH produces a corresponding increase in D-lactate oxidation, and transport approaches an upper limit that is similar to the specific transport activity of wild-type membrane vesicles. However, the quantity of enzyme required to achieve maximum initial rates of transport varies somewhat with different transport systems.

Binding of 2-(N-dansyl)aminoethyl-β-D-thiogalactoside (DG_2) to membrane vesicles containing the *lac* transport system is dependent upon D-lactate oxidation, and this fluorescent probe can be utilized to quantitate the number of *lac* carrier proteins in the membrane vesicles (see previous discussion). When *dld-3* membrane vesicles are reconstituted with increasing amounts of D-LDH, there is a corresponding increase in the binding of DG_2. Assuming that each *lac* carrier protein molecule binds one molecule of DG_2, it can be estimated that there is at least a sevenfold to eightfold excess of *lac* carrier protein relative to functional D-LDH in reconstituted *dld*$^-$ vesicles. A similar determination can be made for wild-type vesicles. These vesicles contain approximately 0.07 nmole of D-LDH per milligram membrane protein (based on the specific activity of the homogeneous enzyme preparation) and about 1.1 nmoles of *lac* carrier protein per milligram membrane protein, yielding a ratio of about 15 for *lac* carrier protein relative to D-LDH.

Although the rate and extent of transport increases dramatically with reconstitution, the rate and extent of labeling of *dld*$^-$ vesicles with radioactive vinylglycolate remains constant. As discussed above, this compound is transported via the lactate transport system, and oxidized to a reactive product by D- and L-LDH's on the inner surface of the vesicle membrane. The observation that reconstituted *dld*$^-$ membranes do not exhibit enhanced labeling by vinylglycolate suggests that bound D-LDH is present on the outer surface of the vesicles. In this case, the reactive product released from D-LDH would be diluted into the external medium, whereas if the enzyme were on the inner surface of the vesicle membrane, the rate of labeling would be expected to increase with reconstitution, since the reactive product should accumulate within the vesicles to higher effective concentrations. This conclusion is consistent with recent experiments of KONINGS (submitted for publication) and SHORT, KABACK, and KOHN (submitted for publication), demonstrating that reduced 5-N-methylphenazonium-3-sulfonate (HAUSKA, 1972), an impermeable electron carrier, drives transport as well as reduced phenazine methosulfate, its lipophilic analog.

The suggestion that D-LDH is localized on the outer surface of reconstituted *dld*$^-$ membrane vesicles, as opposed to the inner surface of native ML 308-225 vesicles, has received strong support based on results of recent experiments with antibody prepared against D-LDH (SHORT, KABACK, KOHN, submitted for publication; SHORT et al., submitted for publication). Incubation of ML 308-225 membrane vesicles with anti-D-LDH does not inhibit D-LDH activity (assayed by tetrazolium dye reduction, oxygen uptake, and/or D-lactate-dependent transport) unless the vesicles are disrupted physically or spheroplasts are lysed in the presence of antibody. In striking contrast, treatment of reconstituted *dld*$^-$ vesicles with anti-D-LDH results in marked inhibition of D-LDH activity. The titration curves obtained with reconstituted *dld-3* membrane vesicles are almost identical quantitatively to that obtained with the homogeneous preparation of D-LDH. In addition to providing information about the localization of D-LDH in native and reconstituted vesicles, the results with the native vesicle preparations are consistent with other experiments that demonstrate that essentially all of the vesicles catalyze active transport (cf. above) and therefore cannot be inverted or sufficiently damaged to allow access of anti-D-LDH to D-LDH.

In view of these findings with antibody against D-LDH, it is pertinent to discuss recent experiments carried out with antiserum prepared against ATPase (SHORT, KABACK, KOHN, submitted for publication). It has been reported (FUTAI, 1974a; VAN THIENEN, POSTMA, 1973) that incubation of membrane vesicles with antiserum prepared against this enzyme results in 30% to 50% inhibition of membrane-bound ATPase activity, leading to the conclusion that a significant number of vesicles are inverted or damaged, or that ATPase becomes dislocated to the outer surface of the vesicle membrane during lysis (ALTENDORF, STAEHELEN, 1974; FUTAI, 1974a). It should be emphasized that most of the studies referred to were carried out at pH 9.0 (i.e., the pH optimum for the membrane-bound ATPase), and that D-LDH activity is not inhibited by anti-D-LDH at this pH. Moreover, when ML 308-225 membrane vesicles are treated with antiserum against ATPase at pH 8.0 or below, there is no significant inhibition of ATPase activity. FUTAI (1974) has stated, however, that he does observe inhibition of ATPase activity by antiserum at pH 8.0 and below. The reason for this apparent discrepancy may be related to differences in the methods used to prepare membrane vesicles (SHORT, KABACK, KOHN, submitted for publication). Thus, homogenization of membrane vesicles at pH 8.0 or below sensitizes ATPase activity to antiserum, although D-lactate-dependent oxygen consumption and active transport in the homogenized vesicles are still completely resistant to the effects of anti-D-LDH γ globulin. Finally, it has been demonstrated (SHORT, KABACK, KOHN, submitted for publication) that ATPase is not firmly bound to the membrane and becomes easily dissociated during vesicle preparation. When the specific activities of ATPase, cytochrome b_1, and D-LDH are compared in sonicated cells, spheroplasts, and membrane vesicles, that of ATPase remains constant, while the specific activities of cytochrome b_1 and D-LDH increase fivefold and threefold, respectively. It is apparent, therefore, that 60% to 80% of the ATPase activity of the cell is lost during the preparation of membrane vesicles. Taken as a whole, these findings indicate that ATPase is loosely bound to the inner surface of the vesicle membrane, and that it may be disclosed from the inside to the outside surface or even lost entirely under relatively mild conditions. In any case, it seems apparent that ATPase activity *per se* or its inhibition by antibody is not a very useful means of determining membrane sidedness.

The flavin moiety of the holoenzyme appears to be critically involved in binding D-LDH to the membrane (SHORT, KABACK, KOHN, 1974). Treatment with [1-^{14}C]hydroxybutynoate leads to the inactivation of D-LDH by modification of the flavin adenine dinucleotide coenzyme bound to the enzyme (cf. Figure 1). Enzyme labeled in this manner does not bind to *dld*$^-$ membrane vesicles. The findings suggest that the flavin coenzyme itself may mediate binding or, alternatively, that covalent inactivation of the flavin may result in a conformational change that does not favor binding. It is tempting to speculate on the relevance of this finding to the synthesis of membrane-bound dehydrogenases in the intact cell. Possibly, the apoprotein moiety of D-LDH is synthesized on cytoplasmic ribosomes but is not inserted into the membrane until coenzyme is bound. If this does happen, D-LDH mutants that are defective in the flavin binding site should exhibit soluble material that cross-reacts immunologically with native D-LDH.

MECHANISM OF ACTIVE TRANSPORT

It is not known how energy released by oxidation of D-lactate or other electron donors is coupled to transport. There are several contending theories that attempt to provide an answer to this question, but experimental evidence to date does not allow a clear choice to be made among them. The evidence currently available reveals a necessity for bringing together aspects of theories that have been looked upon heretofore as mutually exclusive.

There are certain fundamental observations that any proposed mechanism for energy-coupling must take into account:

(1) There is no correlation between rates of oxidation of various electron donors and their relative effects on transport. Succinate and NADH, for instance, are oxidized more rapidly than D-lactate by *E. coli* membrane vesicles, yet D-lactate is by far the most effective energy source for transport.

(2) The cytochrome chain of *E. coli* vesicles is completely reduced by D-lactate, L-lactate, succinate, or NADH. Since D-lactate stimulates transport so much more effectively than the other electron donors, the site at which energy is coupled to transport must lie within a segment of the electron transfer chain between D-LDH and the common cytochrome chain. This conclusion is supported by a number of other experimental approaches (KABACK, 1972; HONG, KABACK, 1973; KABACK, 1973).

(3) D-Lactate-dependent transport is completely blocked by uncouplers such as carbonylcyanide *m*-chlorophenylhydrazone (CCCP) or 2,4-dinitrophenol (DNP), although these agents do not inhibit D-lactate oxidation.

(4) The Michaelis constant (K_m) for efflux is much higher than the K_m for uptake; the V_{max} for influx and efflux are identical.

(5) Although all inhibitors of D-lactate oxidation block uptake, only those that act after the site of energy-coupling in the electron transfer chain cause efflux when added to preloaded vesicles.

(6) There is a large excess of carriers relative to D-LDH.

(7) Valinomycin-induced potassium efflux results in uptake of lactose and other solutes in the absence of D-lactate oxidation.

An initial model proposed by KABACK and BARNES (1971) depicted the carriers as electron transfer intermediates in which a change from the oxidized to the reduced state results in translocation of the carrier-substrate complex to the inner surface of the membrane and a concomitant decrease in the affinity of the carrier for substrate. This model was intended merely to provide a working hypothesis that could account for much of the data available at that time within the framework of common biochemical concepts. Aside from its basis in experimental observation, the model did not require the membrane to exhibit any special characteristics other than its function as a diffusion barrier. Its major defects are that it fails to account adequately for the behavior of electron-transfer-coupling mutants (HONG, KABACK, 1972) and other mutants (SIMONI, SHALLENBERGER, 1972; ROSEN, 1973a, 1973b; GUTNICK, this volume) which exhibit normal electron transfer properties but are defective in D-lactate-dependent transport, for the inhibitory action of uncoupling agents, for the excess of carriers relative to D-LDH, and for certain carriers that are insensitive to sulfhydryl reagents (SHERRIS, KERWAR, KABACK, manuscript in preparation; KABACK, KERWAR, unpublished information).

A very different type of hypothesis, one that emphasizes the positioning of respiratory chain components within the matrix of the membrane, was proposed by MITCHELL (MITCHELL, 1973; MITCHELL, 1963; MITCHELL, 1966; MITCHELL, 1967). As visualized by this chemiosmotic model, oxidation of electron donors is accompanied by expulsion of protons into the external medium, leading to a pH gradient and/or electrical potential across the membrane. This electrochemical gradient is postulated to be the driving force for inward movement of transport substrates by way of passive diffusion in the case of lipophilic cations such as dibenzyldimethylammonium ion (BOKEEVA et al., 1970), via facilitated diffusion in the case of positively charged substrates such as lysine or potassium ions (in the presence of valinomycin) or via coupled movement of protons with a neutral substrate such as lactose or proline (i.e., "symport"). In instances where sodium efflux is observed (LOMBARDI, REEVES, KABACK, 1973), the chemiosmotic model invokes the concept of sodium-proton "antiport," which is postulated to catalyze electroneutral exchange of internal sodium with external protons and vice versa (WEST, MITCHELL, 1974). Moreover, the inhibitory effects of uncoupling agents on transport are attributed to the ability of these compounds to conduct protons across the membrane, thus short-circuiting the "proton-motive force" that drives transport (MITCHELL, MOYLE, 1967).

A review by HAROLD (1972) lucidly discusses the chemiosmotic model and summarizes much of the evidence that supports it. More recent experiments (ROSEN, 1973a; ROSEN, 1973b; HIRATA, ALTENDORF, HAROLD, 1973; KASHKET, WILSON, 1973; ALTENDORF, HAROLD, SIMONI, 1974) have provided more direct support for this hypothesis, and some of these findings have been corroborated and extended in this laboratory. During D-lactate oxidation, lipophilic cations such as dimethyldibenzylamine (in the presence of tetraphenylboron) (LOMBARDI, REEVES, KABACK, 1973; HIRATA, ALTENDORF, HAROLD, 1973; ALTENDORF, HAROLD, SIMONI, 1974), safranine (SCHULDINER, KABACK, manuscript in preparation), and triphenylmethylphosphonium (SCHULDINER, KABACK, manuscript in preparation) are accumulated. Moreover, there is a quantitative correlation between the steady-state level of accumulation of rubidium (in the presence of valinomycin), triphenylmethylphosphonium and of dibenzyldimethylammonium (in the presence of tetraphenylboron). These observations are consistent with the interpretation that D-lactate oxidation generates a membrane potential that is positive outside. In addition, when potassium-loaded membrane vesicles are treated with valinomycin (a potassium-specific ionophore) and rapidly diluted into media lacking potassium, lactose and other transport substrates are taken up (HIRATA, ALTENDORF, HAROLD, 1973; KASHKET, WILSON, 1973; ALTENDORF, HAROLD, SIMONI, 1974; LOMBARDI et al., 1974), indicating that a positive membrane potential can drive lactose uptake in the absence of electron flow. Finally, binding of dansylgalactosides by vesicles containing the *lac* carrier protein can be induced by potassium or thiocyanate diffusion gradients in the absence of D-lactate oxidation (SCHULDINER et al., in press), suggesting that the membrane potential may cause the binding site in the carrier to become accessible on the exterior surface of the membrane. Although the interpretation of some of these observations can be questioned (LOMBARDI, REEVES, KABACK, 1973; LOMBARDI et al., 1974; KABACK et al., 1974) and the chemiosmotic hypothesis does not readily explain all of the observations discussed above, it appears clear that the basic mechanism of respiration-dependent transport involves the generation of a membrane potential (outside positive) by means of proton extrusion and/or charge separation within the membrane.

REFERENCES

ABRAMS, A., SMITH, J. B.: Biochem. Biophys. Res. Commun. 44, 1488 (1971).
ALTENDORF, K., HAROLD, F. M., SIMONI, R. D.: J. Biol. Chem. 249, 4587 (1974).
ALTENDORF, K. H., STAEHELEN, L. A.: J. Bacteriol. 117, 888 (1974).
BARNES, E. M., Jr., KABACK, H. R.: J. Biol. Chem. 246, 5518 (1971).
BERGER, E. A.: Proc. Nat. Acad. Sci. U.S.A. 70, 1514 (1973).
BOKEEVA, L. E., GRINIUS, L. L., JASAITIS, A. A., KULIENE, V. V., LEIRTSKY, D. O., LIBERMAN, E. A., SEVERNA, I. I., SKULACHEV, V. P.: Biochim. Biophys. Acta 216, 13 (1970).
BUTLIN, J. D.: Ph.D. Thesis, Canberra City, Australia: Australia National University 1973.
DEVOR, K. A., SCHAIRER, H. R., RENZ, D., OVERATH, P.: Eur. J. Biochem. 45, 451 (1974).
FLEET, G. W. J., PORTER, R. R., KNOWLES, J. R.: Nature 224, 511 (1969).
FUTAI, M.: Biochemistry 12, 2468 (1973).
FUTAI, M.: J. Membrane Biol. 15, 15 (1974a).
FUTAI, M.: Biochemistry 13, 2327 (1974b).
GUTNICK, D. L.: A genetic approach to the study of oxidative phosphorylation in bacteria (this volume).
HAROLD, F. M.: Bact. Rev. 36, 172 (1972).
HAUSKA, G.: FEBS Letters 28, 217 (1972).
HIRATA, H., ALTENDORF, K .H., HAROLD, F. M.: Proc. Nat. Acad. Sci. U.S.A. 70, 1804 (1973).
HONG, J.-s, KABACK, H. R.: Proc. Nat. Acad. Sci. U.S.A. 69, 3336 (1972).
HONG, J.-s., KABACK, H. R.: In: CRC Critical Reviews in Microbiology (eds. A. I. LASKIN, H. LECHAVALIER), p. 333. Cleveland: Chemical Rubber Co. 1973.
KABACK, H. R.: In: Current Topics in Membranes and Transport (eds. F. BRONNER, A. KLEINZELLER), p. 36. New York: Academic Press 1970a.
KABACK, H. R.: Ann. Rev. Biochem. 39, 561 (1970b).
KABACK, H. R.: Biochim. Biophys. Acta 265, 367 (1972).
KABACK, H. R.: In: Bacterial Membranes and Walls (ed. L. LEIVE), p. 241. New York: Marcel Dekker 1973.
KABACK, H. R.: Science, in press.
KABACK, H. R., BARNES, E. M., Jr.: J. Biol. Chem. 246, 5523 (1971).
KABACK, H. R., KERWAR, G. K.: (unpublished information).
KABACK, H. R., REEVES, J. P., SHORT, S. A., LOMBARDI, F. J.: Arch. Biochem. Biophys. 160, 215 (1974).
KABACK, H. R., SHORT, S. A., KACZOROWSKI, G., FISHER, J., WALSH, C. T., SILVERSTEIN, S. C.: Fed. Proc. 33, 1394 (1974).
KASHKET, E. R., WILSON, T. H.: Proc. Nat. Acad. Sci. U.S.A. 70, 2866 (1973).
KIEFER, H., LUNDSTROM, J., LENNOX, E. S., SINGER, S. J.: Proc. Nat. Acad. Sci. U.S.A. 67, 1688 (1973).
KLEIN, W. L., BOYER, P. D.: J. Biol. Chem. 247, 7257 (1972).
KOHN, L. D., KABACK, H. R.: J. Biol. Chem. 248, 7012 (1973).
KONINGS, W. N.: Arch. Biochem. Biophys. (submitted for publication).
KONINGS, W. N., BISSCHOP, A., VOENHUIS, M., VERMEULEN, C. A.: J. Bacteriol. 116, 1456 (1973).
KONINGS, W. N., KABACK, H. R.: Proc. Nat. Acad. Sci. U.S.A. 70, 3376 (1973).
LOMBARDI, F. J., KABACK, H. R.: J. Biol. Chem. 247, 7844 (1972).
LOMBARDI, F. J., REEVES, J. P., KABACK, H. R.: J. Biol. Chem. 248, 3551 (1973).

LOMBARDI, F. J., REEVES, J. P., SHORT, S. A., KABACK, H. R.: Ann. N.Y. Acad. Sci. 227, 312 (1974).
MITCHELL, P.: Biochem. Soc. Symp. 22, 142 (1963).
MITCHELL, P.: Biol. Rev. 41, 445 (1966).
MITCHELL, P.: Fed. Proc. 26, 1370 (1967).
MITCHELL, P.: J. Bioenergetics 4, 163 (1973).
MITCHELL, P., MOYLE, J.: Biochem. J. 104, 588 (1967).
OR, A., KANNER, B. I., GUTNICK, D. L.: FEBS Letters 35, 217 (1973).
PARNES, J. R., BOOS, W.: J. Biol. Chem. 248, 4429 (1973).
RADDA, G. K., VANDERKOOI, J.: Biochim. Biophys. Acta 265, 509 (1972).
REEVES, J. P., HONG, J.-s., KABACK. H. R.: Proc. Nat. Acad. Sci. U.S.A. 70, 1917 (1973).
REEVES, J. P., LOMBARDI, F. J., KABACK, H. R.: J. Biol. Chem. 247, 6204 (1972).
REEVES, J. P., SHECHTER, E., WEIL, R., KABACK, H. R.: Proc. Nat. Acad. Sci. U.S.A. 70, 2722 (1973).
ROSEN, B. P.: J. Bacteriol. 116, 1124 (1973a).
ROSEN, B. P.: Biochem. Biophys. Res. Commun. 53, 1289 (1973b).
RUDNICK, G., WEIL, R., KABACK, H. R.: J. Biol. Chem. 250, 1371 (1975).
RUDNICK, G., WEIL, R., KABACK, H. R.: J. Biol. Chem. (in press).
RUOHO, A. E., KIEFER, H., ROEDER, P. E., SINGER, S. J.: Proc. Nat. Acad. Sci. U.S.A. 70, 2567 (1973).
SCHAIRER, H. U., HADDOCK, B. A.: Biochem. Biophys. Res. Commun. 48, 54 (1972).
SCHULDINER, S., KABACK, H. R.: (manuscript in preparation).
SCHULDINER, S., KERWAR, G. K., WEIL, R., KABACK, H. R.: J. Biol. Chem. (in press).
SCHULDINER, S., KUNG, H.-f., WEIL, R., KABACK, H. R.: J. Biol. Chem. (submitted for publication).
SCHULDINER, S., SPENCER, R. D., WEIL, R., WEBER, G., KABACK, H. R.: (manuscript in preparation).
SHAW, L., GRAU, F., KABACK, H. R., HONG, J.-s., WALSH, C. T.: J. Bacteriol. (submitted for publication).
SHECHTER, E., GULIK-KRZYWICKI, T., KABACK, H. R.: Biochim. Biophys. Acta 274, 466 (1972).
SHERRIS, D. I., KERWAR, G. K., KABACK, H. R.: (manuscript in preparation).
SHORT, S. A., HAWKINS, T., KOHN, L. D., KABACK, H. R.: J. Biol. Chem. (submitted for publication).
SHORT, S. A., KABACK, H. R., KACZOROWSKI, G., FISHER, J., WALSH, C. T., SILVERSTEIN, S. C.: Proc. Nat. Acad. Sci. U.S.A. (in press).
SHORT, S. A., KABACK, H. R., KOHN, L. D.: Proc. Nat. Acad. Sci. U.S.A. 71, 1461 (1974).
SHORT, S. A., KABACK, H. R., KOHN, L. D.: J. Biol. Chem. (in press).
SHORT, S. A., WHITE, D. C., KABACK, H. R.: J. Biol. Chem. 247, 7452 (1972).
SIMONI, R. D., SHALLENBERGER, M. K.: Proc. Nat. Acad. Sci. U.S.A. 69, 2663 (1972).
VAN THIENEN, G., POSTMA, P. W.: Biochim. Biophys. Acta 323, 429 (1973).
WALSH, C. T., ABELES, R. H., KABACK, H. R.: J. Biol. Chem. 247, 7858 (1972).
WALSH, C. T., KABACK, H. R.: J. Biol. Chem. 248, 5456 (1973).
WALSH, C. T., KABACK, H. R.: N.Y. Acad. Sci. 235, 519 (1974).
WALSH, C. T., SCHONBRUNN, A., LOCKRIDGE, O., MASSEY, V., ABELES, R. H.: J. Biol. Chem. 247, 6004 (1972).
WEINER, J. H.: J. Membrane Biol. 15, 1 (1974).
WEST, I. C., MITCHELL, P.: Biochem. J. 144, 87 (1974).

On the Diversity of Links between Transport and Metabolism in Bacteria

Franklin M. Harold

Division of Research, National Jewish Hospital and Research Center, Denver, Colorado 80206; and Department of Microbiology, University of Colorado School of Medicine, Denver, Colorado 80220

INTRODUCTION

In bacteria, the primary organelle of energy transduction is the cytoplasmic membrane. As in all cells, the membrane is the locus of the permeability barrier and of transport systems; but it also houses the respiratory chain and ancillary enzymes of oxidative phosphorylation, photosynthetic pigments, the flagellar motor, and other energy-linked functions. The recent surge of research into the mechanisms of bacterial energy coupling can be attributed to both technical and conceptual advances. Among the technical advances are the application of ionophores, the isolation of mutants defective in energy coupling, and especially the development of membrane vesicles into a major research tool. Chief among the conceptual advances is the growing recognition that Peter Mitchell's chemiosmotic theory provides a rational and comprehensive framework upon which the experimental observations can be assembled. A guide to some and a goad to others, chemiosmotic theory has lent direction to the revolution that is now transforming bioenergetics.

The principles of chemiosmotic coupling and the evidence upon which the application of these principles rests have been presented in detail both by Mitchell himself (MITCHELL, 1966, 1970a, 1970b, 1973) and by others (GREVILLE, 1969; SKULACHEV, 1971; HAROLD, 1972; HAMILTON, in press). Suffice it here to recall the four basic postulates: (1) The cytoplasmic membrane must be largely impermeable to protons and to ions generally, except for specialized transport systems. (2) A number of vectorial metabolic pathways translocate protons across the membrane in an electrogenic manner (four such pathways have now been documented in bacteria: the respiratory chain, the photosynthetic apparatus, the DCCD-sensitive ATPase, and bacteriorhodopsin). (3) The translocation of protons generates a proton-motive force and sets up a circulation of protons across the membrane. Finally, there exist various molecular devices that link the proton circulation to the performance of useful work including the synthesis of ATP by oxidative phosphorylation, the accumulation of nutrients, the generation of reducing power, and even the rotation of flagella (LARSEN et al., 1974).

Not all the proposed coupling mechanisms are fully established and others may well exist. But the principle that bacterial membranes separate electrical charges by vectorial metabolic reactions and employ the recombination of these charges to perform various kinds of work now appears solidly founded. It is the basis from which I would like to examine more closely the relationship of the proton circulation to the uptake of metabolites by bacteria.

SECONDARY COUPLING OF TRANSPORT TO METABOLISM

According to the chemiosmotic theory, transport systems are linked to the metabolic machinery not by covalent chemical intermediates, nor by energized conformational states of membrane proteins, but via the components of the proton-motive force, the pH gradient (ΔpH), and the membrane potential ($\Delta\Psi$). Charged metabolites can respond to these directly; uncharged metabolites are thought to be translocated by symport with protons so as to render the overall process electrogenic. Secondary coupling (MITCHELL, 1967) refers to the transmission of the driving force via secondary bonds (including ionic ones) rather than by the covalent bonds with which biochemists are accustomed to deal. Experimental tests of this controversial proposal have been conducted chiefly with two complementary systems. The streptococci are bacteria that ordinarily lack the respiratory chain and rely upon ATP produced by substrate-level phosphorylation to drive transport processes. Conversely, membrane vesicles from aerobic bacteria such as *Escherichia coli* respire and accumulate various metabolites but do not normally carry out oxidative phosphorylation; the coupling of respiration to the transport systems is thus effected without the intermediacy of ATP (KABACK, HONG, 1973; KABACK, in press). In both systems, strong evidence has been obtained that certain transport systems are indeed linked to the major energy-yielding pathways via the proton-motive force.

Existence of a Proton-Motive Force

Glycolyzing streptococci generate both membrane potential and pH gradient, apparently by proton extrusion through the ATPase (HAROLD,

PAVLASOVA, BAARDA, 1970; HAROLD, PAPINEAU, 1972a, b). A pH gradient of up to one unit and a $\Delta\Psi$ of -150 to -200 mV (interior negative) correspond to a total proton-motive force of over 200 mV, sufficient to support concentration gradients of at least three orders of magnitude. Respiring vesicles apparently do not maintain a large pH gradient (KABACK, HONG, 1973), but do generate a membrane potential of the order of -150 mV, probably by proton expulsion (HIRATA, ALTENDORF, HAROLD, 1973; ALTENDORF, HAROLD, SIMONI, 1974, in press). These potentials were calculated from the distribution of lipid-soluble cations, which are known to disrupt energy coupling and might thus report an artifact rather than a physiologically functional charge imbalance. Such misgivings lend weight to recent measurements based on the quenching of certain fluorescent dyes (LARIS, PERSHADSINGH, 1974; KASHKET, WILSON, 1974) which do not dissociate transport from metabolism. We may thus possess a nondestructive measure of potential though one must keep in mind that neither the manner of dye binding nor the mechanism of quenching is fully understood.

Accumulation in Response to an Artificial Potential Gradient

An artificial membrane potential drives the accumulation of β-galactosides and of a series of neutral amino acids by streptocci deprived of metabolic substrate (ASGHAR, LEVIN, HAROLD, 1973; KASHKET, WILSON, 1972, 1973). In analogous experiments with *E. coli* vesicles, uptake of β-galactoside, proline, glycine, lysine, and several other substrates was coupled to the potential gradient (HIRATA, ALTENDORF, HAROLD, 1973, 1974). Control experiments were performed to ensure that transport was mediated by the physiological transport systems and did not involve participation of either the ATPase or the respiratory chain. Under appropriate conditions, transport can also be driven by a pH gradient (WEST, MITCHELL, 1972; KASHKET, WILSON, 1973; HAMILTON, in press). Interpretation of these findings in chemiosmotic terms is most secure for the elegant study of KASHKET and WILSON (1973), in which the time-course of thiomethyl-galactoside uptake was quantitatively correlated with the rise and decay of the proton-motive force.

Symport with Protons

Movement of protons together with sugars or amino acids has been observed in several systems including *S. lactis* (KASHKET, WILSON, 1974) and *E. coli* (WEST, MITCHELL, 1972, 1973). For the uptake of β-galactosides by *E. coli*, the stoichiometry was found to be one to one.

Mutants defective in the coupling of β-galactoside transport to metabolism still translocated sugar but not protons (WEST, WILSON, 1973), nor did they respond to an artificial potential gradient (HIRATA, ALTENDORF, HAROLD, 1974). Since these mutations map in the Y gene, it is likely that both proton and substrate associate with a single polypeptide. Uptake of glucose by *Neurospora* was also accompanied by influx of an equivalent amount of H^+ in an electrogenic manner (SLAYMAN, SLAYMAN, 1974).

Inhibition by Ionophores

It has been recognized for some years that proton conductors, valinomycin, and other ionophores dissociate transport from metabolism (reviewed by MITCHELL, 1966, 1970b; HAROLD, 1972; HAMILTON, in press). Most of the results support the original proposal that these reagents dissipate the proton-motive force (HAROLD, BAARDA, 1968; ASHGHAR, LEVIN, HAROLD, 1973; HIRATA, ALTENDORF, HAROLD, 1973, 1974; ALTENDORF, HIRATA, HAROLD, in press; WEST, MITCHELL, 1972) though doubts persist in some cases. For example, HATEFI and his associates have reported that certain uncouplers bind stoichiometrically to a specific protein in the mitochondrial membrane (HANSTEIN, HATEFI, 1974) and it has been suggested that a similar interaction may underlie the inhibition of transport by CCCP (KABACK et al., 1974).

Defective ATPase as a Proton-Conducting Channel

Mutants of *E. coli* deficient in ATPase can still utilize the respiratory chain as an energy donor for transport. Unexpectedly, however, membrane vesicles prepared from the mutant cells proved deficient in this coupling option as well. This was explained by the discovery (ALTENDORF, HIRATA, HAROLD, 1974; ROSEN, 1973) that such vesicles are exceedingly permeable to protons and cannot maintain a membrane potential, whether metabolic or artificial. It appears that the defective ATPase complex is readily lost, thereby exposing a proton-conducting element in the membrane. Dicyclohexylcarbodiimide seals the proton channel and restores both transport and the capacity to establish a potential gradient.

Respiration and ATP as Alternatives

By now there is overwhelming evidence that in aerobic bacteria, respiration and ATP are alternative energy donors for many transport processes (reviewed by HAROLD, 1972, 1974; BOYER, KLEIN, 1972; HAMIL-

TON, in press). This is not in itself specific support for any one theory of energy coupling, but it is more readily accommodated by the chemiosmotic theory than by any alternative proposed so far.

A massive body of evidence has thus been built up, mostly during the past five years. Data not easily reconciled with energy coupling via the proton-motive force have also been collected, notably by Kaback and his associates (KABACK, HONG, 1973; KABACK, in press; LOMBARDI et al., 1974; BOYER, KLEIN, 1972). A recent addition to the list is the report that a fluorescent derivative, dansylgalactoside, enters the hydrophobic phase of membrane vesicles only when they are allowed to respire (REEVES et al., 1973); whether this warrants the inference that translocation itself is energy linked remains to be determined. It is perhaps still possible to accommodate the data within the framework of conformational coupling of transport to respiration or ATP (BOYER, KLEIN, 1972); this alternative will only be rigorously challenged when a purified carrier protein is incorporated into a lipid bilayer membrane and asked to carry out "active transport" in response to an artificial potential gradient. But for the transport systems we have considered so far, energy coupling via the proton-motive force is at once the most comprehensive and the most economical explanation available.

PRIMARY SYSTEMS FOR METABOLITE TRANSPORT

Transport systems that are linked to the proton circulation can be recognized by quite simple criteria--inhibition by proton conductors, for instance--whether the ultimate energy donor be ATP or the respiratory chain. By the same token, not all transport systems fit the pattern expected from the chemiosmotic theory. It now seems clear that bacteria exploit a variety of coupling mechanisms and that the proton circulation is not the only link between metabolism and transport.

Nutrient Uptake by Group Translocation

The phosphotransferase system for sugar transport remains by far the best known example of a vectorial metabolic pathway (ROSEMAN, 1972). The work of transport and accumulation is in this case effected by the sugar-specific Enzyme II which mediates transfer of a phosphoryl group from the (soluble) Factor III to the incoming sugar molecule (SIMONI, ROSEMAN, 1973) but nothing is known of the mechanism by which the sugar molecule is enabled to traverse the barrier. MITCHELL (1973) has suggested that the sugar may be translocated together with protons so that once again the proton circulation could provide the driving force. However, the well-known failure of proton-conducting uncouplers to inhibit sugar uptake by the phosphotransferase system (HOFFEE, ENGELSBERG, LAMY, 1964; PAVLASOVA, HAROLD, 1969) makes this unlikely. Other apparent cases of group translocation driven by a donor molecule of high group-transfer potential are the uptake of purines and pyrimidines by combination with phosphoribosylpyrophosphate and that of fatty acids by reaction with acetyl Coenzyme A.

Binding Proteins and the Role of ATP

Uptake of many substances by *E. coli* and other gram-negative bacteria requires the participation of periplasmic binding proteins that can be released by osmotic shock. Among the metabolites so transported are sulfate, phosphate, galactose, histidine, ribose, and many others (OXENDER, 1972; HEPPEL, ROSEN, 1973). Membrane vesicles lack binding proteins and the transport functions that require their presence, but retain carriers that are more firmly associated with the membrane, including those for proline, glycine, and β-galactosides. There thus appear to be two classes of transport systems that differ in their dependence on dissociable binders; in some cases shockable as well as nonshockable transport systems have been described for a single substrate.

BERGER, HEPPEL, and their associates (BERGER, 1973; BERGER, HEPPEL, in press) have now demonstrated that these two classes differ also with respect to metabolic coupling. In *E. coli*, uptake of proline, serine, phenylalanine, glycine, and cysteine is mediated by transport systems resistant to osmotic shock. Accumulation of these metabolites was inhibited by dinitrophenol and other proton conductors whether respiration or glycolysis was the source of energy. In a mutant lacking ATPase, respiration, but not glycolysis, supported accumulation. These findings are consistent with energy coupling via the proton-motive force, as discussed above. By contrast, the uptake of glutamine, diaminopimelic acid, arginine, histidine, and ornithine--all of which require periplasmic binding proteins--was specifically dependent on ATP (or on an unidentified metabolite derived from it). These transport systems could not be energized by respiration in absence of the ATPase but were supported by glycolysis. By the same token, all were resistant to proton conductors. The identity of the actual energy donor and, for that matter, the role of the binding protein have not yet been established. But the limited information at hand already suggests the existence of a large class of transport systems independent of the proton circulation and linked to the metabolic machinery in a manner yet to be defined. From the studies by BOOS (1974) on the methyl-galactoside carrier of *E. coli*, it appears that energy coupling may be effected at the entry step.

Dependence of Ion Transport on Both ATP and the Proton Circulation

Extension and modification of the original chemiosmotic theory is also required to accommodate recent observations on ion transport by *S. faecalis*. It will be recalled that Mitchell originally envisaged accumulation of K^+ and other cations in response to the electrical potential, uptake of anions as the protonated species in response to the pH gradient, and neutral antiport carriers such as Na^+/H^+ exchange to maintain the osmotic balance (MITCHELL, 1966). There is mounting evidence that ion movements in mitochondria can be so rationalized (KLINGENBERG, 1970; COTY, PEDERSEN, 1974; DOUGLAS, COCKRELL, 1974), but the bacteria do not quite conform to the model.

Consider the accumulation of K^+ by *S. faecalis*, a process strictly dependent upon concurrent glycolysis. Studies with mutants, ionophores, and other inhibitors led to the conclusion that the primary process is the electrogenic extrusion of protons, mediated by the DCCD-sensitive ATPase. K^+ accumulates in response to the electrical potential, whereas Na^+ exits from the cell by electroneutral exchange for protons (HAROLD, PAVLASOVA, BAARDA, 1970; HAROLD, PAPINEAU, 1972a, b). However, as we pointed out at the time, K^+ is not simply in equilibrium with the electrical potential. In retrospect, the most telling points are that $^{42}K^+/K^+$ exchange, just like net uptake, depends on concurrent glycolysis; that DCCD and other inhibitors of the ATPase block net uptake but not the homologous exchange (HAROLD et al., 1969); and that none of the inhibitors elicit efflux of K^+ from the cells down the concentration gradient, unless valinomycin is present as well (HAROLD, BAARDA, PAVLASOVA, 1970; HAROLD, PAPINEAU, 1972b). Clearly, the properties of the K^+-transport system are quite unlike those of valinomycin or those to be expected of a secondary porter. The simplest scheme capable of accommodating all the data would be one in which ATP is required both to generate the electrical imbalance and to confer "mobility" upon the "carrier" (a channel with appropriate gating may be equally acceptable). The Na^+/H^+ antiporter also is quite unlike monensin in its properties, since glycolysis is required both for $^{22}Na^+/Na^+$ exchange and for Na^+/H^+ exchange (HAROLD, BAARDA, PAVLASOVA, 1970;

HAROLD, PAPINEAU, 1972b). The results suggest that ATP may be required for Na^+/H^+ exchange but that the ATPase is not.

The accumulation of phosphate and of arsenate by *S. faecalis* displays yet another complex relationship to the proton-motive force. Uptake of these anions is strictly unidirectional, dependent on ATP generation, and attains a concentration gradient of 400 and more. The pH gradient in these cells is at most one unit (interior alkaline) and often much less, which is too small to make more than a minor contribution to the driving force for anion accumulation. Indeed, under carefully controlled conditions, it proved possible to collapse the entire proton-motive force, both ΔpH and $\Delta\Psi$, without any inhibition of phosphate or arsenate uptake. All the results suggest that ATP or some unidentified derivative, rather than the proton-motive force, is the energy donor for the accumulation of phosphate (unpublished results). Yet under other conditions, proton conductors had been found to be severely inhibitory (HAROLD, BAARDA, 1968). The apparent discrepancy has now been resolved by the recognition that uptake of phosphate and of arsenate requires that the cytoplasmic pH be at least neutral; it is not the existence of a pH gradient that matters, but the absolute internal pH (unpublished results), and this in turn generally depends on the capacity of the cells to extrude protons. A tentative interpretation is that uptake of arsenate and phosphate is an ATP-driven electroneutral exchange for OH^-. Potassium, long known to be required for phosphate uptake, serves only to balance the anion by exchanging for protons.

The diversity of energy-coupling mechanisms in bacteria encourages one to speculate about their evolutionary origins. During the initial, anaerobic phase of cellular evolution, the accumulation of metabolites from the primordial soup may have been accomplished both by group translocation and by transport systems linked to an ATP-driven proton circulation. DCCD-sensitive ATPases are probably ubiquitous among prokaryotic creatures including mitochondria and chloroplasts, suggesting that they belong to an ancient family whose structure and function have been conserved. With the appearance of alternative means of generating the proton-motive force by trapping light or by respiration, the ATPase could assume its modern function of mediating ATP synthesis during oxidative and photosynthetic phosphorylation. Now, direct utilization of ATP as the energy donor for transport seems to make adaptive sense since bacteria (unlike, say, mitochondria) are free-living organisms exposed to a dilute environment and fluctuating energy supplies. Primary-coupled transport systems, which lend themselves more readily to strictly unidirectional operation, may thus have been favored by selection, particularly for such vital metabolites as K^+ and phosphate. Perhaps we may infer from the more restricted distribution of the binding proteins that these represent a later development, superimposed upon the earlier pattern of carriers linked to the proton circulation.

SUMMARY

According to the chemiosmotic theory, the linkage between transport carriers and the major metabolic pathway of bacteria is an indirect one, effected entirely via the proton-motive force. Studies in many laboratories with membrane vesicles and with intact cells of *Escherichia coli* and *Streptococcus faecalis* have demonstrated the following main points: utilization of respiration and of ATP as alternative energy donors for transport; generation of a proton-motive force of sufficient magnitude both by respiration and by glycolysis; accumulation of substrates in response to an artificial proton-motive force; symport of sugars and of some amino acids with protons; inhibition of accumulation by reagents and defects that dissipate the proton-motive force. Thus for the transport of many metabolites by bacteria, energy coupling via the proton-motive force is well established.

However, it now appears that bacteria possess a variety of transport systems, some of which are linked to the metabolic machinery by mechanisms other than the proton circulation. In gram-negative bacteria, transport systems that require dissociable binding proteins appear to be energized by ATP (or by a metabolic derivative). Recent data suggest that in *S. faecalis*, the same is true for the uptake of phosphate and of glutamate, and perhaps for that of K^+ as well. The significance and evolutionary origin of the diversity of bacterial transport mechanisms is considered.

REFERENCES

ALTENDORF, K., HAROLD, F. M., SIMONI, R. D.: Impairment and restoration of the energized state in membrane vesicles of a mutant of *Escherichia coli* lacking adenosine triphosphatase. J. Biol. Chem. 249, 4587-4593 (1974).

ALTENDORF, K., HIRATA, H., HAROLD, F. M.: Accumulation of lipid-soluble ions and of rubidium as indicators of the electrical potential in membrane vesicles of *Escherichia coli*. J. Biol. Chem. (in press).

ASGHAR, S. S., LEVIN, E., HAROLD, F. M.: Accumulation of neutral amino acids by *Streptococcus faecalis*: Energy coupling by a proton motive force. J. Biol. Chem. 248, 5225-5233 (1973).

BERGER, E. A.: Different mechanisms of energy coupling for the active transport of proline and glutamine in *Escherichia coli*. Proc. Nat. Acad. Sci. U.S.A. 70, 1514-1518 (1973).

BERGER, E. A., HEPPEL, L. A.: Different mechanisms of energy coupling for the shockable and non-shockable amino acid permeases of *Escherichia coli*. J. Biol. Chem. (in press).

BOOS, W.: Pro and contra carrier proteins: Sugar transport via the periplasmic galactose-binding protein. Curr. Topics Membranes and Transport 5, 51-136 (1974).

BOYER, P. D., KLEIN, D. W.: Energy-coupling mechanisms in transport. In: Membrane Molecular Biology (ed. C. F. FOX, A. D. KEITH), pp. 323-344. Stamford, Conn.: Sinauer Associates 1972.

COTY, W. A., PEDERSEN, P.: Phosphate transport in rat liver mitochondria: Kinetics and energy requirements. J. Biol. Chem. 249, 2593-2598 (1974).

DOUGLAS, M. G., CICKRELL, R. S.: Mitochondrial cation-hydrogen ion exchange: Sodium-selective transport by mitochondria and submitochondrial particles. J. Biol. Chem. 249, 5464-5471 (1974).

GREVILLE, G. D.: A scrutiny of Mitchell's chemiosmotic hypothesis. Curr. Topics Bioenergetics 3, 1-78 (1969).

HAMILTON, W. A.: Energy coupling in microbial transport. Advances in microbial physiology (in press).

HANSTEIN, W. G., HATEFI, Y.: Characterization and localization of mitochondrial uncoupler binding sites with an uncoupler capable of photoaffinity labeling. J. Biol. Chem. 249, 1356-1362 (1974).

HAROLD, F. M.: Conservation and transformation of energy by bacterial membranes. Bacteriol. Rev. 36, 172-230 (1972).

HAROLD, F. M.: Chemiosmotic interpretation of active transport in bacteria. Ann. N.Y. Acad. Sci. 227, 297-311 (1974).

HAROLD, F. M., BAARDA, J. R.: Inhibition of membrane transport in *Streptococcus faecalis* by uncouplers of oxidative phosphorylation and its relationship to proton conduction. J. Bacteriol. 96, 2025-2034 (1968).

HAROLD, F. M., BAARDA, J. R., BARON, C., ABRAMS, A.: Inhibition of membrane-bound adenosine triphosphatase and cation transport in *Streptococcus faecalis* by N,N'-dicyclohexylcarbodiimide. J. Biol. Chem. 244, 2261-2268 (1969).

HAROLD, F. M., BAARDA, J. R., PAVLASOVA, E.: Extrusion of sodium and hydrogen ions as the primary process in potassium ion accumulation by *Streptococcus faecalis*. J. Bacteriol. 101, 152-159 (1970).

HAROLD, F. M., PAPINEAU, D.: Cation transport and electrogenesis by *Streptococcus faecalis*. I. The membrane potential. J. Membrane Biol. 8, 27-44 (1972a).

HAROLD, F. M., PAPINEAU, D.: Cation transport and electrogenesis by *Streptococcus faecalis*. II. Proton and sodium extrusion. J. Membrane Biol. 8, 45-62 (1972b).

HAROLD, F. M., PAVLASOVA, E., BAARDA, J. R.: A transmembrane pH gradient in *Streptococcus faecalis*: Origin and dissipation by proton conductors and N,N'-dicyclohexylcarbodiimide. Biochim. Biophys. Acta 196, 235-244 (1970).

HEPPEL, L. P., ROSEN, B.: Binding proteins released by osmotic shock. In: Bacterial Membranes and Walls (ed. L. LEIVE), pp. 209-239. New York: Marcel Dekker 1973.

HIRATA, H., ALTENDORF, K., HAROLD, F. M.: Role of an electrical potential in the coupling of metabolic energy to active transport by membrane vesicles of *Escherichia coli*. Proc. Nat. Acad. Sci. U.S.A. 70, 1804-1808 (1973).

HIRATA, H., ALTENDORF, K., HAROLD, F. M.: Energy coupling in membrane vesicles of *Escherichia coli*: Accumulation of metabolites in response to an electrical potential. J. Biol. Chem. 249, 2939-2945 (1974).

HOFFEE, P., ENGELSBERG, E., LAMY, F.: The glucose permease system in bacteria. Biochim. Biophys. Acta 79, 337-350 (1964).

KABACK, H. R.: Transport studies in bacterial membrane vesicles. Science (in press).

KABACK, H. R., HONG, J.-S.: Membranes and transport. C.R.C. Critical Reviews in Microbiology 22, 333-376 (1973).

KABACK, H. R., REEVES, J. P., SHORT, S. A., LOMBARDI, F. J.: Mechanisms of active transport in isolated bacterial membrane vesicles. XVII. The mechanism of action of carbonylcyanide m-chlorophenylhydrazone. Arch. Biochem. Biophys. 160, 215-222 (1974).

KASHKET, E. R., WILSON, T. H.: Galactoside accumulation associated with ion movements in *Streptococcus lactis*. Biochem. Biophys. Res. Commun. 49, 615-620 (1972).

KASHKET, E. R., WILSON, T. H.: Proton-coupled accumulation of galactosides in *Streptococcus lactis* 9762. Proc. Nat. Acad. Sci. U.S.A. 70, 2866-2869 (1973).

KASHKET, E. R., WILSON, T. H.: Proton motive force in fermenting *Streptococcus lactis* 7962 in relation to sugar accumulation. Biochem. Biophys. Res. Commun. 59, 879-886 (1974).

KLINGENBERG, M.: Metabolite transport in mitochondria: An example for intracellular membrane function. Essays Biochem. 6, 119-159 (1970).

LARIS, P. C., PERSHADSINGH, H. A.: Estimations of membrane potentials in *Streptococcus faecalis* by means of a fluorescent probe. Biochem. Biophys. Res. Commun. 57, 620-626 (1974).

LARSEN, S. H., ADLER, J., GARGUS, J. J., HOGG, R. W.: Chemomechanical coupling without ATP: The source of energy for motility and chemotaxis in bacteria. Proc. Nat. Acad. Sci. U.S.A. 71, 1239-1243 (1974).

LOMBARDI, F. J., REEVES, J. P., SHORT, S. A., KABACK, H. R.: Evaluation of the chemiosmotic interpretation of active transport in bacterial membrane vesicles. Ann. N.Y. Acad. Sci. 227, 312-327 (1974).

MITCHELL, P.: Chemiosmotic coupling in oxidative and photosynthetic phosphorylation. Biological Reviews 41, 445-502 (1966).

MITCHELL, P.: Translocations through natural membranes. Advances in Enzymology 29, 33-87 (1967).

MITCHELL, P.: Reversible coupling between transport and chemical reactions. In: Membranes and Ion Transport (ed. E. E. BITTAR) Vol. I, pp. 192-256. New York: Interscience Publishers 1970a.

MITCHELL, P.: Membranes of cells and organelles: Morphology, transport and metabolism. In: Organization and Control in Prokaryotic and Eukaryotic Cells. XXth Symp. Soc. Gen. Microbiol., pp. 121-166. London: Cambridge University Press 1970b.

MITCHELL, P.: Performance and conservation of osmotic work by proton-coupled solute porter systems. Bioenergetics 4, 63-91 (1973).

OXENDER, D. L.: Membrane transport. Ann. Rev. Biochem. 41, 477-814 (1972).

PAVLASOVA, E., HAROLD, F. M.: Energy coupling in the transport of β-galactosides by *Escherichia coli*: Effect of proton conductors. J. Bacteriol. 98, 198-204 (1969).

REEVES, J. P., SHECHTER, E., WEIL, R., KABACK, H. R.: Dansyl-galactoside, a fluorescent probe of active transport in bacterial membrane vesicles. Proc. Nat. Acad. Sci. U.S.A. 70, 2722-2726 (1973).

ROSEMAN, S.: Carbohydrate transport in bacterial cells. In: Metabolic Transport (ed. L. E. HOKIN) Vol. VI, pp. 42-91. New York: Academic Press 1972.

ROSEN, B.: β-Galactoside transport and proton movements in an adenosine triphosphatase-deficient mutant of *Escherichia coli*. Biochem. Biophys. Res. Commun. 53, 1289-1296 (1973).

SIMONI, R. D., ROSEMAN, S.: Sugar transport. VII: Lactose transport in *Staphylococcus aureus*. J. Biol. Chem. 298, 966-976 (1973).

SKULACHEV, V. P.: Energy transformations in the respiratory chain. Current Topics Bioenergetics 4, 127-190 (1971).

SLAYMAN, C. L., SLAYMAN, C. W.: Depolarization of the plasma membrane of *Neurospora* during active transport of glucose: Evidence for a proton-dependent cotransport system. Proc. Nat. Acad. Sci. U.S.A. 71, 1935-1939 (1974).

WEST, I., MITCHELL, P.: Proton coupled β-galactoside translocation in nonmetabolizing *Escherichia coli*. Bioenergetics 3, 445-462 (1972).

WEST, I., MITCHELL, P.: Stoichiometry of lactose-H^+ symport across the plasma membrane of *Escherichia coli*. Biochem. J. 132, 587-592 (1973).

WEST, I., WILSON, T. H.: Galactoside transport dissociated from proton movement in mutants of *Escherichia coli*. Biochem. Biophys. Res. Commun. 50, 551-558 (1973).

Mechanisms of Membrane Electron and Proton Transfer

P. Leslie Dutton, Katie M. Petty, Roger C. Prince, and Richard J. Cogdell
Johnson Research Foundation, University of Pennsylvania, Philadelphia, Pennsylvania 19174

INTRODUCTION

This report describes the current status of our work on determining how the energy made available during electron transfer processes may be handled and conserved in biological membrane systems. For this work, four principal approaches have been followed: (a) a systematic search was conducted to determine and characterize all electron transfer components associated with the energy-conserving membrane; (b) an estimation was made of the location of each component, with respect to the containing membrane; (c) the thermodynamic properties of each component were measured to determine the tendency of each component to receive or donate electrons and to discover whether a component is capable of proton transfer coupled to the oxidation-reduction processes; and (d) a determination was made of whether the electron and proton transfer reactions of each component occur at rates commensurate with an involvement in the energy supply system of the cell.

The results of experiments with the simple membranes of the photosynthetic bacterium *Rhodopseudomonas sphaeroides* reported here (and referred to in earlier studies as *Rps. spheroides*) provide some general information on the underlying forces governing the effective operation of carriers that handle electrons, or electrons and protons, within and across the membrane. Such information has extended our concepts of the primary factors governing energy conservation mechanisms and cellular economy.

THE PHOTOSYNTHETIC MEMBRANES OF *RPS. SPHAEROIDES* AS A SYSTEM FOR THE STUDY OF ELECTRON AND PROTON TRANSFER

The membranes from the bacterium are obtained from the parent cell by the usual cell breakage techniques (French press, sonication, etc.). The particles obtained are derived from the cytoplasmic membrane and are called chromatophores. These are vesicles of average diameter, ∿500 Å, and their suspensions are optically clear. In some ways, chromatophores are analogous to submitochondrial particles, in

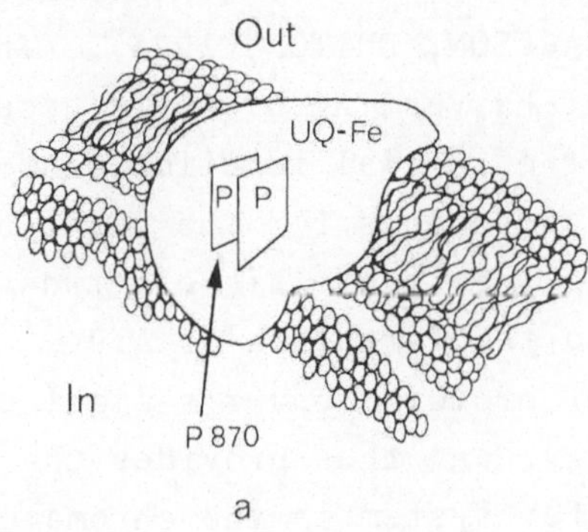

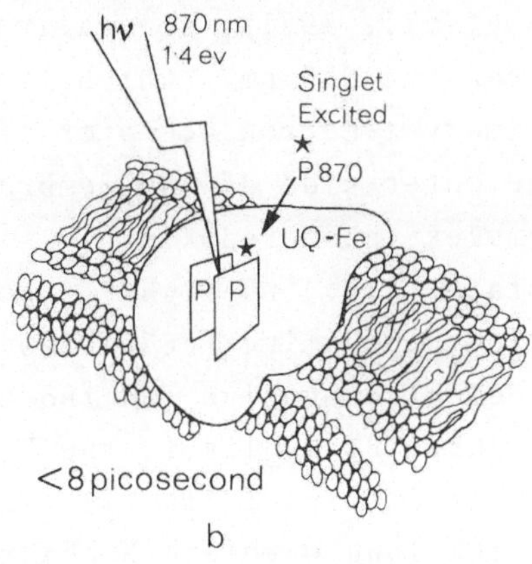

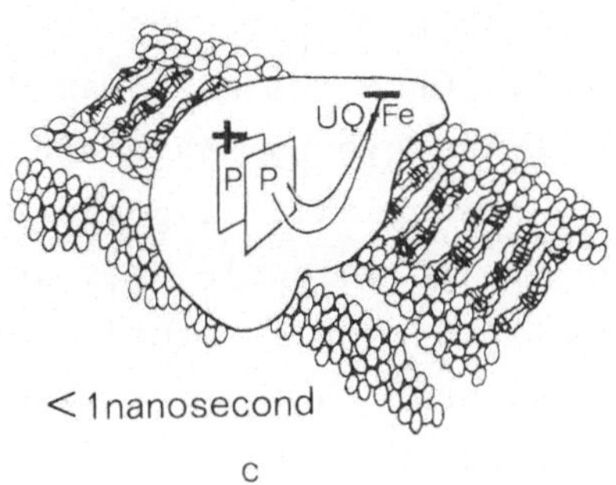

Fig. 1. The primary photoreactants of *Rps. sphaeroides*. See text for details

that they are inside-out with respect to the parent cytoplasmic membrane contained within the cell wall. Thus, for instance, cytochrome c_2, which in the intact organism occupies the periplasmic space between the cell membrane and the cell wall (i.e., outside the cell membrane), is found on the inner side of the chromatophore membrane (PRINCE et al., 1975; JACKSON, DUTTON, 1973). The chromatophore membranes contain all light-harvesting pigments necessary to transfer absorbed light energy to the special reaction center bacteriochlorophyll-protein and many redox components for the light-driven electron transfer system. The principal part of this electron transfer chain is cyclic and is coupled to the formation of ATP.

The reaction center protein converts light energy into electrochemical potential energy, and thus provides the redox driving force for each electron transfer system in the chromatophore. Figure 1 shows a section of a chromatophore, indicating how the reaction center is thought to lie across the membrane. The "P-P" represents the magnesium porphyrin dimer (NORRIS et al., 1971), generally called P870 (870 for its longest absorption band at 870 nm), which lies near the inner side of the vesicle. The primary electron acceptor of the reaction center appears to be nearer the outer side of the membrane and is thought to be a ubiquinone-iron complex, photoredoxin.

In the unexcited state (i.e., no light), the half-reduction potential of P870 is quite electropositive ($P870/P870^+$, E_m + 450 mV), while that of photoredoxin is electronegative (Pd^-/Pd E_m - 50 mV at pH 7.0). Following excitation (Figure 1B) by light, the P870 assumes an excited singlet state (and possibly a triplet state). The energy of the incident light at 870 nm, the long wavelength absorbance of the reaction center, is about 1.4 electron volts (≡ 32 Kcal/mole). In the excited state, the E_m of the P870 can be regarded as being very negative and easily capable of reducing photoredoxin. Figure 1C shows the situation after the electron transfer from P870 to photoredoxin; this occurs in less than 1 nsec (10^{-9} sec). Once the P870 has delivered an electron to photoredoxin, it resumes its ground state electropositive character, and hence acts as a fairly strong oxidizing agent; photoredoxin at the other end of the reaction center molecule acts as a fairly negative reducing agent. To prevent the simple return of the electron, the reaction center is organized such that it is kinetically unfavorable for the electron to go directly from Pd^- to $P870^+$ (a possible molecular mechanism for this has been suggested by LEIGH et al., 1974).

Thus, the light energy is converted into two forms of electrical energy: the redox energy of $P870^+$ and Pd^- (ΔE_m ∿500 mV), and the electrostatic field alterations caused by the physical separation of charges within the protein arranged across the membrane (15 to 30 Å). The pulsed generation of this state in such a rapid time (<1 nsec) is an important experimental reason why we use chromatophores of *Rps. sphaeroides*. Another reason is that the "dark" electron transfer system that exists to handle the recombination of the electron on Pd (Pd^-) and its positive "hole" on P870 ($P870^+$) is cyclic and thus is regenerative. There are also important similarities of this system with the mitochondrial ubiquinone-cytochrome *b*-cytochrome *c* region (see DUTTON and WILSON, 1974). The components that have been found to be kinetically active in the potential span between Pd^- and $P870^+$ are introduced in Figure 2. On the inside of the chromatophore, there are two cytochrome c_2 molecules per reaction center. Cytochrome c_2 is water soluble and interacts with $P870^+$ at the aqueous-membrane interface. At the negative end of the reaction center are 10 to 20 ubiquinone-$_{10}$ molecules (REED, 1969) ($E_m \simeq + 20$ mV at pH 7.0) and these receive an electron from Pd^- and take a proton from the external aqueous phase. The ubiquinone then reacts with the cytochrome *b*. We have no direct evidence for the location of the *b*-cytochrome, although from its hydrophobic nature it would be expected to be embedded in the membrane lipoprotein matrix. Some indications that it is in functional contact with the inner aqueous phase will be presented here. The electron

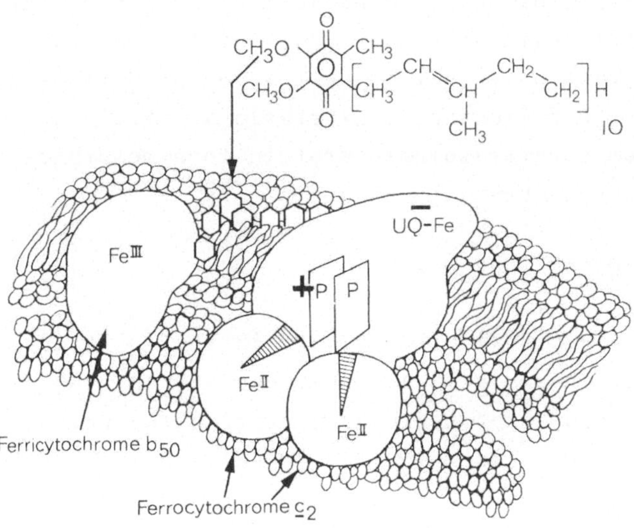

Fig. 2. Electron transfer components of the cyclic photosynthetic electron transport chain of *Rps. sphaeroides*. The diagrammatic symbolism remains constant through all the remaining figures of this type

transfer cycle is completed when the electron leaves the cytochrome *b* (E_m + 50 mV at pH 7.0) and arrives at cytochrome c_2 (E_m + mV, pH independent). The chemical and physical details of this region of the chain remain obscure, as is also the case in the analogous mitochondrial region.

During the redox processes described above, energy is converted into chemical potential energy that can ultimately be used to form ATP. The ATPase enzymes on the chromatophores appear in negatively stained electron micrographs as knobs protruding from the outer side of the membrane in a pattern similar to that of submitochondrial particles (REED, RAVEED, 1972). The ATPase has also been localized on the outside of the vesicle-using antibodies (PRINCE et al., 1975).

THE CYTOCHROME c_2 TO REACTION CENTER P870 ELECTRON TRANSFER REACTION ACROSS THE AQUEOUS-MEMBRANE INTERFACE

The two times shown in Figure 3 indicate that the electron transfer from cytochrome c_2 to flash-oxidized P870 ($P870^+$) is biphasic. These biphasic kinetics can be monitored experimentally either by the course of cytochrome c_2 photooxidation or $P870^+$ reduction; both give the same results, as would be expected for the direct ferrocytochrome c_2 to $P870^+$ reaction. Figure 4 shows the time course for oxidation of the first cytochrome c_2 oxidized and also the time course of the second oxidation following a second activating flash, while the first cytochrome is still oxidized. This possibility of examining the exact kinetics of the two individual cytochrome c_2 oxidations has provided information for the following considerations pertaining to the delivery of electrons from a water-soluble protein to a membrane-bound protein across the aqueous-membrane interface.

The Biphasic Nature of the Electron Transfer

The fast phase of oxidation occurs in the 10 to 40 μsec time range; the slower phase is usually between 300 and 400 μsec in half time, but can be slowed markedly by preparing the chromatophores in low ionic strength media. (Normally 100 mM KCl and 50 mM buffer are present.)

The biphasic nature of the cytochrome c_2 oxidation/$P870^+$ reduction is indicative of two physical states in equilibrium, with perhaps a "binding site" on the membrane's inner side. Thus, the fast ∿40 μsec phase may be oxidation of those cytochrome c_2 molecules oriented prox-

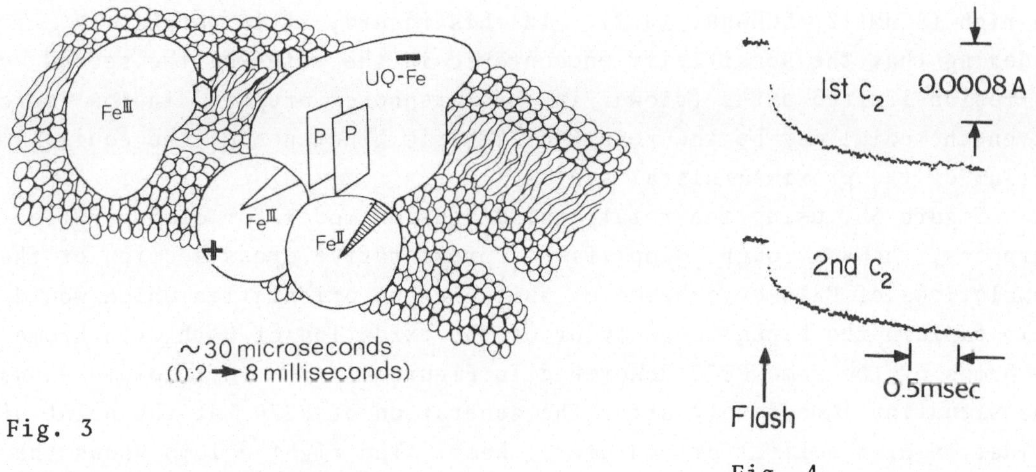

Fig. 3

Fig. 4

Fig. 3. Cytochrome c_2 photooxidation. The reduction of P^+ by reduced cytochrome c_2 is biphasic, as discussed in the text. The half times of the two phases are ∿30 μs and 0.2 to 8 ms respectively, the latter being experimentally variable depending on the ionic strength of the suspending medium. The symbols are explained in Figure 2

Fig. 4. Kinetics of the photooxidation of the first and second cytochrome c_2 monitored at 550 minus 540 nm. The first cytochrome was monitored by averaging 64 single flashes separated sufficiently to allow complete re-reduction between flashes, the second by flashing at a rate sufficient to keep the first cytochrome fully oxidized. These flash rates were 10/min and 240/min, respectively. Chromatophores of *Rps. sphaeroides* Ga (BChl 22.6 μM) were poised at an ambient redox potential (E_h) of 240 mV with 3 μM Diaminodurene (DAD) and Phenazine methosulfate (PMS) as mediators. Antimycin (2 μM) was present to slow the re-reduction of cytochrome c_2

imally to (or "on") the membrane at the time of the flash. The variable slow phase could be the rate at which those cytochrome c_2 molecules, initially not favorably oriented distally to (or "off") the membrane, achieve a proximal orientation that permits their rapid oxidation.

Two easily visualized alternatives are shown in Figure 5a. The upper one describes an on/off equilibrium situation, the ratio of the fast and slow phases being related to the equilibrium constant and the rate of the slow phase giving the rate constant. The alternative shown lower in the figure suggests that the time spent on the binding site is very large compared to the time off under physiological conditions, but that a proximal/distal equilibrium exists on a basis of molecular rotational diffusion of cytochrome c_2. The rate of the slow phase of oxidation in this case would be governed by the rotational diffusion rates. The sensitivity of reaction rates to an angular deviation from the juxtaposition of two "active" points on reactants has been shown to be high (SCHMITZ, SCHURR, 1972). In this regard, it is worth considering that the sensitivity encountered in the rates of the second diffusion-limited phase (slower in chromatophores prepared in low ionic strength media) may be the result of kinetic hindrance of the equilibrium of the proximal/distal states.

Figure 5b, using the rotational diffusion model for demonstration purposes, shows a rough, simplified representative cross-section of the populations of P870-cytochrome c_2 units in the orientation which would help explain the biphasic behavior of the oxidation of both cytochrome c_2 hemes by the same P870 described in Figure 4. The left column shows the situation immediately after the generation of $P870^{\dot{+}}$ at the point of oxidation of the first cytochrome c_2 heme. The right column shows the same situation, after the oxidation of the first heme, at the point of oxidation of the second cytochrome c_2 heme. The equilibrium constant K for c_2 (distal) $\rightleftharpoons c_2$ (proximal) in the figure is 0.25. This would

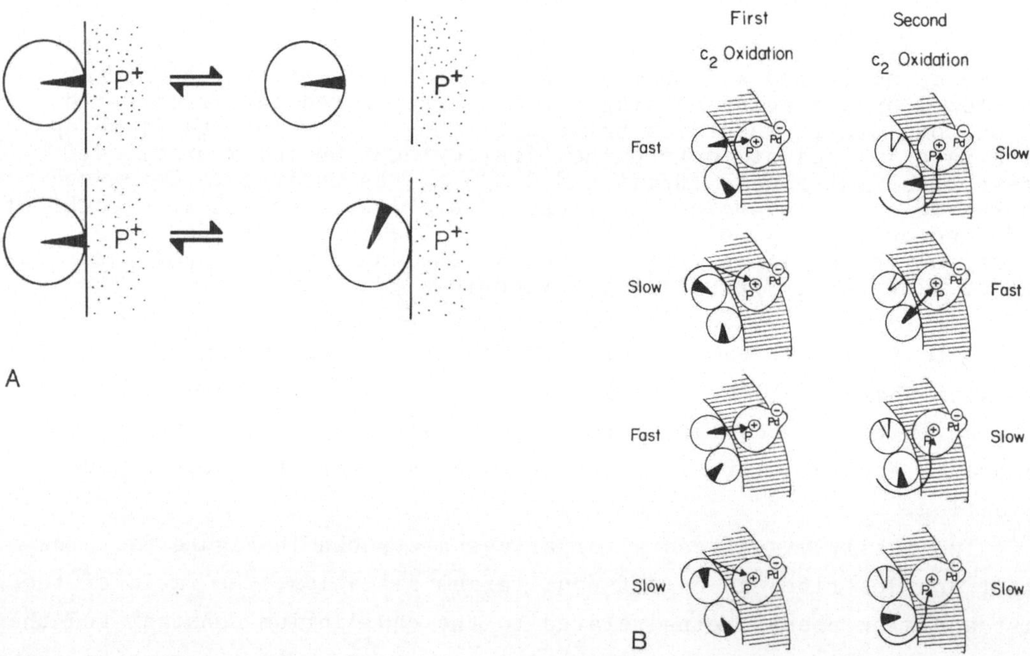

Fig. 5. Physical relationship of cytochrome c_2 with the membrane and reaction center. (A) Two alternatives for the proximal/distal equilibrium of cytochrome c_2 with the membrane-bound reaction center. The top scheme is an on/off diffusional equilibrium; the bottom alternative a rotational model. The black wedge symbolizes a site on the cytochrome where the electrons enter and leave. The reality of the equilibrium is probably composed of both alternatives. (B) A simplified model to explain the experimentally observed biphasic kinetics of the oxidation of both cytochrome c_2 molecules. The proximal/distal equilibrium constant is chosen as 0.25

produce a fast-to-slow ratio on the first cytochrome c_2 oxidation of 1:1 and a ratio of 1:3 on the second.

In experimental terms, Figure 4 reveals an approximate fast-to-slow ratio of 1:1 for the first heme and a 1:1.5 for the second, which, although a trend in the right direction, is relatively less in slow phase for the second heme oxidation than expected from the model. The apparent loss of the slow phase, however, could conceivably lie in the usual diminished total extent of the second heme oxidized on the second flash (i.e., only 70% to 80% of ideal is, in fact, oxidized on the second flash, the remainder being picked up by the third flash). If the diminished total extent is due entirely to slow phase loss, then the fast-to-slow phase ratio would be nearer to 1:3. More work on this matter is needed. In this regard, investigations to try to mimic in vivo cytochrome c_2 kinetics are under way, and pertinent preliminary results are discussed in the next section.

Model Systems

Figure 6 depicts results from exploratory work with soybean phospholipid vesicles as the supporting medium for purified isolated reac-

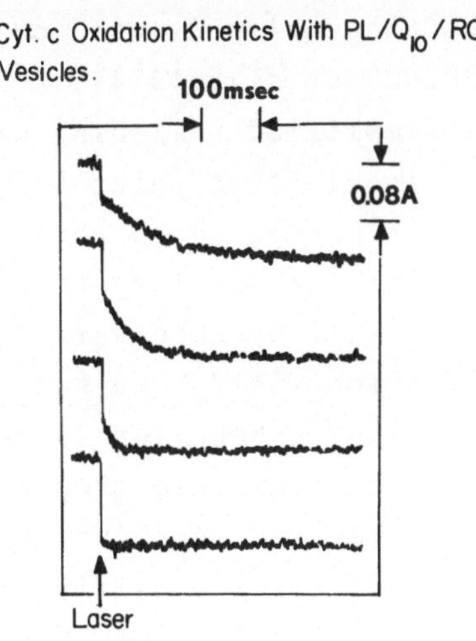

Fig. 6. Laser-induced cytochrome *c* oxidation by reaction centers in phospholipids. The reaction center protein was incorporated into soybean phospholipids, using an unpublished method devised by Drs. R. G. Cogdell and C. A. Wraight, after the detergent lauryl dimethylamine oxide (LDAO) was removed by dithionite treatment (P. L. DUTTON, unpublished method). A large excess of UQ was present to slow down the return of the electron from Pd^- to oxidized cytochrome *c*. The model system was composed of 250 mg of soybean phospholipids in 3.0 ml 10 mM Tris Cl (pH 8), 1 mM UQ_{10}, and 5 μM reaction centers. The experiment shown was performed at a reaction center concentration of 2 μM and the four traces represent the photooxidation of 3, 10, 30, and 40 μM cytochrome *c* respectively

tion center protein from *Rps. sphaeroides*. Little or no detergent is present. Kinetics of laser-induced cytochrome oxidation were taken after addition of mammalian cytochrome *c* in different amounts ranging from a cytochrome *c* and P870 ratio of 1:1 to 20:1 (see Figure 6 for details). At lower concentrations, the kinetics of the cytochrome oxidation are clearly biphasic and are consistent with a proximal/distal relationship of cytochrome *c* with the reaction center. Thus, as the cytochrome *c* concentration is increased, (a) the number of the cytochromes in the proximal position with respect to the reaction center at any time is increased, as indicated by a greater fast-to-slow ratio; and (b) the rate of the diffusional, collision-dependent course of the slower phase of cytochrome *c* oxidation is enhanced. Identical experiments with *Rps. sphaeroides* cytochrome c_2 so far have yielded no biphasic kinetics of oxidation, the course of cytochrome c_2 oxidation being governed in rate by collisional factors (first-order basis, with the amount of ferrocytochrome c_2 added), indicating no significant binding or proximal orientation.

Thus, on a simple overall electrostatic basis, mammalian cytochrome *c* (isoelectric point 10; see KE et al., 1970) behaves in a predictable manner from its electropositive character, relative to that of either the electronegative phospholipids used or the reaction center (isoelectric point 6.2; PRINCE et al., 1974).

The same perhaps oversimplified reasoning explains why the electronegative cytochrome c_2 (isoelectric point 5.5; BARTSCH, 1971) did not indicate any binding. It may be added that the half-reduction potential of cytochrome c_2 in this model system was that of the free form (E_m 345 mV); perhaps it will be necessary to produce an environment for the cytochrome c_2 which yields the natural membrane E_m of 295 mV, before the in vivo kinetic behavior is achieved. It is conceivable that presumed binding areas (electropositive?) are more susceptible in cytochrome c_2 to masking (perhaps by minor amounts of detergent?) than in the overall electropositive mammalian cytochrome *c*; but even in the latter case, indications of a membrane-reaction center binding complex are ultimately obliterated by detergent concentrations in the 1% to 2% range (PRINCE et al., 1974).

General View of Physiological Function of *c*-Type Cytochromes

Figure 7 shows how the respective *c*-type cytochromes of photosynthetic bacteria like *Rps. sphaeroides*, *Rps. capsulata*, and *R. rubrum*, and mitochondria from various sources may react with their membrane electron donors and acceptors. The operation of two (as opposed to one) cytochrome *c* molecules in each electron transfer system would obviously be expected to serve to increase the electron transfer rates in a situation where cytochrome *c* is operating between two large and perhaps relatively immobile proteins. The existence of two cytochrome *c* molecules per electron transfer system may well be general. This appears to be the case in mitochondria ranging from sea urchin sperm (WILSON, EPEL, 1968) to rat liver (ESTABROOK, HOLOWINSKI, 1961) and in photosynthetic bacteria ranging from *Chromatium D* (CASE, PARSON, 1971) to the above-named species. The finding in the case of *Rps. sphaeroides* (see also CASE and PARSON, 1971) that both cytochrome c_2 molecules are equally available for oxidation by $P870^+$ allows us (generalizing in Figure 7) to place the two cytochrome *c* molecules in parallel between the "reductase" and "oxidase," and rule out an "in series" model, i.e., one in which one cytochrome interacts with the oxidase and the other interacts with the reductase, with net electron transfer from reductase to oxidase occurring via intercytochrome *c* electron transfer). Each cytochrome heme must therefore be capable of handling reductase-to-oxidase electron transfer alone. The distal/proximal equilibria discussed in this paper provide good evidence for both the necessary mobility of the cytochrome *c*-type molecule and also the existence of appropriate constraints on that mobility, which are provided by the membrane and/or the adjacent membrane-bound reductase and oxidase, that keep the cytochrome fairly close. It is encouraging that the distal/proximal equilibrium constant of the two cytochrome c_2 molecules

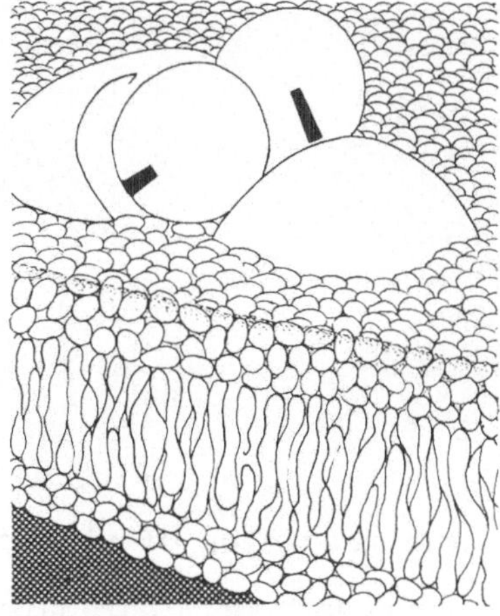

Fig. 7. General picture of membrane cytochrome *c* located between reductase and oxidase

operating with respect to one P870 is of appropriate magnitude to keep the time spent by cytochrome c_2 proximal to either the P870 ("oxidase"), or the reductase, roughly the same. Thus, we anticipate in simplistic terms that when a cytochrome *c* is distal to its oxidase, it may be proximal to the reductase. A final physiological role for the mobility observed in the case of the cytochrome c_2 molecule could be that, although the binding sites on the cytochrome *c* for the reductase and the oxidase may be in different positions on the molecule, the movement of the cytochrome *c* increases the possibility that electrons enter and leave the molecule at a single place.

THE EXTERNAL MEMBRANE PROTONATION OF REDUCED UBIQUINONE

Subsequent to the reduction of photoredoxin in <1 nsec following flash activation of P870, the electron is passed to the secondary electron acceptor, ubiquinone (UQ) (see HALSEY and PARSON, 1974), and a proton is picked up from the external aqueous phase (Figure 8), as first discovered by CHANCE et al. (1970). This proton uptake causes a change in the pH of the external medium, which can be monitored spec-

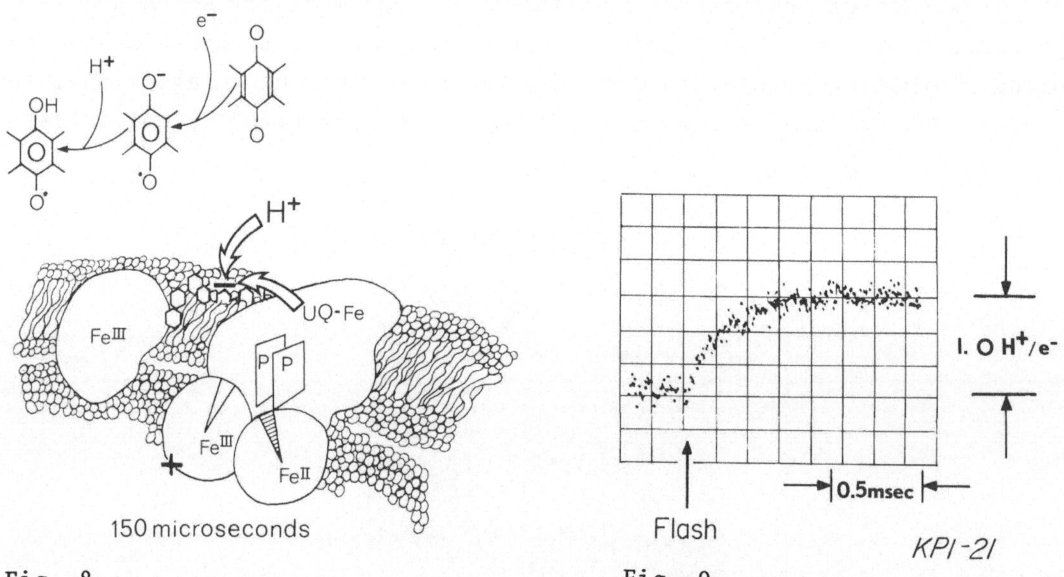

Fig. 8 Fig. 9

Fig. 8. Proton uptake on the reduction of ubiquinone

Fig. 9. The kinetics of proton uptake. Chromatophores (18 µM BChl) were suspended in 10 mM KCl, pH 8.0, 5 µM phenol red, 4 µM DAD, 2 µM antimycin (E_h = 200 mV). This is an average of 256 single scans spaced far enough apart that the charge completely decayed between flashes

trophotometrically using pH-indicator dyes (bromocresol purple and chlorphenol red, pH 5.2 to 6.8; phenol red, pH 6.8 to 8.2; cresol red, pH 7.2 to 8.8).

At pH 7.0, 1.0 ± 0.05 H^+ is taken up (half rise time 150 μsec) for every electron received by UQ (Figure 9). The extent and rate of this H^+ uptake is affected neither by the membrane proton conducting uncouplers nor agents like valinomycin. The rate of H^+ uptake is increased only about fivefold over a 1,000-fold increase in hydrogen ion concentration (pH 8.3 to 5.3), and the addition of 50% ethylene glycol (a tenfold increase in viscosity at 20°C) is without effect. The insensitivity of the rate of membrane H^+ uptake indicates that the rate-limiting step in the membrane UQ reduction and protonation is the transfer of the electron from Pd^- to UQ. The activation energy of this event, as measured by H^+ uptake over the 4°C to 40°C temperature range, is 10.5 Kcal/mole (independent of pH).

The lack of any discernible break(s) ("transition point[s]") in the Arrhenius plot over this temperature range (bacterial cell growth temperature, 30°C) adds to our failure to reveal any constraints by the physical state of the membrane and environment (apart from temperature) on the UQ reduction and protonation.

THE INTRAMEMBRANE PROTON AND ELECTRON TRANSFER FROM UBIQUINONE TO CYTOCHROME *b*

The reduction of ubiquinone is followed by its oxidation and the concomitant reduction of a *b*-type cytochrome (Figure 10). This reaction process can best be monitored in the presence of antimycin, which prevents the reoxidation of the cytochrome. Under such conditions, the half rise time of reduction of the cytochrome can be measured (Figure 11); under the conditions shown, the half time is of the order of 2 msec.

An Arrhenius plot of the temperature dependence of this reduction indicates an activation energy of about 10 Kcal/mole, with no noticeable break in the slope between 4°C and 40°C. Such a result indicates that the reduction of cytochrome *b*, like that of the ubiquinone reduction, is not influenced by the "fluidity" of the membrane lipids. We are taking this result as a hint that the ubiquinone complement (10 to 20 per electron transfer system) is not swimming "free" in the membrane matrix with this kind of diffusional process governing intramembrane electron and proton flow. We are currently investigating the attractive hypothesis that electrons and protons are carried to cytochrome *b*

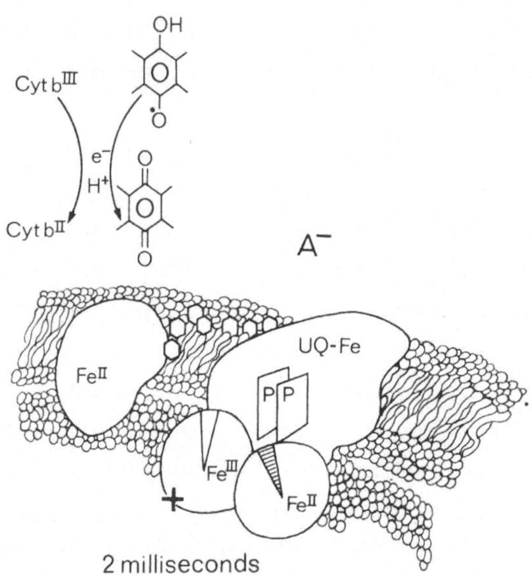

Fig. 10. The reduction of cytochrome *b*

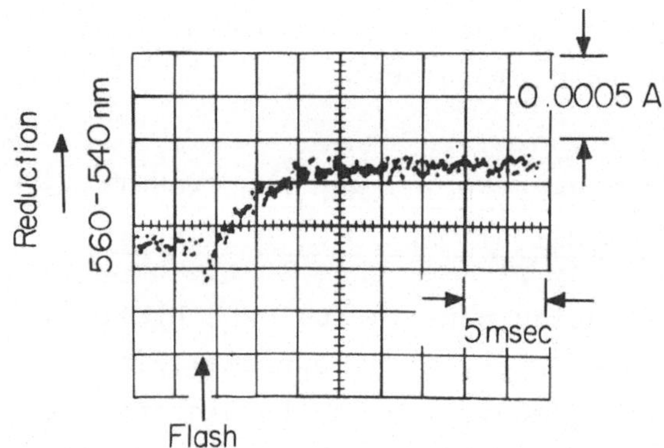

Fig. 11. The kinetics of cytochrome *b* reduction. Chromatophores (16.6 μM BChl) were suspended in 100 mM KCl, 20 mM MOPS, pH 7.0, with 2 μM antimycin and 4 μM (DAD (E_h = 190 mV). Single scans averaged 256

from photoredoxin (and succinate dehydrogenase) in the form of a ubiquinoprotein complex. We envisage this complex to be comprised of a number of quinones anchored in the protein by means of the long isoprenoid chains, with the aromatic head groups suitably aligned for effective, sequential electron/proton transfer through the membrane to a suitable point of entry on the *b*-cytochrome.

A FUNCTIONAL pK ON CYTOCHROME *b* AND THE FATE OF THE PROTON

Steady-state redox potentiometry (see DUTTON and WILSON, 1974) of the cytochrome *b* involved in cyclic electron flow indicates (Figure 12) that below pH 7.4 the midpoint potential of the cytochrome varies by -60 mV per pH unit, while above pH 7.4, the midpoint potential is independent of pH. This finding tells us that below pH 7.4, the reduction of the cytochrome involves not only the addition of an electron to the oxidized form but also of a proton, while above pH 7.4, the reduction requires only an electron. A question now arises: Is this pK functional in the time scale of electron flow (i.e., in milliseconds) as well as in the time scale of redox potentiometry (i.e., in tens of seconds)? If it is a pK functional in the rapid time scale, it might have important consequences related to energy coupling, since at pH values below pH 7.4, the proton picked up by the reduced semiquinone would be handed on to the cytochrome and would not be released from the electron transfer system until the *oxidation* of the *b*-cytochrome at the earliest (Figure 13A). In contrast, at pH values greater than 7.4, the proton would be released from the electron transfer system upon the *reduction* of the cytochrome *b* (Figure 13B).

To determine whether the pK *is* functional, we have examined the rate of proton release (after its single flash-induced uptake) from chromatophores in the presence of the uncoupling agent FCCP. The rate of reappearance of the proton in the outer aqueous phase will be a function of the proton permeability of the membrane. The uncoupling agent FCCP renders membranes permeable to protons (HENDERSON et al.,

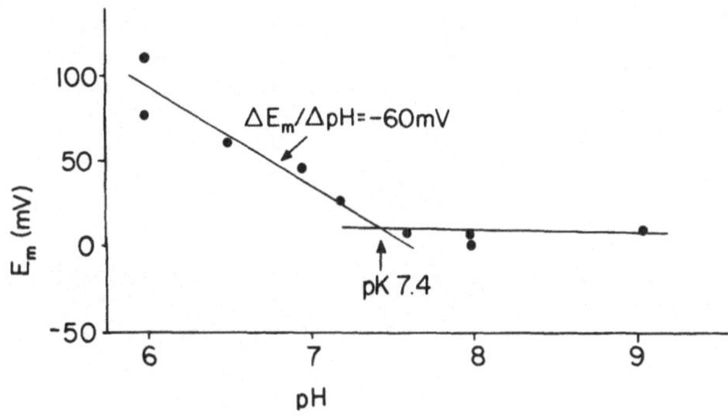

Fig. 12. The oxidation-reduction midpoint potential of cytochrome *b*. Chromatophores (~100 μM BChl) were suspended in 100 mM KCl, 50 mM buffer, at the appropriate pH (using MES, MOPS, TES, and Tris) with 50 μM DAD, 20 μM PMS, Phenazine ethosulfate, duroquinone, 3 μM pyocyanine, and 40 μM 2-hydroxyl-1,4-naphthaquinone as mediators

1969), so that in the presence of saturating amounts of this compound, the rate of reappearance of the proton in the outer aqueous phase becomes a function of the rate of proton release from the electron transfer chain. In other words, we define an uncoupler-sensitive proton as one that has been released from the chain. If the pK of the cytochrome *b* is functional in the time scale of electron transport, at pH

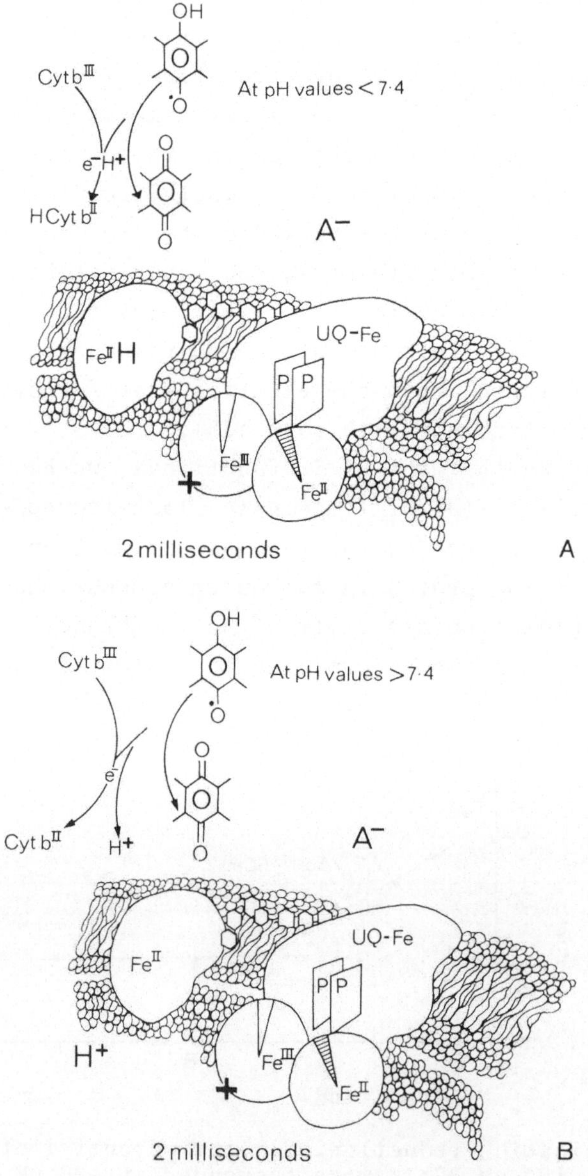

Fig. 13. The involvement of the electron and proton in the reduction of cytochrome *b*

values below pH 7.4, the proton will not be released within the chromatophore until the cytochrome is oxidized by the next carrier in the sequence. Inhibitors of this oxidation, such as antimycin, will thus inhibit proton release. However, at pH values above pH 7.4, the proton will be released before the reduction of the cytochrome *b*, since at these values of pH the cytochrome reduction does not require a proton. Under single turnover conditions at these high values of pH, therefore, the rate of proton release will be independent of the rate of cytochrome *b* reoxidation and hence antimycin-independent.

From the experiment shown in Figure 14, this indeed seems to be the case; the rate of reappearance in the external medium of the protons taken up for the protonation of UQ, in the presence of FCCP, is antimycin-sensitive below pH 7.4, and insensitive above this pH. The marked synergistic inhibition by antimycin plus FCCP on proton release at low pH is unexplained at this time.

Another approach to determining the functionality of the cytochrome *b* pK has been to trap pH indicator dyes *inside* chromatophores in order to directly monitor internal proton release. As we have discussed, our usual experimental conditions for proton measurements have the pH indicator dyes added outside the chromatophores (COGDELL et al., 1972). However, if chromatophores are soaked in a concentrated (20 mM)

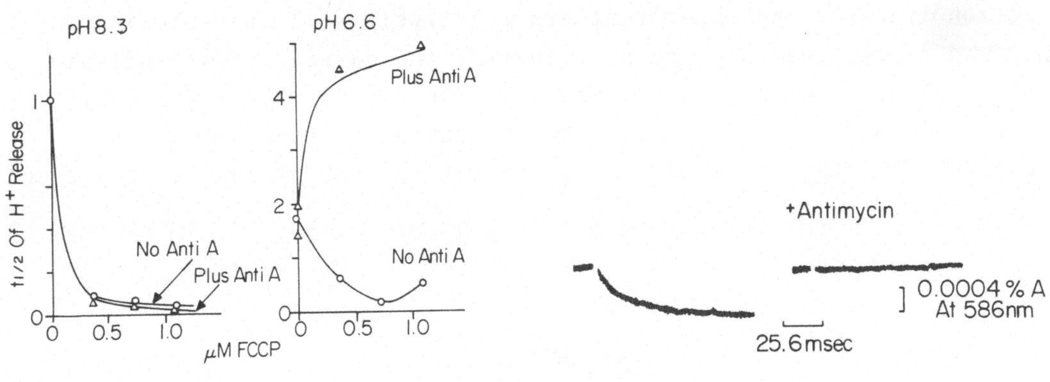

Fig. 14 Fig. 15

Fig. 14. The effect of antimycin on proton release. The conditions were essentially those of Figure 9, with the pH as indicated. The half time of proton release is measured in milliseconds

Fig. 15. The kinetics of proton release inside the chromatophores. Chromatophores (17.2 μM BChl) with chlorphenol red trapped inside the vesicles were suspended in 100 mM KCl, pH 6.01 with 5 μM DAD and 2 μM antimycin where indicated (E_h = 200 mV). These traces are the average of 16 flashes spaced far enough apart for the charge to have completely decayed between flashes and to have had a baseline (recorded in the absence of indicator) subtracted from each

alkaline (pH 9 to 10 to dissolve the indicator) aqueous solution of the dyes for 36 hr, some indicator penetrates inside the vesicles. The external indicator is then removed on a Sephadex G-25 column. To date, we have not found a suitable indicator for pH values above pH 7.4, but we have been able to get consistent results using chlorphenol red (pK = 6.0). An experiment is shown in Figure 15, where it is evident that at pH 6.0, antimycin inhibits the release of the proton within the chromatophore. Of course, this experiment could be explained in other ways, since the location of the indicator is essentially unknown, although we do know that it is not outside the vesicles. However, taken with the data of Figure 14, there is good evidence that the pK of the reduced form of the cytochrome *b* detected by steady-state redox potentiometry is indeed functional on the time scale of electron flow, and that the proton is released into the inner aqueous phase. As far as we know, this is the first time a pK on a cytochrome has been shown to be relevant to its function.

AN OVERALL PICTURE OF ENERGY CONVERSION AND THE CYCLIC ELECTRON TRANSFER SYSTEM

Figure 16 shows the current model. The solid arrows represent electron transfer seqûences that are relatively well characterized. The open arrows indicate proton transfer; the dashed arrows indicate the two alternative release points from the system, before cytochrome *b* reduction (pH values <7.4) and after cytochrome *b* reoxidation (pH values <7.4). In the latter case, the fate of the proton is at present

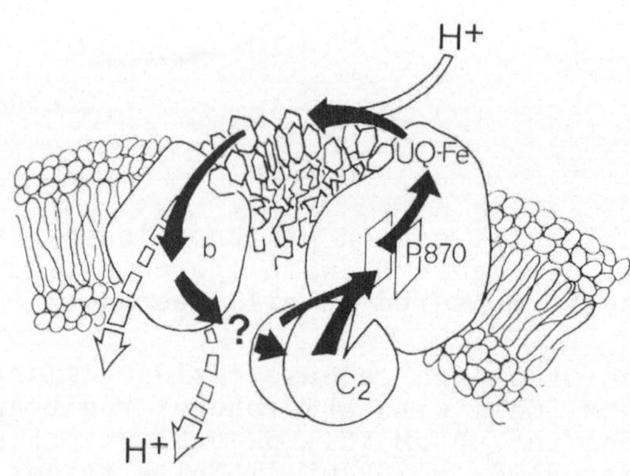

Fig. 16. Our current model for electron and proton flow in the cyclic, photosynthetic system of *Rps. sphaeroides*

unknown, although it does appear to be ultimately released from the electron transport system at a point within the chromatophore. This could be at the reduction of cytochrome c_2, since its reduction does not require a proton (E_m is pH-independent from pH 5 to 9). However, in contrast to the considerable detail now known for the cytochrome c_2 to cytochrome *b* part of the cycle, we do not know any detailed kinetics of the cytochrome *b* to c_2 electron flow beyond the knowledge that the electron does leave cytochrome *b* and arrive at cytochrome c_2.

In its simplest form, the results are not inconsistent with the chemiosmotic view that electron transfer serves to effect the vectorial transfer of protons across the membrane. However, chemical and physical information concerning the large redox potential span from cytochrome *b* (E_m 50 mV at pH 7.0) to cytochrome c_2 (E_m 295 mV at pH 7.0) is not yet available, leaving questions concerning the role of the cytochrome *b* pK. Does it have a ΔpH regulatory function in the system, or is it part of a proton-linked chemical or conformational change on the cytochrome *b* driven by the redox potential drop to cytochrome c_2, or is it entirely fortuitous? The answers await investigations into the exact sequences of events regarding protons and electrons between the *b* and c_2 cytochromes, and determination of the possible involvement of intermediary carriers that may operate there.

REFERENCES

BARTSCH, R. G.: Methods in Enzymology (ed. A. SAN PIETRO) Vol. XXIII, pp. 344-363. New York-London: Academic Press 1971.

CASE, G. D., PARSON, W. W.: Biochim. Biophys. Acta 253, 187-202 (1971).

CHANCE, G., CROFTS, A. R., NISHIMURA, M., PRICE, B.: Eur. J. Biochem. 13, 364-374 (1970).

COGDELL, R. J., JACKSON, J. B., CROFTS, A. R.: J. Bioenerg. 4, 412-429 (1972).

DUTTON, P. L., WILSON, D. F.: Biochim. Biophys. Acta 346, 165-212 (1974).

ESTABROOK, R. W., HOLOWINSKY, A.: J. Biophys. Biochem. Cytol. 9, 19-27 (1961).

HALSEY, Y. D., PARSON, W. W.: Biochim. Biophys. Acta 347, 404-418 (1974).

HENDERSON, P. J. F., McGIVAN, J. D., CHAPPELL, J. B.: Biochem. J. 111, 521-535 (1969).

JACKSON, J. B., DUTTON, P. L.: Biochim. Biophys. Acta 325, 102-113 (1973).

KE, B., CHANEY, T. H., REED, D. W.: Biochim. Biophys. Acta 216, 373-383 (1970).

LEIGH, J. S., NETZEL, T. L., DUTTON, P. L., RENTZEPIS, P. M.: FEBS Letters 48, 136-140 (1974).

NORRIS, J. R., UPHAUS, R. A., CRESPI, H. L., KATZ, J. J.: Proc. Nat. Acad. Sci. U.S.A. 68, 625-628 (1971).

PRINCE, R. C., BACCARINI-MELANDRI, A., HAUSKA, G. A., MELANDRI, B. A., CROFTS, A. R.: Biochim. Biophys. Acta (in press).
PRINCE, R. C., COGDELL, R. J., CROFTS, A. R.: Biochim. Biophys. Acta 347, 1-13 (1974).
REED, D. W.: J. Biol. Chem. 244, 4936-4941 (1969).
REED, D. W., RAVEED, D.: Biochim. Biophys. Acta 283, 79-91 (1972).
SCHMITZ, K. S., SCHURR, J. M.: J. Phys. Chem. 76, 534-545 (1972).
WILSON, D. F., EPEL, D.: Arch. Biochem. Biophys. 126, 83-90 (1968).

The Relation of Proton Gradients to Energy Conservation in Isolated Chloroplasts

M. Avron
Biochemistry Department, Weizmann Institute of Science, Rehovot, Israel

INTRODUCTION

Chemiosmotic coupling (MITCHELL, 1966; JAGENDORF, 1967) predicts that oxidation reduction energy is converted via proton movements into energy stored in the form of a proton gradient and a membrane potential. These secondary forms can, in turn, serve as the driving force for ATP formation.

During the past few years, we have concentrated our efforts on designing and utilizing techniques that have enabled us to measure accurately the size of these proton gradients and membrane potentials under a variety of conditions (ROTTENBERG et al., 1971, 1972; SCHULDINER et al., 1972a). It was found that under conditions that are optimal for ATP synthesis, proton concentration gradients as large as 10,000 (ΔpH = 4) could be measured, but only insignificant membrane potentials were observed. This finding seemed in agreement with our earlier results (KARLISH et al., 1968, 1969), which indicated that agents that might be expected to completely abolish a membrane potential, had one existed, were ineffective in decreasing the efficiency of ATP production by isolated chloroplasts.

In this chapter, I will describe some of our recent work on the ability of chloroplasts to utilize externally imposed membrane potentials as the driving force for ATP formation, and a more quantitative study of the relation of ATP formation to the size of ΔpH across the thylakoid membrane.

RESULTS AND DISCUSSION

The elegant experiments of JAGENDORF and collaborators (see JAGENDORF, 1967) have clearly illustrated that chloroplasts are capable of utilizing energy in the form of proton gradients as a driving force for ATP formation in the absence of any other independent energy source. Our initial attempts to utilize energy stored in the form of a membrane potential as a driving force for ATP formation have resulted in failure. However, when such an externally produced membrane potential was superimposed on a suboptimal proton gradient, chloroplasts clearly could use the membrane potential energy as a driving force for ATP formation (SCHULDINER et al., 1972b, 1973). As can be seen in Table 1, it was immaterial whether the suboptimal proton gradient was achieved by lowering the light intensity or by changing the pH of the dark stage. In either case, the imposition of a membrane potential by including KCl + valinomycin in the dark stage resulted in a marked increase in the observed yield of ATP. It should be noted (Table 1) that this increase in yield was not observed when the proton gradient was maximal. Similar observations were also reported by URIBE and LI (1973). To determine whether these stimulations are indeed due to the imposition of a membrane potential, we have run two types of control experiments. In one, the chloroplasts were preincubated with KCl, so that no significant K^+ movement would be expected during the light-to-dark transition and therefore no membrane potential should develop. Under these conditions, no increase in ATP formation was observed (SCHULDINER et al., 1972b). Alternatively, a permeable anion, picrate,

Table 1. Membrane potential as a driving force for ATP formation by postillumination ATP synthesis in chloroplasts[a]

pH		Light intensity	ATP formed (nmoles x mg chlorophyll^{-1})	
Light stage	Dark stage	(ergs x cm^{-2} x sec^{-1})	Choline chloride	KCl + valinomycin
6.5	8.5	3.5 x 10^5	154	160
6.5	8.5	9 x 10^3	9	29
6.5	7.5	3.5 x 10^5	61	144
6.5	7.5	9 x 10^3	3	26

[a](From SCHULDINER et al., 1972b.) The reaction mixture of the light stage contained, in a final volume of 1.2 ml: Sorbitol, 100 mM; choline-chloride, 66 mM; pyocyanine, 15 µM; phenylene-diamine-di HCl, 5 mM, at pH 6.5; MES-Tris, pH 6.5, 5 mM; and chloroplasts containing 160 µg chlorophyll. In the dark stage, the volume was 1.5 ml (final volume 2.7 ml) and the final concentration (in 2.7 ml): $Mg(H_2PO_4)_2$, 0.8 mM (containing 1.6 x 10^7 cpm ^{32}P); ADP, 1 mM; Tris-HEPES, 5 mM; choline chloride or KCl, 100 mM; valinomycin, where indicated, 3 µM Final pH was adjusted by addition of Tris base. Chloroplasts were illuminated 2 minutes at the indicated intensity, before injection into the dark reaction mixture. The reaction was stopped 15 seconds later by the addition of trichloroacetic acid to a final concentration of 3%, and the contents were assayed for the ATP formed.

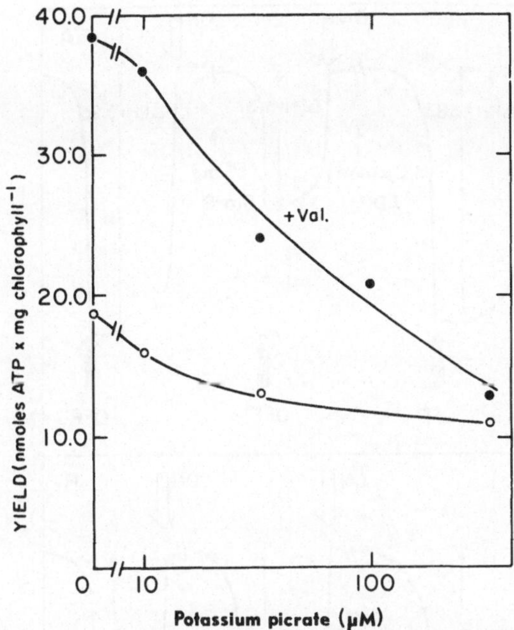

Fig. 1. Inhibition of the membrane-potential-driven ATP formation by picrate. (From SCHULDINER et al., 1973.) Details are as described in Table 1, except for the addition of picrate at the dark stage, at the indicated final concentrations

was introduced together with the K^+. Under these conditions, one would expect the movement of potassium picrate, rather than K^+, during the light-dark transition and therefore no membrane potential should be formed. As can be seen in Figure 1, the introduction of a sufficient concentration of picrate indeed fully inhibited the increase in ATP formation due to the K^+ + valinomycin (SCHULDINER et al., 1973).

Thus, we can conclude that chloroplasts have the capability to use energy stored either in the form of proton gradients or membrane potentials as a driving force for ATP formation.

We can now turn our attention to the relation between the magnitude of these potentials and the ability to synthesize ATP.

First, we may ask, does phosphorylation vary the magnitude of the measured ΔpH values? If the energy stored in the proton gradient does drive ATP formation, it would be expected that ATP formation would create an additional drain on the size of ΔpH and so decrease its magnitude. Figure 2 indicates that this is clearly the case (PICK et al., 1973). When ADP is added to an otherwise complete photophosphorylating system, a marked decrease in the magnitude of ΔpH is observed. This decrease is reversed or prevented by the post-(Figure 2A) or pre- (Figure 2B) addition of the energy transfer inhibitor Dio-9.

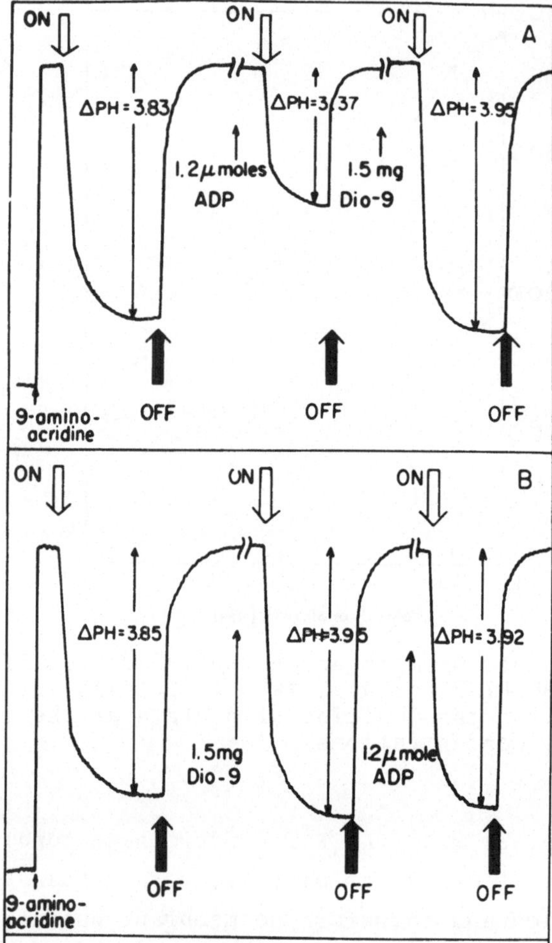

Fig. 2. Decrease in the magnitude of ΔpH by photophosphorylation. (From PICK et al., 1973.) The reaction mixture contained, in a total volume of 3.0 ml: Tricine, pH 8.6, 50 mM; KCl, 20 mM; $MgCl_2$, 5 mM; arsenate, 1 mM; pyocyanine, 5 μM; 9-aminoacridine, 1 μM; and chloroplasts containing 30 μg of chlorophyll; where indicated, ADP, 0.4 mM, and Dio-9, 0.5 mg/ml, were added

Second, we may ask, does the rate of ATP formation depend on the magnitude of ΔpH and what is the form of that dependence? Figure 3A illustrates the effect of two uncouplers on the rate of ATP formation and the magnitude of ΔpH. Figure 3B shows these data plotted to indicate the rate of ATP formation as a function of ΔpH (PICK et al., 1974). It is evident that with all three uncouplers tested, two phases can be clearly distinguished: (a) in the range of ΔpH up to a threshold value of about 2.5, no ATP formation was observed, and (b) above this threshold value, there was a sharp dependence of the rate of ATP formation on the magnitude of ΔpH. The data obtained by varying sev-

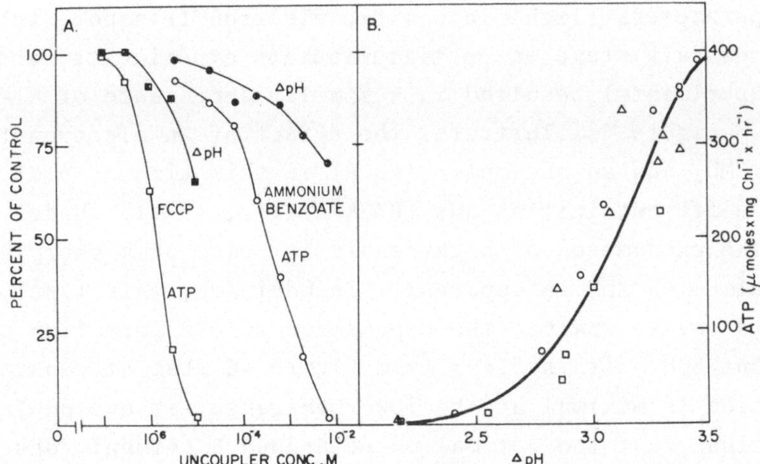

Fig. 3. Dependence of the rate of ATP formation on the magnitude of ΔpH varied by uncouplers. (From PICK et al., 1974.) The reaction mixture contained in a total volume of 3.0 ml: Tricine, pH 8.0, 60 mM; magnesium acetate, 2 mM; inorganic phosphate, 1 mM (containing 6×10^6 cpm ^{32}P); ADP, 1 mM; pyocyanine, 30 μM; 9-aminoacridine, 1 μM; chloroplasts containing 20 μg of chlorophyll, and the uncouplers at the concentrations indicated. ATP formation and ΔpH were determined. In Figure 3B, the rate of phosphorylation is plotted as a function of ΔpH for ammonium benzoate (0), FCCP (□), and dianemycin (Δ)

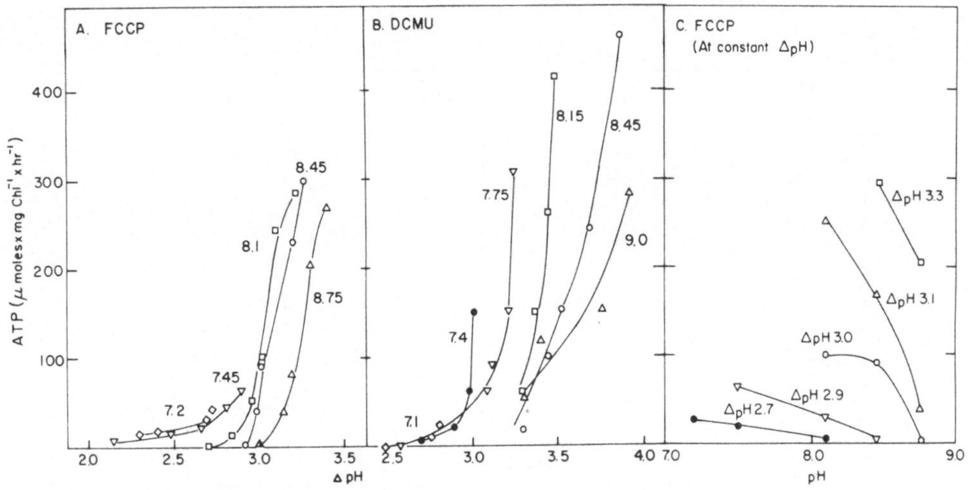

Fig. 4. Dependence of the rate of ATP formation on the magnitude of ΔpH at different external pH values. (From PICK et al., 1974.) The reaction mixture medium was essentially as described for Figure 3, but contained diquat, 15 μM, and NaN_3, 1 mM, in place of pyocyanine. Photophosphorylation and ΔpH were controlled by the additions of FCCP, 0.1 to 2 μM (A) or DCMU, 3 to 600 nM (B). In Figure 4C, the dependence of the rate of photophosphorylation on external pH at constant ΔpH is plotted. The data are taken from Figure 4A

eral other parameters (light intensity, electron transport inhibitors, pH of light or dark stage in postillumination experiments, and ΔpH in acid base experiments) resulted in a similar dependence of ATP formation on ΔpH. Figure 4 illustrates the effect of an electron transport inhibitor (DCMU) and an uncoupler (FCCP) in this kind of plot when measured at different initial pHs (PICK et al., 1974). Under all the tested pHs, the phenomenon of a threshold followed by a sharp dependence of ATP formation on ΔpH is apparent. In addition, this type of experiment permits one to analyze the dependence of ATP formation on medium pH at constant ΔpH. It is clear from Figure 4C that at constant ΔpH, phosphorylation is maximal at the lower pH range, around pH 7. Thus, one may conclude that the optimal pH of around 8 commonly observed for photophosphorylation is the result of two opposing effects: (a) the optimum of the phosphorylation reaction at constant ΔpH which is around pH 7 and (b) the maximal ΔpH which is observed above pH 9 (see SCHULDINER et al., 1972a).

We can now return to the effect of a membrane potential on the yield of ATP formation and ask: How does the effect influence the ΔpH dependence of ATP formation? Figure 5 shows such an analysis, where ΔpH was varied by varying the pH of the light stage in a postillumination experiment. In Figure 5B, the data are plotted to show the yield of ATP as a function of ΔpH (SCHULDINER et al., 1973). As can be seen,

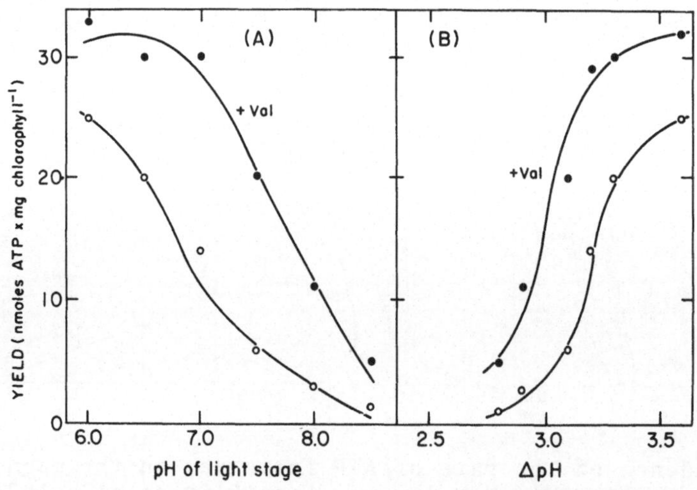

Fig. 5. The effect of a membrane potential on the ΔpH dependence of ATP formation. (From SCHULDINER et al., 1973.) Details are as described for Table 1, except for the pH of the light stage which was as indicated and the light intensity which was 900 ergs x cm^{-2} x sec^{-1}. The ΔpH plotted equals the difference between the chloroplasts' internal pH in the light stage and the pH of the dark stage (8.6)

when a membrane potential was superimposed (+ val.), the curve shifted toward lower ΔpH values, resulting in a lower threshold value. The superimposition of a membrane potential lowers the ΔpH value required for a given rate of ATP synthesis.

It was previously reported (SAKURAI et al., 1964) that at very low light intensities, it takes some time before the rate of phosphorylation attains its steady state rate. Figure 6A demonstrates that at pH 7.2 and at very low light intensities, the rate of phosphorylation shows an intensity-dependent time lag. However, no such lag can be seen in the buildup of ΔpH. In Figure 6B, the rate of phosphorylation at any time point is plotted as a function of the ΔpH at that point. Clearly, a threshold value is again observed (2.4 in this case). Thus, the time lag may be a function of the time needed to build up a minimal ΔpH value.

Finally, the question should be asked, are the values of ΔpH measured sufficient to drive ATP synthesis by the proposed chemiosmotic mechanism (MITCHELL, 1966)? To form a molecule of ATP by the transfer of 2 protons, ΔpH must exceed 5.5, when no membrane potential exists

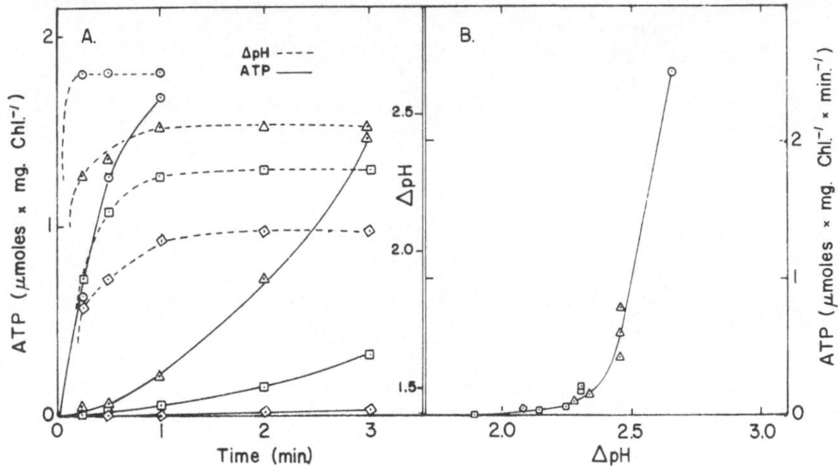

Fig. 6. The relation of the time lag in photophosphorylation at low light intensities to the magnitude of the proton gradient. (From PICK and AVRON, 1975.) The reaction medium contained in 3 ml: Na-tricine maleate, pH 7.2, 30 mM; sorbitol, 3 mM; $MgCl_2$, 1 mM; ADP, 1 mM; inorganic phosphate, 1 mM containing P^{32} (3×10^7 cpm/μmol); ATP μM; pyocyanine, 25 μM; 9-aminoacridine, 0.5 μM; and chloroplasts containing 75 μg chlorophyll. Samples of 0.2 ml were withdrawn to determine ATP^{32} produced during illumination. Light intensities (red light) were in erg x cm^{-2} x sec^{-1}: o-o 4.5×10^5, Δ-Δ 6.0×10^4, □-□ 4×10^4, ◇-◇ 1.5×10^4. In Figure 6B, rate of phosphorylation at any time point is plotted as a function of the magnitude of ΔpH at that time point

and if we accept the measured phosphate potential in chloroplasts of 17 Kcal/mole (see ROTTENBERG et al., 1972). Since the latter is a maximal value, and most of the reported experiments were performed under a lower phosphate potential, we designed an experiment to test whether the ΔpH dependence of ATP formation is sensitive to changes in phosphate potential. Figure 7 illustrates the results of such an experiment, where ΔpH was controlled by varying the light intensity and the phosphate potential was changed by a factor of 400 by varying the ADP/ATP ratio (PICK et al., 1974). It is clear that the dependence of the rate of ATP formation on the magnitude of ΔpH was only marginally, if at all, dependent on the phosphate potential during the reaction. We must conclude, therefore, that the ΔpH values of 3 to 4, which permit measurable rates of ATP formation, are thermodynamically insufficient for a movement of 2 protons to provide sufficient energy for the synthesis of an ATP molecule. Nevertheless, it is amply clear that the energy stored in the form of proton gradients or membrane potentials can be used by chloroplasts to drive ATP formation.

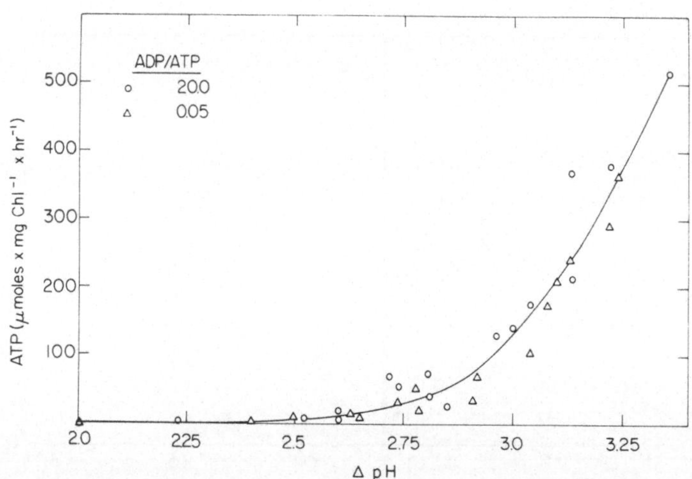

Fig. 7. Insensitivity of the phosphorylation dependence of ΔpH to changes in the phosphate potential. (From PICK et al., 1974.) The reaction medium contained in 3 ml: KCl, 30 mM; Na-tricine, pH 8.3, 30 mM; $MgCl_2$, 5 mM; inorganic phosphate, 1.33 mM (containing P^{32} 6 x 10^6 cpm/μmol); pyocyanine 30 μM; 9-aminoacridine, 1 μM; and chloroplasts containing 14 μg chlorophyll. The initial concentrations of ADP and of ATP were 2.0 mM and 0.1 mM or 0.1 and 2 mM, respectively. The rate of phosphorylation and the magnitude of the ΔpH were varied by changing the light intensity (red light) from 1.3 x 10^3 - 3.5 x 10^5 erg x cm^{-2} x sec^{-1}. Maximal changes in ADP/ATP ratios during the reaction in the light were between 17 to 20 and 0.04 to 0.05, respectively

The answer to this dilemma may lie either in introducing a variation in the chemiosmotic hypothesis, which will permit the coupled transfer of more protons per electron per site and so a higher proton-to-ATP ratio, or in suggesting that the proton gradients and membrane potentials are energy storage devices in equilibrium with another, yet undefined, high-energy intermediate state, which directly drives the synthesis of ATP.

REFERENCES

JAGENDORF, A. T.: Acid base transitions and phosphorylation by chloroplasts. Fed. Proc. 26, 1361-1369 (1967).

KARLISH, S. J. D., AVRON, M.: Dinitrophenol and valinomycin as uncouplers in isolated chloroplasts. FEBS Letters 1, 21-24 (1968).

KARLISH, S. J. D., SHAVIT, N., AVRON, M.: On the mechanism of uncoupling in chloroplasts by ion permeability inducing agents. Eur. J. Biochem. 9, 291-298 (1969).

MITCHELL, P.: Chemiosmotic coupling in oxidative and photosynthetic phosphorylation. Bodwin: Glynn Research, Ltd. 1966.

PICK, U., AVRON, M.: Proton gradients, proton concentrations and photophosphorylation. In: Proc. 3rd International Congress on Photosynthesis (ed. M. AVRON). Amsterdam: Elsevier 1975.

PICK, U., ROTTENBERG, H., AVRON, M.: Effect of phosphorylation on the size of the proton gradient across chloroplast membranes. FEBS Letters 32, 91-94 (1973).

PICK, U., ROTTENBERG, H., AVRON, M.: The dependence of photophosphorylation in chloroplasts on ΔpH and external pH. FEBS Letters (1974).

ROTTENBERG, H., GRUNWALD, T., AVRON, M.: Direct determination of ΔpH in chloroplasts and its relation to the mechanism of photoinduced reactions. FEBS Letters 13, 41-44 (1971).

ROTTENBERG, H., GRUNWALD, T., AVRON, M.: Determination of ΔpH in chloroplasts. I. Distribution of (^{14}C)-methylamine. Eur. J. Biochem. 25, 54-63 (1972).

SAKURAI, H., NISHIMURA, M., TAKAMIYA, A.: Studies on photophosphorylation. I. Two step excitation kinetics of photophosphorylation. Plant and Cell Physiol. b., 309-324 (1965).

SCHULDINER, S., ROTTENBERG, H., AVRON, M.: Determination of ΔpH in chloroplasts. II. Fluorescent amines as a probe for the determination of ΔpH in chloroplasts. Eur. J. Biochem. 25, 67-70 (1972a).

SCHULDINER, S., ROTTENBERG, H., AVRON, M.: Membrane potential as a driving force for ATP synthesis in chloroplasts. FEBS Letters 28, 173-176 (1972b).

SCHULDINER, S., ROTTENBERG, H., AVRON, M.: Stimulation of ATP synthesis by a membrane potential in chloroplasts. Eur. J. Biochem. 39, 455-462 (1973).

URIBE, E. G., LI, B. C. Y.: Stimulation and inhibition of membrane dependent ATP synthesis in chloroplasts by artificially induced K^+ gradients. J. Bioenergetics 4, 435-444 (1973).

Light Energy Transduction in *Halobacterium halobium**

Walther Stoeckenius, Roberto A. Bogomolni, and Richard H. Lozier
Cardiovascular Research Institute and Department of Biochemistry and Biophysics, University of California, San Francisco, California 94143; and Ames Research Center, NASA, Moffett Field, California 94035

Halobacteria depend on high concentrations of NaCl for growth and survival. They occur naturally in environments such as salt lakes, where the salt concentration due to evaporation is near or at saturation. Typically, the temperature and solar radiation density are high in such locations, at least during the day, and the O_2 concentration in the brine must be low unless O_2-producing halophilic algae are present. Most halobacteria do not contain chlorophyll, and it was rather puzzling to observe that many of them do not ferment sugars either and appear to rely entirely on oxidative phosphorylation as their energy source (for a review, see LARSEN, 1963, 1967). The solution to this puzzle has been provided by the observation that halobacteria can use light energy to drive metabolic processes through a chlorophyll-independent mechanism.

PHOTOPHYSIOLOGY OF *HALOBACTERIUM HALOBIUM*

When cells of *Halobacterium halobium* are suspended in a salt solution without nutrients and kept under anaerobic conditions in the dark, the ATP content of the cells drops rapidly to a low level where it remains relatively constant. Either aeration or illumination of such a suspension with white light under anaerobic conditions will cause a rise in intracellular ATP to the original level or higher. The ATP level remains at that value as long as the cells are illuminated or have access to oxygen. When returned to the dark and anaerobiosis, the ATP content drops to the same low level established before the illu-

---

*This research was supported by NHLI Program Project Grant HL-06285 and NASA Life Scientist Grant NGL 05-025-014. We thank Arlette Danon and San-Bao Hwang for permission to reproduce data that were obtained in collaboration.

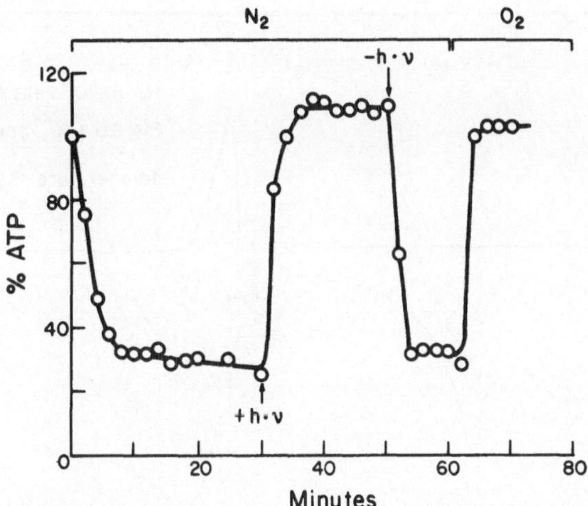

Fig. 1. Changes in the ATP concentration of cells incubated in salt solution under anaerobic and aerobic conditions. The initial ATP concentration before the cells are made anaerobic is taken as 100%. Either light or oxygen can restore the ATP concentration in the cells to the original level

mination (Figure 1) (DANON, STOECKENIUS, 1974). The light effect is insensitive to known inhibitors of the electron transport chains of either respiration (KCN, NQNO, antimycin) or photosynthesis (DCMU), all of which abolish the O_2 effect on the cell suspension. The effect of both light and O_2 is, however, inhibited by uncouplers of oxidative phosphorylation such as DNP, FCCP, CCCP, which are thought to act as proton ionophores (MITCHELL, 1972), and DCCD and Dio-9--the inhibitors of mitochondria and chloroplast ATPases.

These observations are most readily explained by Mitchell's chemiosmotic theory of energy coupling, if we assume that in these cells not only O_2 but also light, even in the absence of a chlorophyll-linked electron transport chain, can generate an electrochemical proton gradient. This effect is indeed easily demonstrated (OESTERHELT, STOECKENIUS, 1973). Light causes an acidification of the medium, which is sensitive to the uncouplers but not to the ATPase inhibitors. When the acidification is generated by O_2, it is blocked by the electron transport chain inhibitors, but when it is generated by light, the electron transport chain inhibitors have no effect (Figure 2).

These findings strongly suggest that light and O_2 act as two alternative energy sources for these cells and that the energy from either is first converted into an electrochemical proton gradient, which in turn can drive the synthesis of ATP. If both energy sources use an ATP-synthesizing system located in the same membrane, we should

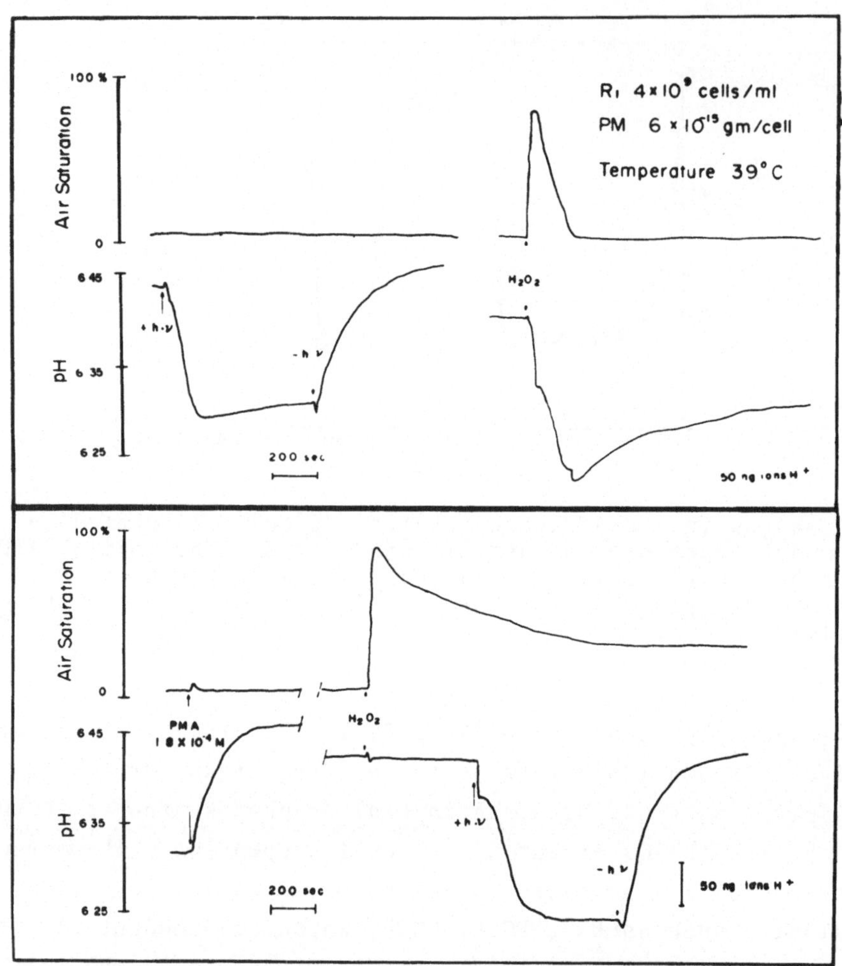

Fig. 2. Inhibition of respiratory proton translocation by phenylmercuric acetate (PMA). Continuous recordings of pH and oxygen tension changes in an anaerobic cell suspension. Acidification of the cell suspension induced by light or by oxygen from an H_2O_2 pulse is shown in the upper traces. Addition of 1.8 x 10^{-4} M PMA causes an alkalinization of the suspension and results in a complete inhibition of the H_2O_2-induced acidification. The light-driven proton ejection is not affected

expect that illumination inhibits respiration. This is indeed observed, as shown in Figure 3 (OESTERHELT, STOECKENIUS, 1973; OESTERHELT, KRIPPAHL, 1973).

We have described three effects of light on these cells: ATP synthesis, proton ejection, and inhibition of respiration. Action spectra for all three effects are identical: they show a broad maximum around 590 nm, which, except for a slight red shift, corresponds very well to the absorption spectrum of a membrane fraction--the purple

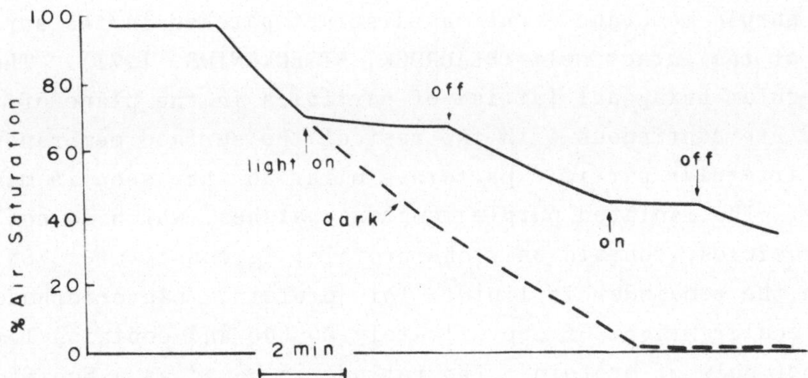

Fig. 3. Registration with a Clark electrode of O_2 consumption of *H. halobium* R_1 cells in a closed chamber. Light strongly inhibits respiration. 5 x 10^9 cells/ml; 1.5 x 10^5 ergs cm^{-2} sec^{-1}

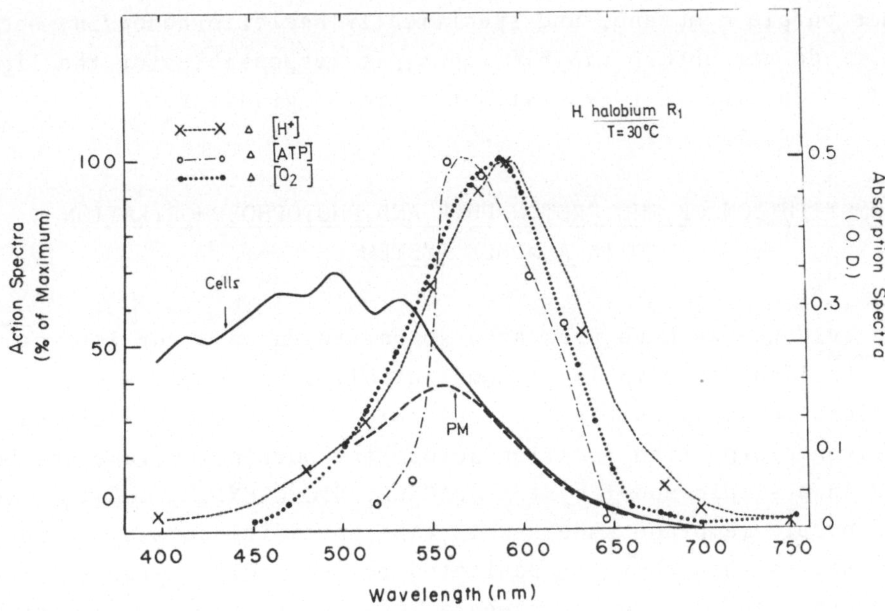

Fig. 4. Action spectra for the light-induced effects in *H. halobium* R_1 cells. The absorption spectrum of the cell suspension and the absorbance due to the purple membrane are shown for comparison

membrane--that we have isolated from these cells (STOECKENIUS, KUNAU, 1968; OESTERHELT, STOECKENIUS, 1971). The red shift of the action spectrum apparently arises from the overlapping absorption of the main carotenoid, α-bacterioruberine, which is typically present in high concentrations (Figure 4).

The purple membrane occurs as distinct patches in the surface membrane of the intact cell (BLAUROCK, STOECKENIUS, 1971). The patches show a regular hexagonal lattice of particles in the plane of the membrane and are continuous with the rest of the surface membrane, which shows an irregular particle pattern similar to that seen in many other membranes. The isolated purple membrane patches, which do not close up to form vesicles, contain only one protein; it constitutes 75% of their mass, and the remainder is lipid. This protein, "bacteriorhodopsin," has a molecular weight of approximately 26,000 and contains 1 mole of retinal per mole of protein. The retinal is bound as a Schiff base to a lysine residue of the protein (OESTERHELT, STOECKENIUS, 1973). The retinylidene-lysine chromophore is stabilized and its absorption maximum is red-shifted from the near ultraviolet to 570 nm by protonation and complexation with aromatic amino acid residues of the protein (LEWIS, 1973; LEWIS et al., in preparation). The action spectra show that this purple membrane, and specifically bacteriorhodopsin, because the lipids do not absorb visible light, is responsible for the light energy transduction that generates the proton gradient.

RECONSTITUTION OF THE PROTON PUMP AND PHOTOPHOSPHORYLATION IN A MODEL SYSTEM

The evidence we have presented so far to support our interpretation of the light effects is circumstantial; other explanations for our observations remain possible. Direct evidence for the postulated light-driven proton translocation across the purple membrane has been obtained in a simple model system (RACKER, STOECKENIUS, 1974). The isolated purple membrane sheets will take up additional lipid and form closed vesicles when they are sonicated together with natural or synthetic phospholipids. Such a preparation of purple membrane-containing lipid vesicles, when illuminated with light absorbed by bacteriorhodopsin, shows a reversible change of pH (Figure 5). This change can be demonstrated either with a glass electrode that monitors the pH of the suspending medium or with pH indicators enclosed in the vesicles. The pH decreases in the vesicle interior and increases in the suspending medium. Compared with intact cells, the direction of the pH gradient is therefore in the opposite direction and the preferential orientation of the bacteriorhodopsin in the vesicles must also be the opposite of that found in intact cells. This has been confirmed by freeze-fracture electron microscopy (HWANG, STOECKENIUS, unpublished). The pH effect

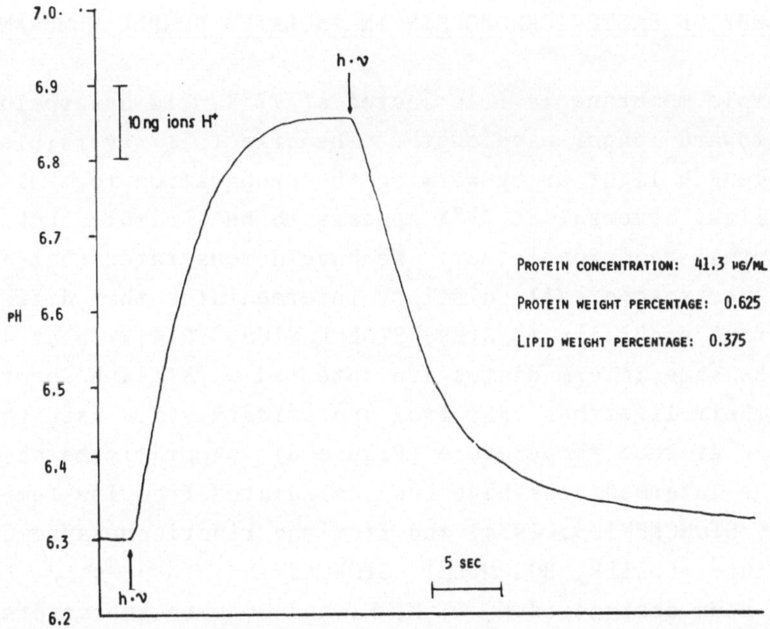

Fig. 5. Record of pH changes during illumination in a suspension of phospholipid vesicles with incorporated purple membrane. A 250-watt quartz iodine lamp with heat filter and Schott OG 5 filter were used, yielding $\sim 2 \times 10^6$ ergs cm^{-2} sec^{-1}

is sensitive to uncouplers and the potential associated with the proton gradient has also been demonstrated with tetraphenylboron in this system (RACKER, HINKLE, 1974). We have further extended the model system by including an ATP-synthesizing component. Mitochondrial ATPase (coupling factor F_1) and hydrophobic protein, when incorporated into the purple membrane vesicles, yield a preparation that uses light energy absorbed by the purple membrane to synthesize ATP from ADP and inorganic phosphate (RACKER, STOECKENIUS, 1974).

These experiments unequivocally demonstrate that bacteriorhodopsin can use absorbed light energy to establish and maintain an electrochemical gradient across the membrane; in other words, it functions as a light-driven proton pump. Preliminary evidence indicates that the quantum yield is one proton per photon absorbed and that at saturating light intensities 200 to 500 protons per second are translocated by one bacteriorhodopsin molecule (BOGOMOLNI et al., in preparation). These findings indicate that a fast cyclic photoreaction must take place in bacteriorhodopsin. It should be accompanied by absorption changes, which might be detected either with flash spectrophotometry at physiological temperatures or with conventional spectrophotometry at low temperatures.

SPECTROSCOPY OF BACTERIORHODOPSIN IN ISOLATED PURPLE MEMBRANE

When purple membrane is illuminated at 77°K, its absorption spectrum shifts toward longer wavelength. The effect is reversible either by long-wavelength light or by warming the preparation to > 213°K. Whereas the light reversal at 77°K appears to be a simple back reaction, the thermal reaction is not. We have demonstrated that it occurs via several spectroscopically distinct intermediates that differ widely in their thermal stability (LOZIER, STOECKENIUS, 1974). What are apparently the same intermediates are detected with flash spectrophotometry; their lifetimes vary from approximately 1.0 μsec to approximately 5 msec at room temperature (Figure 6). Approximate absorption spectra of the intermediates have been calculated from low-temperature data (LOZIER, STOECKENIUS, 1974) and from the kinetic measurements at room temperature (LOZIER, BOGOMOLNI, STOECKENIUS, in press). The spectra have been designated K, L, M, N, and O, with subscripts indicating their absorption maxima. The initial complex is termed R_{570}; in the dark, it is slowly converted to still another form, R_{560}, which in

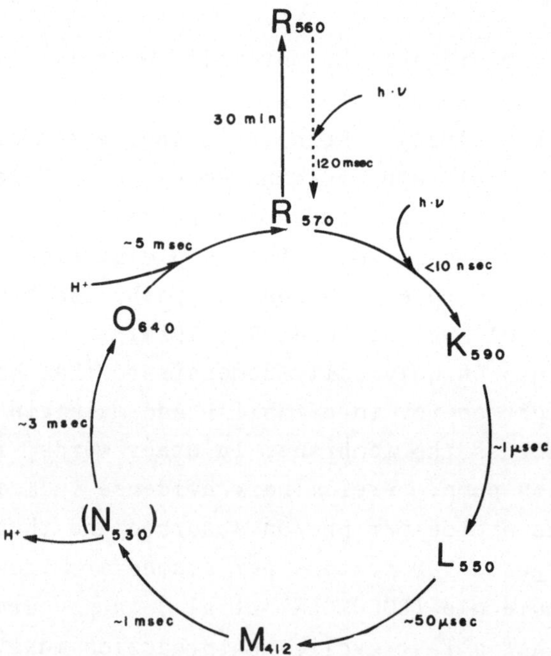

Fig. 6. Reaction cycle of bacteriorhodopsin. Times indicated are approximate half times for the reactions at 20°C and neutral pH (see text for details). Inclusion of intermediate N_{530} is provisional. The $R_{560} \rightarrow R_{570}$ transition is not a direct transition; its reaction time is light-intensity-dependent

the light at room temperature returns to the R_{570} complex within a few seconds. This transition will not be considered here, because the $R_{570} \rightarrow R_{560}$ reaction is too slow to allow appreciable concentration of R_{560} to accumulate under the conditions of the experiments described. The only light reaction in the R_{570} photoreaction cycle is the $R_{570} \rightarrow K_{590}$ transition; all other reactions occur in the dark. As one would expect, the R_{570} photoreaction cycle is accompanied by changes in the protonation of bacteriorhodopsin. This cycle was first investigated in a preparation of purple membrane modified by the presence of ether at high NaCl concentrations (OESTERHELD, HESS, 1973). It also occurs in the absence of ether and at low NaCl concentrations.

When bacteriorhodopsin reaches the M_{412} intermediate, one proton is released, and one proton is bound when M_{412} returns to the R_{570} complex (LOZIER, BOGOMOLNI, STOECKENIUS, in press). This observation provides an explanation for the light-driven proton translocation across the purple membrane. Electron microscopy and X-ray diffraction studies have shown that bacteriorhodopsin is oriented uniformly across the purple membrane (BLAUROCK, STOECKENIUS, 1971). Let us assume that the release and uptake of protons involves different groups on the molecule, which are located on opposite sides of the membrane, and let us assume that a transfer of protons between these groups occurs through the membrane. Bacteriorhodopsin will undergo one reaction cycle for every proton absorbed by R_{570}. In strong, continuous light, it will therefore cycle continuously and will eject one proton on one side of the membrane and absorb one proton on the other side in every cycle, thus generating an electrochemical gradient across the membrane, provided that rapid back reactions in and back diffusion through the membrane are prevented. The transfer of protons between the two groups on opposite sides of the membrane presumably occurs through a series of proton exchanging groups in the bacteriorhodopsin molecule. We have obtained evidence in support of this proposed mechanism.

Resonance Raman spectroscopy has shown that the retinal-lysine Schiff base linkage in R_{570} is protonated and that it is unprotonated in M_{412} (LEWIS et al., 1974). However, this linkage is relatively inaccessible from the outside, because it is buried in a hydrophobic environment in the interior of the membrane (OESTERHELT, STOECKENIUS, 1971; LEWIS et al., in preparation). One would, therefore, not expect it to exchange protons rapidly with the aqueous medium, and the pH changes observed in the medium should not involve the Schiff base directly but other groups located on the surface of the membrane. Indeed, when purple membrane is suspended in D_2O and kept in the dark, exchange of the proton on the Schiff base for a deuteron occurs only slowly; in the light, however, the exchange is very rapid. The exchange is detected by a characteristic shift in the Raman band of the C=N stretching vibration. This finding suggests that the Schiff base linkage may be one of the groups involved in the transfer of protons through the membrane (LEWIS et al., 1974). Alternatively, the Schiff base could become directly accessible to the medium at one or several of the intermediate stages in the photoreaction cycle. This, however, does not seem to be the case. When purple membrane suspended in D_2O is exposed to a series of light flashes, each of which is short enough to cause only a single turnover of bacteriorhodopsin, we do not observe a uniform increase in the concentration of deuteronated Schiff bases with each flash, but rather an abrupt increase after the fourth flash (LEWIS et al., in preparation). This is clearly a strong argument in favor of the participation of the Schiff base as one link in the assumed chain of proton transfer groups thought to span the membrane.

A reversible change in the pK of one group in such a chain could drive the proton translocation along the chain, provided back reactions are prevented, and it is tempting to speculate that the excited Schiff base plays this role; large pK changes are known to occur upon excitation of molecules containing acidic groups (WELLER, 1961). Alternatively, light-induced changes in the chromophore could induce further conformational changes in the protein, resulting in effective pK changes of the Schiff base and/or some other group(s) in the protein. Such changes have been, of course, postulated for the light reaction in visual pigments. While at least the early intermediates in the photocycle of bacteriorhodopsin are similar to the intermediates in the photolysis of visual pigments, no isomerization of the retinal has been detected in bacteriorhodopsin so far. The chromophore in both the R_{570} and the M_{412} complex appears to be all-transretinal (JAN, 1974). In any event, conformational changes in the protein clearly do occur and are probably necessary to prevent back reactions. The first energy transduction step must be the excitation of the chromophore; how the energy of the excited molecule is conserved long enough to drive proton translocation is a fascinating question.

REFERENCES

BLAUROCK, A. E., STOECKENIUS, W.: Structure of the purple membrane. Nature New Biology 233, 152-155 (1971).

BOGOMOLNI, R. A., BAKER, R. A., LOZIER, R. H., STOECKENIUS, W.: Light-driven proton translocation in *H. halobium* (in preparation).

DANON, A., STOECKENIUS, W.: Photophosphorylation in *Halobacterium halobium*. Proc. Nat. Acad. Sci. U.S.A. 71, 1234-1238 (1974).

JAN, L.: Investigations on rhodopsin and bacteriorhodopsin. Thesis, California Institute of Technology, Pasadena, May, 1974.

LARSEN, H.: Halophilism. In: The Bacteria (ed. I. C. GUNSALES, R. Y. STANIER) Vol. VI, pp. 297-342. New York, London: Academic Press 1963.

LARSEN, H.: Biochemical aspects of extreme halophilism. Advan. Microbiol. Physiol. 1, 97-132 (1967).

LEWIS, A.: Tunable laser resonance Raman spectroscopy of the visual process. I. The spectrum of rhodopsin. J. Raman Spectroscopy 1, (1973).

LEWIS, A., SPOONHOWER, J., BOGOMOLNI, R. A., LOZIER, R. H., STOECKENIUS, W.: Tunable laser resonance Raman spectroscopy of bacteriorhodopsin. Proc. Nat. Acad. Sci. U.S.A. 71, 4462-4466 (1974).

LOZIER, R. H., BOGOMOLNI, R. A., STOECKENIUS, W.: Bacteriorhodopsin: A light-driven proton pump in *Halobacterium halobium*. Biophys. J. (in press).

LOZIER, R. H., STOECKENIUS, W.: The photochemical reaction cycle of bacteriorhodopsin in isolated purple membrane. Fed. Proc. 33, 1408, Abstract (1974).

MITCHELL, P.: Chemiosmotic coupling in energy transduction: A logical development of biochemical knowledge. J. Bioenergetics 3, 5-24 (1972).

OESTERHELT, D., HESS, B.: Reversible photolysis of the purple complex in the purple membrane of *Halobacterium halobium*. Eur. J. Biochem. 37, 316-326 (1973).

OESTERHELT, D., KRIPPAHL, G.: Light inhibition of respiration in *Halobacterium halobium*. FEBS Letters 36, 72-76 (1973).

OESTERHELT, D., STOECKENIUS, W.: Rhodopsin-like protein from the purple membrane of *Halobacterium halobium*. Nature New Biology 233, 149-152 (1971).

OESTERHELT, D., STOECKENIUS, W.: Functions of a new photoreceptor membrane. Proc. Nat. Acad. Sci. U.S.A. 70, 2853-3857 (1973).

RACKER, E., HINKLE, P. C.: Effect of temperature on the function of a proton pump. J. Membrane Biol. 17, 181-188 (1974).

RACKER, E., STOECKENIUS, W.: Reconstitution of purple membrane vesicles catalyzing light-driven proton uptake and adenosine triphosphate formation. J. Biol. Chem. 249, 662-663 (1974).

STOECKENIUS, W., KUNAU, W. H.: Further characterization of particulate fractions from lysed cell envelopes of *Halobacterium halobium* and isolation of gas vacuole membranes. J. Cell Biol. 38, 337-357 (1968).

WELLER, A.: Fast reactions of excited molecules. In: Progress in Reaction Kinetics (ed. G. PORTER) Vol. I, pp. 187-214. Oxford: Pergamon Press 1961.

Rhodopsin in Experimental Membranes: An Approach to Elucidate Its Role in the Process of Phototransduction

M. Montal

Departamento de Bioquimica, Centro de Investigación y de Estudios Avanzados del Instituto Politécnico Nacional, Apartado Postal 14-740, México 14, D. F., México

INTRODUCTION

Photoreceptor cells in both vertebrate and invertebrate retinas behave as single quantum detectors. The chemical basis for this light sensitivity has been accounted for by the presence, in these specialized cells, of a chromophore-bearing protein generically known as rhodopsin (WALD, 1968). In the rod cell of the vertebrate retina, rhodopsin has been purified and characterized to a great extent (cf. KROPF, 1972; ABRAHAMSON, FAGER, 1973): rhodopsin is a lipoglycoprotein which constitutes over 80% of the membrane protein in the cell outer segment (ROBINSON, GORDON-WALKER, BOWNDS, 1972; HEITZMAN, 1972; HELLER, LAWRENCE, 1970; STEINEMANN, STRYER, 1973). Its chromophore, retinaldehyde, is covalently attached as a Schiff-base to the ε-amino group of a lysine in the protein moiety, traditionally called opsin (AKHTAR et al., 1967; BOWNDS, 1967). It is well established that the absorption of a light quantum leads to the immediate isomerization of the polyene chromophore from the 11-cis to the transisomeric form (HUBBARD, KROPF, 1958; HUBBARD, BOWNDS, YOSHIZAWA, 1965) and that this primary photochemical event is followed by a sequence of chemical changes (dark thermal reactions) that have been characterized by spectral changes and lifetimes at various temperatures. The final reaction is the hydrolysis of the Schiff base. The problem, therefore, is to establish how these photochemical processes lead to cellular excitation (WALD, BROWN, GIBBONS, 1963; HAGINS, 1965, 1972).

It has been demonstrated that photon absorption by vertebrate photoreceptors displaces the resting membrane potential to more negative values. This hyperpolarizing receptor potential appears to be the connecting signal between the receptor cell and a network of several types of retinal neurons, which eventually results in the generation of nerve impulses in the fibers of the optic nerve (cf. TOMITA, 1970).

The work of HAGINS, PENN, and YOSHIKAMI (1970) and PENN and HAGINS (1972), using external microelectrodes to map the current distribution around rods in the retina of rats, has shown that the receptor cell maintains a large, resting *dark* current (of the order of 10^{-10} A/rod) entering the cell along the outer segment where the rhodopsin-containing discs are located. The current leaves the cell at the inner segment by an active process, presumably involving a Na^+/K^+ exchange pump located near the junction of the inner and outer segments (KORENBROT, CONE, 1972; KORENBROT, 1973; ZUCKERMAN, 1973). This flow of dark current from the inner to the outer segment would have K^+ as the dominant charge carrier along the cell's inner segment and Na^+ along the outer segment. Absorption of light by rhodopsin leads to large changes in this ionic current: the Na^+-conductance decreases (TOYODA, NOSAKI, TOMITA, 1969; BAYLOR, FUORTES, 1970; but see also LASANSKI and MARCHIAFAVA, 1974) to such an extent that excitation of a single rhodopsin molecule stops transiently ($\sim$1 sec) about 10^7 Na^+ ions from flowing across the plasma membrane of the outer segment (KORENBROT, CONE, 1972). The amplification exerted by the closure, presumably of a few Na^+ channels per photon absorbed, would be sufficient to trigger a voltage change at the synapse between the photoreceptor and higher order neurons (HAGINS, 1972). KORENBROT and CONE (1972) measured the osmotic behavior of isolated rod outer segments (frog and rat), and also established that light reduces the influx of Na^+ ions but not K^+ ions. The mechanism by which the excited rhodopsin molecule, located in the free-floating inner discs, achieves such a large reduction in ionic flux across the receptor cell membrane is not understood.

Considerable attention has been recently paid to the suggestion of YOSHIKAMI and HAGINS (1971) and HAGINS (1972), that light could release Ca^{++} ions accumulated inside the discs: the increase in the cytoplasmic concentration of Ca^{++} would block the plasma membrane Na^+ channels.

Several recent communications appear to support the Ca^{++}-hypothesis (SZUTZ, CONE, 1974; POO, CONE, 1973; CONE, in press; ABRAHAMSON, FAGER, MASON, 1974; LIEBMAN, 1974; HENDRICKS, DAEMEN, BONTING, 1974). In essence, they report that the Ca^{++} content of rod disc membranes decreases upon illumination. In addition, HAGINS and YOSHIKAMI (1974) showed that in the presence of the Ca^{++}-ionophore X-537A (PRESSMAN, 1972; CELIS, ESTRADA-O., MONTAL, 1974), the concentration of Ca^{++} in the bathing medium required to block the Na^+-channels was reduced by more than 3 orders of magnitude relative to the control retina (see also BROWN and PINTO, 1974).

On the other hand, large invertebrate photoreceptors such as those of the ventral eye of *Limulus* have been impaled with microelectrodes and then studied with conventional electrophysiological techniques. In these cells rhodopsin is located in the cell membrane and the absorption of light leads to an increase in Na^+ conductance and consequently to a depolarizing receptor potential (cf. FUORTES and O'BRYAN, 1972a).

It seems clear, therefore, that in both vertebrate and invertebrate photoreceptors, the excited rhodopsin molecules modify the cell membrane conductance. A simple mechanism by which rhodopsin may do so is by being itself a light-triggered ion-translocator (WALD, BROWN, GIBBONS, 1963; CONE, 1972, in press; WALD, 1974). A direct way to test this possibility would be to incorporate native rhodopsin into a membrane system amenable to the experimental methods required to investigate the postulated transport properties.

In the present work, an attempt is being made to assemble a "functionally active" rhodopsin membrane. For this purpose, we have devised new methods of bilayer formation (MONTAL, MUELLER, 1972; MONTAL, 1973, 1974, and in press) and chemical manipulation of membrane proteins (GITLER, MONTAL, 1972a, b; MONTAL, KORENBROT, 1973; MONTAL, in press). We are far from being satisfied with the results and from claiming any particular mode of action for rhodopsin, but the results obtained so far encourage us to pursue this approach as a potentially powerful one. Thus, we present a summary of our results as a progress report.

LIPID PROTEIN ASSEMBLY

Planar Lipid Bilayers

Planar bilayers can be assembled from two lipid monolayers at the air-water interface by the hydrophobic apposition of their hydrocarbon tails through an aperture in a hydrophobic partition which separates the two monolayers (cf. MONTAL, 1974). Three significant advantages of these bilayers over the black lipid films (cf. MUELLER and RUDIN, 1969) can be recognized.

1. Defined chemical composition: the bilayer is composed exclusively of lipid, hence the adverse effect of the presence of hydrocarbon solvents in black films previously used in functional reconstitution studies is overcome.

2. Asymmetric bilayers: these are readily formed by apposing two monolayers of different chemical composition (MONTAL, 1973, 1974).

This unique feature allows the exploration of the role of the sidedness, not only of the lipid but also of the protein, in determining the vectorial aspects of the process under investigation.

3. Incorporation of proteins: it is possible, in principle, to form a bilayer from monolayers initially composed of both lipid and protein. To have the protein as a constituent of the organized interface, this may be allowed to penetrate from the aqueous phase into the expanded monolayer, or alternatively it could be spread together with the lipid as a monolayer at the air-water interface. We have been unable so far to incorporate rhodopsin into bilayers by following the first approach. The second outlined strategy requires the lipid protein complex to be "soluble" in an organic solvent but still native and active. This is achieved by enhancing the partition of rhodopsin lipoprotein vesicles into an apolar solvent following ion-pair formation between Ca^{++} and the hydrophilic groups in the lipoprotein, thus rendering the overall complex neutral (GITLER, MONTAL, 1972a, b; MONTAL, KORENBROT, 1973; MONTAL, in press).

Rhodopsin-Lipid Complex

Removal of Detergent and Lipoprotein Vesicle Formation

Rhodopsin has been solubilized from rod outer segment membranes and purified in a variety of ionic and nonionic detergents (WALD, BROWN, 1952; HELLER, 1968; SHICHI et al., 1969; HONG, HUBBELL, 1972, 1973; APPLEBURRY et al., 1974). To extract rhodopsin into a solvent, the detergent must be removed. In addition, it is known that photobleached rhodopsin cannot be regenerated by supplementing it with 11-cis retinaldehyde when solubilized in most detergents (SNODDERLY, 1967; SHICHI et al., 1969; HONG, HUBBELL, 1972, 1973; HUBBELL, in press). Because the function of rhodopsin is not known, it is desirable to meet with the high regenerability that characterizes rhodopsin in vivo.

Several approaches have been successfully applied to remove detergent from rhodopsin.

(a) Dialysis. The technique involves mixing of detergent solutions of delipidated rhodopsin and lipids and removing the detergent slowly by dialysis (HONG, HUBBELL, 1972, 1973; CHEN, HUBBELL, 1973; HUBBELL, in press; APPLEBURRY et al., 1974) as has been applied in other membrane reconstitution studies (MARTONOSI, 1968; cf. RAZIN, 1972; cf. KAGAWA, 1974). This method yields rhodopsin lipoprotein vesicles with high regenerability (HONG, HUBBELL, 1973) and enables the vesicles to perform the transition from Metarhodopsin $I_{480\,nm}$ to Metarhodopsin $II_{380\,nm}$ with comparable kinetic competence relative to that in the native disc membrane (APPLEBURRY et al., 1974). CHEN and HUBBELL (1973) submitted these vesicles to the freeze-fracture procedure and found a fracture face covered with particles of 80 to 110 Å in diameter; these particles are absent in pure lipid vesicles and have been associated with the presence of the protein in the hydrophobic interior of the bilayer (BRANTON, 1966; PINTO DA SILVA, BRANTON, 1972; HUBBELL, in press; MONTAL, in press; VAIL, PAPAHADJOPOULOS, MOSCARELLO, 1974).

(b) Sonication in salt media. RACKER (1973) has shown that functionally active vesicles are formed by sonicating lipids with protein (in a small volume of detergent) in salt media. Among the membrane proteins reported to be assembled into a lipid vesicle with this method is bacteriorhodopsin of *Halobacterium halobium* (RACKER, HINKLE, 1974).

(c) Bio-Beads SM-2 (Bio-Rad Laboratories). HOLLOWAY (1973) reported that Bio-Beads SM2 (a neutral porous styrene-divinylbenzene copolymer) have a high affinity for Triton X-100 and thus could be used to remove detergent from protein samples. Detergent removal is not complete; the residual amount may be near the critical micelle concentration after an incubation of 1 hr. Thus, mainly detergent monomers remain bound to the protein. FELDBERG (1974) applied Holloway's technique to remove Triton as well as cetyltrimethylammonium bromide (CTAB) from rhodopsin. He assayed for the regeneration yield and found that while in CTAB the regeneration was less than 10%, it was complete after an incubation of 1 hr with Bio-Beads.

We have followed FELDBERG's procedure and recombined the resultant rhodopsin with lipids by sonication. The rhodopsin-lipid recombinants appear under the electron microscope as homogeneous vesicles of about 2,500 Å in diameter. The fracture faces of these vesicles show a homogeneous distribution of particles of approximately 100 Å in diameter. Thus, the recombinants formed by this procedure display a similar structure to those formed by dialysis (HONG, HUBBELL, 1972, 1973; CHEN, HUBBELL, 1973).

Alberto Darszon, in my laboratory, has recently found that when the sonication step is performed in the presence of radioactively labeled Ca^{++} or Na^{+} ions, the labeled ions are retained by the vesicles (cf. BANGHAM, HILL, MILLER, 1974). In addition, light causes the release of significant quantities of both ions. The experiment is performed as follows: After sonication, the vesicles are collected by centrifugation at 105,000 x *g* for 30 min and the pellet is washed by resuspension and centrifugation at least two times. The supernatant is centrifuged again and yields another pellet that is subsequently washed. The two sets of recombinants are divided in three aliquots: one is kept in the dark; the second is kept in the dark but is supplemented with an amount of all transretinal equivalent to that present as 11-cis retinal in rhodopsin; and the third is illuminated to bleach the rhodopsin completely. The aliquots are centrifuged, and the radioactivity in both pellet and supernatant is counted. The experiments are still in progress, but the preliminary results suggest that light induces the release of either Ca^{++} or Na^{+} from the vesicles. The ionic content of the pellets is depleted and that of the supernatant is enriched by more than 30% in the bleached preparation, relative to those kept in the dark, in the absence or presence of retinal. The results are even more striking in the set of vesicles that is centrifuged twice, where the bleached versus the dark difference can be as large as 60%.

BONTING and BANGHAM (1968) and DAEMEN and BONTING (1969) had previously shown that retinaldehyde increases the diffusion rate of K^{+} from phosphatidylethanolamine vesicles. Since our vesicles contain this lipid (about 30% of the total lipid), it was necessary to rule out any spurious effect of retinal. The experiments were performed at 4°C, where it is known that the transition from Metarhodopsin II_{380nm} to opsin and free retinal has a $t_{1/2} \cong 160$ min (OSTROY, ERHARDT, ABRAHAMSON, 1966); our experiment was completed within 40 min, so that no major contribution of retinal would be expected. Furthermore, the aliquot kept in darkness but supplemented with transretinal was equivalent to the unsupplemented one, within the experimental error. Experiments are currently under way to establish the temporal sequence of events, and the correlation of the lifetimes of rhodopsin intermediates with the time course of Ca^{++} release, as this is conveniently followed spectrophotometrically (OHNISHI, EBASHI, 1964).

Rhodopsin Proteolipid

Juan Korenbrot and I (MONTAL, KORENBROT, 1973) succeeded in forming a proteolipid of rhodopsin, extracting the dialysed lipoprotein vesicles into hexane by ion-pair formation with Ca^{++}. We noted at that time the turbidity of the hexane phase and attributed it to the formation of an emulsion of the rhodopsin-lipoprotein and water in hexane. Since then, our aim has been to obtain a proteolipid in a transparent hexane phase, either in solution or more likely as a microemulsion (BANGHAM, 1963). The best results have been obtained by extracting the rhodopsin lipoprotein vesicles prepared by removing the detergent with Bio-Beads followed by cosonication with lipid in salt media. Figure 1 illustrates an absorption spectrum of the transparent (no detectable turbidity) hexane phase containing the rhodopsin proteolipid. The curve on the scale labeled 0-0.1 shows a distinct peak of dark rhodopsin with λmax at 498 nm. The curves on the 0-1 scale illustrate the difference between dark (upper trace) and bleached (lower trace) rhodopsin. Several points deserve particular comment:

(a) The ratio of the absorbance at 280 nm to that at 500 nm is a criterion of purity of rhodopsin (KROPF, 1970; ABRAHAMSON, FAGER, 1973). This ratio in the original detergent rhodopsin preparation (not delipidated) was 2.6 to 2.8; after detergent removal and after sonication, it varied between 5 and 8. Thus, considerable denaturation of rhodopsin results from the manipulations required to form the vesicles. The ratio in the proteolipid illustrated in Figure 1 is approximately 10, indicating that the partition into the solvent is not a harmless process.

(b) The dark trace exhibits a shoulder around 360 nm, with the same broad profile as the absorbance at 360 nm that appears in the bleached preparation (lower trace). In hexane, the absorption maxima of 11-cis and all-transretinal are 365 and 368 nm, respectively (HUBBARD, 1956). This is indicative of the presence of free retinal in the unbleached proteolipid and could be attributed to opsin denaturation induced by the manipulations.

(c) The isosbestic point of the dark/bleached traces is around 415 nm; this is comparable to that in detergent-solubilized rhodopsin preparations.

(d) The efficiency of extraction can be as high as 30%. In this particular case, 0.2 mg of rhodopsin (assuming an extinction coefficient for rhodopsin of 40,600 at 498 nm [WALD, BROWN, 1953]) were extracted in 1.0 ml of n-hexane.

These spectral data indicate that a proteolipid of rhodopsin in hexane solution conserves the most characteristic features of rhodopsin in native disc membranes or detergent solutions, but is accompanied by significant amounts of denatured opsin (up to 50%) and retinal. Note that extraction of rhodopsin recombinants prepared by sonication in salt media gives the same type of behavior, but the efficiency of extraction is much lower (less than 10%); however those prepared by dialysis always yield an emulsion.

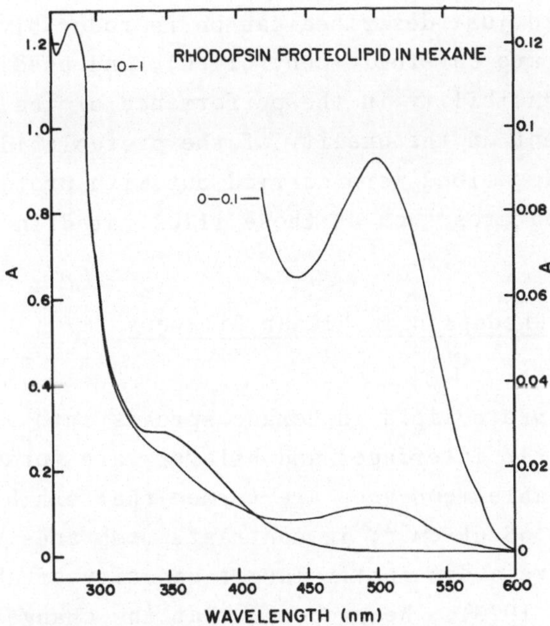

Fig. 1. Absorption spectrum of rhodopsin proteolipid in hexane. Rhodopsin was prepared from rod outer segments, as previously described (MONTAL, KORENBROT, 1973), except that the isolated segments were dialyzed according to RAUBACH, NEMES, and DRATZ (1974), and rhodopsin was solubilized from the membrane with 50 mM CTAB in M/15 phosphate buffer pH 7.0 (SNODDERLY, 1967; HELLER, 1968). All procedures were performed in the dark under dim red light unless otherwise indicated. Phospholipid recombinants were made in the following way: 0.25 ml of detergent-solubilized rhodopsin was incubated with 0.075 g of Bio-Beads SM-2 (washed according to HOLLOWAY, 1973) for 60 min at 20°C, with occasional manual stirring. Thereafter, an aliquot of 0.2 ml of turbid rhodopsin amounting to 1 mg (assuming an extinction coefficient of 40,600 at 498 nm [WALD, BROWN, 1953]) was mixed with a dispersion of phospholipids composed of 7 mg of egg lecithin (Applied Science Laboratories, Inc., State College, Pennsylvania) and 3 mg of *E. coli* phosphatidylethanolamine (Koch-Light Laboratories, Colnbrook Bucks, England) in 1.0 ml of 0.1 M KCl buffered with 0.01 M imidazole at pH 7.0. The mixture was sonicated by immersion of the test tube in a Bransonic ultrasonic cleaner (Heat Systems Ultrasonics, Inc., Plainview, New York; tank dimensions, 3 1/2 x 3 1/2 x 2 5/8 inches; power output, 40 watts) for 5 min at 4°C. The measured molar ratio of lipid to protein in the resulting recombinant was 100:1. Then, 0.1 ml of a 100 mM $CaCl_2$ solution and 1.0 ml of hexane were added to the suspension in rapid succession. The tube was vigorously mixed for 5 min and the two immiscible phases separated by centrifugation in a clinical centrifuge for 1 min. The hexane phase was removed and the absorption spectrum of a 0.5 ml aliquot measured in a Cary 14 recording spectrophotometer (Applied Physics Corporation, Monrovia, California) with the use of a 1-cm path length cuvette at 20°C. The reference was n-hexane. Difference spectra of dark and bleached rhodopsin in hexane (not illustrated) show a symmetric peak with λmax at 498 nm. (See also MONTAL and KORENBROT, 1973.)

The proteolipid just described can be reproducibly prepared (if all the conditions are carefully controlled), and used to form bilayers. The reproducibility in the performance of the bilayer is stringently dependent on the quality of the proteolipid. The bilayer experiments to be described were carried out with proteolipids that exhibit spectral features such as those illustrated in Figure 1.

Rhodopsin in Planar Bilayers

The rhodopsin proteolipid in hexane spreads into a monolayer when placed at an air-water interface, and bilayers are formed by apposing two monolayers. Stable membranes are formed that exhibit an electrical capacity of about 0.65 $\mu F \cdot cm^{-2}$; in contrast, membranes formed from the equivalent lipid have a distinctly larger capacity of about 0.9 $\mu F \cdot cm^{-2}$ (MONTAL, KORENBROT, 1973). We proposed that the change in membrane capacity on incorporation of proteins (see also MONTAL, in press) results predominantly from a change in dielectric thickness.

When the proteolipid emulsion in hexane (MONTAL, KORENBROT, 1973) is used to form bilayers, the membrane conductance in the dark is essentially the same as that of the equivalent lipid membranes, i.e., 10^{-7}-$mho \cdot cm^{-2}$. On illumination, there is a slow development of conductance ($t_{1/2}$ = 1 to 2 min) reaching stable values up to 10^{-4}-$10^{-3} mho \cdot cm^{-2}$ (MONTAL, in press). The conductance is ohmic and symmetric. The performance of this preparation is hard to reproduce, presumably because when rhodopsin is spread from an emulsion it desorbs from the monolayer reforming vesicles in the aqueous phase, thus leaving variable amounts of protein at the interface. To avoid this problem, the preparation shown in Figure 1 was developed, in which the proteolipid is in a macroscopic solution of hexane and is not emulsified.

Figure 2 illustrates that in the dark (left-hand picture) the absolute conductance of rhodopsin bilayers is of the order of 1.5×10^{-9} mho. The conductance fluctuates in discrete steps of $4 \pm 1.5 \times 10^{-10}$ mho. On illumination (right-hand picture), the membrane conductance increases to a stable value of 7.5×10^{-9} mho; it fluctuates between several discrete levels, all of which appear to be multiples of 0.4 nmho. The macroscopic conductance is ohmic up to 100 mV. As illustrated in Figure 3, the conductance of the bleached membrane is symmetric, in this case being 1.1×10^{-8} mho, and the fluctuations are readily discernible even at this low amplification.

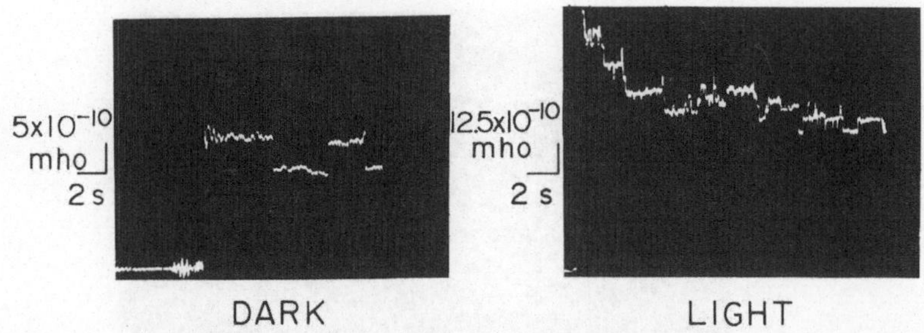

Fig. 2. Membrane currents at constant voltage of rhodopsin-proteolipid bilayers in the presence and absence of illumination. The rhodopsin-proteolipid was prepared as described for Figure 1, except that 10 mg of a partially purified preparation of asolectin (KAGAWA, RACKER, 1971), composed essentially of 37% phosphatidylcholine, 37% phosphatidylethanolamine, and 9% cardiolipin (KAGAWA, RACKER, 1966) were used. Symmetric bilayers were formed by apposing the proteolipid monolayers, as described in detail previously (MONTAL, 1974). Both monolayers were spread by delivering a 10 μl aliquot of the proteolipid in hexane over a clean subphase containing 0.2 M NaCl, 0.01 M Imidazole pH 7, and 1 mM $CaCl_2$. A two-compartment cell was used, the volume of each compartment being 5.0 cc and the area 3.5 cm^2. The area of the aperture where the membrane was formed was 2.8×10^{-4} cm^2. The records illustrate the membrane current in response to a voltage pulse applied across the bilayer. The current measuring device has been described in detail by BAMBERG and LÄUGER (in press). The left-hand photograph illustrates the membrane in the dark: the absolute conductance is of the order of 1.5×10^{-9} mho; the current fluctuates in discrete steps equivalent to $4 \pm 1.5 \times 10^{-10}$ mho. The applied voltage was 19.5 mV. The right-hand photograph presents the membrane 5 min after illumination with white light at an energy of 200 μwatt/cm^2. (Please note the different scale calibration.) The stable membrane conductance fluctuates around a value of 7.5×10^{-9} mho in steps of about 0.4 nmho. The applied voltage was 39 mV. The temperature was 20°C. The macroscopic conductance of this membrane was ohmic up to 100 mV. Above this level, the membrane became very unstable and entered into the region of breakdown. In contrast, bilayers formed with the equivalent lipids (in the absence of rhodopsin) have a stable conductance of about 10^{-11} mho, display small fluctuations (of the order of 10^{-12} mho) only at voltages greater than 100 mV, and break at around 300 mV

At larger amplifications, it is possible to resolve the two distinct types of fluctuations illustrated in Figure 4, where for descriptive purposes the two appear superimposed; two large steps with a lifetime of 2 sec and magnitude of 3.5×10^{-10} mho, and several small steps with a mean lifetime of 0.17 sec and magnitude of $0.45 \pm 0.1 \times 10^{-10}$ mho are shown. The small fluctuations frequently appear to be riding like noisy waves on the large fluctuations at the upper levels of conductance.

As previously noted the hexane solution contains, in addition to spectrally native rhodopsin, denatured opsin and retinal. One obvious possibility is that the conductance and its fluctuations arise from the

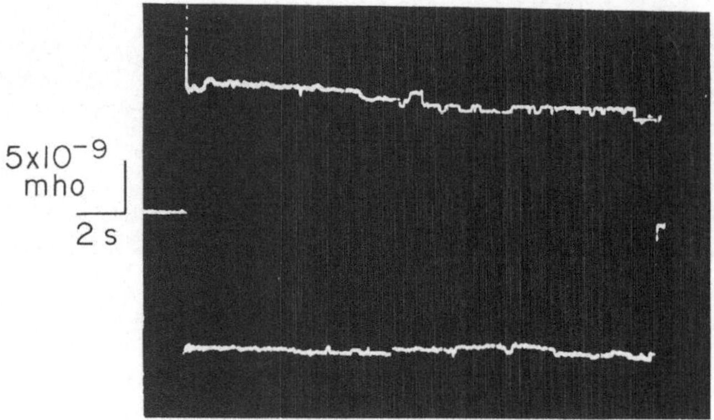

Fig. 3. Membrane currents at constant voltage of bleached rhodopsin-proteolipid bilayers. The conditions are as described for Figure 2. The absolute conductance is 1.1 x 10^{-8} mho, is symmetric, and fluctuates around this value in distinct steps of about 0.5 nmho. The applied voltage was 19.5 mV; temperature, 20°C

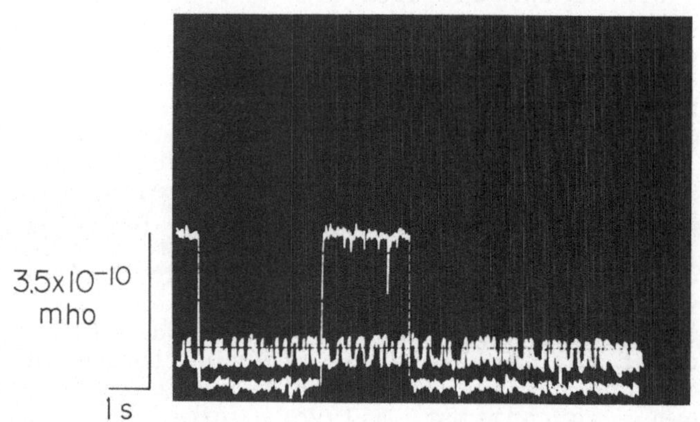

Fig. 4. Current fluctuations of bleached rhodopsin-proteolipid bilayers. The conditions are as described for Figure 2. The large step in the center of the photograph has a magnitude of 0.35 nmho and a lifetime of 2 sec. The small fluctuations have an amplitude of 0.45 ± 0.1 x 10^{-10} mho and a mean lifetime of 0.17 sec. The applied voltage was 10 mV; temperature, 20°C

retinal present in the sample, as well as from its release by illumination from dark rhodopsin. It is well known that retinal is an amphipathic molecule with surfactant capacity (DINGLE, LUCY, 1965; BONTING, BANGHAM, 1968; KOBAMOTO, TIEN, 1971; SCHADT, 1973), and the effects observed could be attributed to local micellization of discrete bilayer domains induced by retinal.

A control experiment shows that when bilayers are formed from monolayers composed only of lipid and an amount of either 11-cis or all-transretinal equivalent to that present in rhodopsin (a molar ratio of 100:1 lipids to retinal, DAEMEN, 1973; SHICHI, 1973), the membrane conductance is 1.5×10^{-10} mho, and the fluctuations are comparable to the small steps detected in proteolipid preparations ($0.45 \pm 0.1 \times 10^{-10}$ mho). In addition, the membranes are not light sensitive. This indicates that the small fluctuations observed in rhodopsin bilayers can be attributed to retinal.

Are the large steps produced by a spurious artifact arising from the manipulations required to extract any protein into a solvent? Control experiments in which a soluble protein, such as cytochrome *c*, is subjected to the manipulations described to obtain the proteolipid of rhodopsin, yield a cytochrome *c* proteolipid (GITLER, MONTAL, 1972b). The bilayers from this proteolipid have a conductance as low as that of equivalent lipids (10^{-11} mho), and only at potentials greater than 50 mV display small fluctuations with an amplitude of 2×10^{-12} mho. It is noteworthy that the large fluctuations in rhodopsin bilayers can be observed even at 10 mV (as in Figure 4), in very stable membranes, so that these cannot be attributed to steps in the breakdown, either spontaneous or dielectric, of the membrane. Thus, the manipulations described do not confer to a lipid-protein complex the activity displayed by the rhodopsin bilayers. As a working hypothesis, we suggest that the 0.4 nmho steps are associated with the presence of opsin; they appear in the dark probably due to the presence of important quantities of opsin; the effect of light is to displace the conductance to higher levels as rhodopsin is bleached.

As mentioned previously, Ca^{++} appears to have a central role in the process of phototransduction. A summary of several experiments is presented in Figure 5. In the absence of Ca^{++} (0.2 M NaCl), the level of conductance is low and does not fluctuate; upon illumination, another level appears and it fluctuates with a mean lifetime of 76 sec. In the presence of 1 mM Ca^{++}, the conductance is larger than in its absence, and it fluctuates in three levels with mean lifetimes going in progressive succession from nonfluctuating to 71 and 25 sec, respectively. On illumination, two additional levels appear, and the two lower levels fuse into a single nonfluctuating one. Thus, it appears that Ca^{++} is not the current carrier, but it does have a modulator effect because the membranes are ostensibly more active.

The most disturbing feature of the results collected to date is the low response time of the membrane. The best experiment obtained so far is presented in Figure 6. A series of five bright flashes was delivered to the membrane and, after a latency of 9 sec after the last

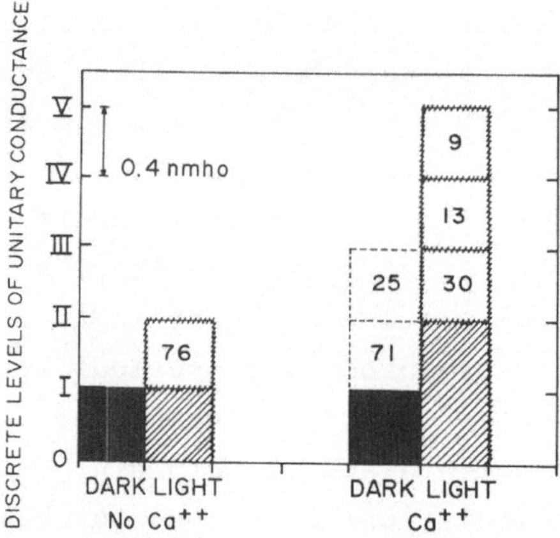

Fig. 5. Current fluctuations of rhodopsin-proteolipid bilayers in the dark and after illumination. The conditions are as described for Figure 2, except that 1 mM $CaCl_2$ was omitted where specified. The data are taken 10 min after the onset of illumination. The unnumbered blocks indicate the nonfluctuating levels of conductance. The numbers indicate the mean lifetime of the fluctuations at that particular level of conductance. The applied voltage was 19.5 mV; temperature, 20°C

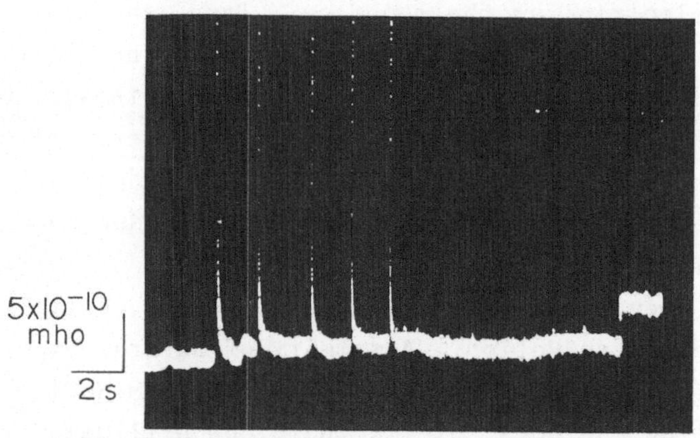

Fig. 6. Membrane current at constant voltage of rhodopsin-proteolipid bilayers in response to five single flashes. The conditions are as described for Figure 2, except that the extraction was performed in 1.0 ml of a mixture of n-hexane (70%) and ethylether (30%). Bright white-light flashes, 1 msec duration, were delivered with a Honeywell Strobonnar 800. The applied voltage was 19.5 mV; temperature, 20°C

flash, a step of 0.4 nmho appeared. The most representative lag time between illumination and onset of conductance changes is 3 min. If the conductance changes observed are related to the mode of action of rhodopsin, there must be a correlation with its bleaching time course. Indeed, we have found that the rhodopsin proteolipid in hexane is bleached in not less than 200 sec at 20°C (white light intensity of 2 mwatt/cm^2 at a distance of 30 cm from the sample). This correlates with the bilayer response time but differs with the bleaching time of rhodopsin in detergent solutions, where it occurs in less than 10 sec at an equivalent light intensity and temperature.

A possible explanation for this discrepancy would be a limited accessibility of the proteolipid in hexane and in the bilayer to H^+ known to be required for the Metarhodopsin I_{480nm} to Metarhodopsin II_{380nm} transition (RADDING, WALD, 1956; WULFF et al., 1958; MATHEWS et al., 1963; OSTROY, ERHARDT, ABRAHAMSON, 1966; FALK, FATT, 1966). This transition appears to be the last step in the dark reactions of rhodopsin, which is fast enough to be involved in the transduction event (HAGINS, 1972).

FUNCTIONAL IMPLICATIONS AND PERSPECTIVE

The experiments described so far are of an exploratory nature and primarily illustrate the primitive stage at which we stand at present. They do show the potential use of the techniques developed and the strategy of our approach in the search of the molecular mechanism of rhodopsin action. At this point, however, a caution is necessary: the values which have been derived from the measurements will need revision when the preparations are improved, when more precise results are obtained, and more refined methods of analysis are applied. Particularly, the analysis of the fluctuations may provide insight into the mechanism. This approach has been recently applied by ZINGSHEIM and NEHER (in press) and KOLB, LÄUGER, and BAMBERG (in press) to study the kinetics of formation of the gramicidin A channel in black lipid films.

The conductance fluctuations observed in the rhodopsin bilayers are reminiscent of those induced in black lipid films and lipid bilayers by molecules presently thought to act by forming transmembrane channels (BEAN et al., 1969; HLADKY, HAYDON, 1970; cf. HAYDON, HLADKY, 1972; EHRENSTEIN et al., 1970, 1974; LA TORRE, EHRENSTEIN, LECAR, 1972; GORDON, HAYDON, 1972; EISENBERG, HALL, MEAD, 1973; BAMBERG, BOHEIM, MONTAL, 1973). Further, the unit conductance of 0.4 nmho indicates a univalent ion flux of 10^8 particles/sec for a 100 mV driving force, which implies a diffusion constant of 10^{-5} cm^2/sec; this is comparable to the diffusion constant of Na^+ in water. The calculated flux is higher by 4 orders of magnitude than the maximal transport rate of an ion carrier such as valinomycin (STARK et al., 1971) but is the expected value for a channel (HILLE, 1970). Let us assume for the sake of discussion that, as suggested by these preliminary experiments, rhodopsin is a transmembrane channel: for a cylindrical pore filled with the bathing electrolyte (0.2 M NaCl) having a length comparable to the membrane thickness (30 Å according to MONTAL and KORENBROT, 1973), its diameter would be about 10 Å (ROBINSON, STOKES, 1959). Such a wide pore would not display, at first approximation, a marked ionic selectivity (cf. EISENMAN, KRASNE, in press). This would be in agreement with the results obtained in both the rhodopsin vesicles and the planar bilayers (see discussion above).

It may not be fortuitous that the conductance steps observed in rhodopsin bilayers have a value of 4 x 10^{-10} mho. To date, the values of the single-channel conductance of various excitable membranes are of the order of 10^{-10} mho: e.g., the chemically activated postsynaptic channel (KATZ, MILEDI, 1972; ANDERSON, STEVENS, 1973; SACHS, LECAR, 1973), the Na^+-channel of nerve (cf. HILLE, 1970; KEYNES, RITCHIE, and ROJAS, 1971), and the black films modified with the excitability-inducing material (BEAN et al., 1969; EHRENSTEIN et al., 1970, 1974).

CHEN and HUBBELL (1973) have prepared rhodopsin vesicles by dialysis with several kinds of phospholipids, and have studied the structure of the recombinant using the freeze-fracture technique: when the frozen recombinants are fractured and etched, the fracture faces display particles on one face and depressions on the complementary face. These findings led CHEN and HUBBELL to conclude that rhodopsin was spanning the entire width of the bilayer. Additional indirect evidence supporting this contention comes from fluorescence energy transfer to 11-cis retinal from different fluorescent donors covalently attached to rhodopsin (WU, STRYER, 1972; STEINEMANN, WU, STRYER, 1973); STRYER et al. inferred from their results that rhodopsin would be an elongated rod with a principal axis of about 70 Å--long enough to traverse the thickness of the membrane (but see also WRIGHT, BROWN, and WALD, 1973 for a recent discussion of the rhodopsin-pore model; also see POO and CONE, 1973).

One may ask the question: Would the opening of a channel provide the amplification characteristic of the transduction process? HAGINS (1972) has elaborated on this point and concluded that "diffusion of Na^+ ions through a single membrane site with a conductance of 20 pmho provides gain enough" (see also CONE, in press).

A second consideration that emerges is that if rhodopsin were a wide pore, why does it appear to allow the equilibration of Na^+ in invertebrate and of Ca^{++} in vertebrate photoreceptors? A possible explanation for this would be that the apparent selectivity of the channel would be a consequence of the predominancy of one ionic species over others in one compartment, this gradient of electrochemical potential arising from the activity of an ion pump. Hence, in the dark, a Ca^{++}-ATPase would concentrate Ca^{++} in the intradiscal space of rod cells and opening of the channel on bleaching would allow its equilibration down its electrochemical gradient. In this regard, BOWNDS et al. (1971) have described the presence of a Ca^{++}-ATPase in highly pure preparation of outer segments, and OSTWALD and HELLER (1972) described a Ca^{++}-ATPase activity in disc membranes and indicated as well that the activity was inhibited by 500 nm of light. Moreover, FALK and FATT (1968, 1972) have described a light-induced change in the conductance of the rod interior (Component II) that could be attributed to an increase in disc membrane conductance of a magnitude such that for each excited rhodopsin molecule, there is a resulting increment in conductance of 0.64×10^{-10} mho.

On the other hand, invertebrate photoreceptors possess a Na^{+}-K^{+} (Strophantidin sensitive) ATPase that in the dark keeps Na^{+} ions out of the cell and thus contributes to the resting membrane potential (BROWN, LISMAN, 1972). On illumination, the channel would open, allowing the influx of Na^{+} with the consequent depolarization of the cell membrane potential (BROWN et al., 1971; HAGINS, 1965, 1972; HAGINS, ADAMS, 1960; HAGINS, ZONANA, ADAMS, 1962; MILLECCHIA, MAURO, 1969). HAGINS (1965) concluded that absorption of single photons by squid photoreceptors resulted in an increase in Na^{+}-conductance of about 0.2×10^{-10} mho.

In *Limulus*, however, the situation is more complex: the response to a single photon can be observed as a transient discrete depolarization, called a quantum bump (cf. FUORTES and O'BRYAN, 1972b). The quantum bumps recorded in the dark-adapted ventral photoreceptor display the same shape of current-voltage curve as the full-size light response. MILLECCHIA and MAURO (1969) have therefore suggested that the increments in conductance associated with the quantum bumps reveal the "unitary" conductances that together elicit the overall light-induced conductance change recorded at higher light levels. The "unitary" conductance, however, is about 5×10^{-8} mho, which is 100 times larger than the single Na^{+}-channel conductance in squid axon (HILLE, 1970). In addition, the kinetics of the photoresponse (cf. FUORTES and O'BRYAN, 1972a) as well as the stochastic properties of single photon responses in *Limulus* (cf. FUORTES and O'BRYAN, 1972b; SREBRO and BEHBEHANI, 1971) can be most adequately fitted by a model in which rhodopsin regulates the plasma membrane conductance by means of a transmitter. Thus, the possibility remains that in invertebrate as in vertebrate photoreceptors, rhodopsin mediates the photoresponse by releasing a transmitter that secondarily modifies the plasma membrane Na^+-channels.

Hence, both in vertebrate (FALK, FATT, 1972) and in invertebrate (HAGINS, 1965) photoreceptors, the absorption of single quantum induces a change in an electrical parameter that can be ascribed to an increase in ion-conductance within an order of magnitude of 10^{-10} mho. Likewise, these photoreceptors seem to be endowed with specific ion pumps required for the dark restoration process by generating the transmembrane electrochemical gradient and perhaps accounting for the selectivity observed. The coupling of downhill ionic movements during activity with uphill movements during recovery is essentially what occurs during the conduction of the nervous impulse (cf. HODGKIN, 1963) and in the Ca^{++}-system of muscle sarcoplasmic reticulum (cf. INESI, 1972). At present, all that can be said is that the data acquired so far are not inconsistent with the model.

Reflecting on the nature of this volume, I have indulged myself with this speculative discussion. One is encouraged, however, to bear in mind the considerations examined, because they are amenable to experimental validation with the distinct approaches hitherto reviewed and the development of more sophisticated ones.

ACKNOWLEDGMENTS

I am indebted to Alberto Darszon and Juan Korenbrot for allowing me to quote from unpublished work, to Ernst Bamberg for the design of the apparatus for the fluctuation measurements, to Adolfo Martínez-Palomo for the electron-micrographs, to G. B. Arden for an informative discussion, and to C. Gómez-Lojero and J. Muñoz for criticism.

REFERENCES

ABRAHAMSON, E. W., FAGER, R. S.: The chemistry of vertebrate and invertebrate visual photoreceptor. In: Current Topics in Bioenergetics (ed. D. R. SANADI) Vol. V, pp. 125-200. New York: Academic Press 1973.

ABRAHAMSON, E. W., FAGER, R. S., MASON, W. T.: Comparative properties of vertebrate and invertebrate photoreceptors. Exp. Eye Res. 18, 51-67 (1974).

AKHTAR, M., BLOSSE, P. T., DEWHURST, P. B.: The active site of the visual protein, rhodopsin. Chem. Commun. 13, 631-632 (1967).

ANDERSON, C. R., STEVENS, C. F.: Voltage clamp analysis of acetylcholine produced end-plate current fluctuations at frog neuromuscular junction. J. Physiol. (London) 235, 655-691 (1973).

APPLEBURRY, M. L., ZUCKERMAN, D. M., LAMOLA, A. A., JOVIN, T. M.: Rhodopsin purification and recombination with phospholipids assayed by the Metarhodopsin I → Metarhodopsin II transition. Biochemistry 13, 3448-3458 (1974).

BAMBERG, E., LÄUGER, P.: Temperature-dependent properties of gramicidin A channels. Biochim. Biophys. Acta (in press).

BAMBERG, E., BOHEIM, G., MONTAL, M.: Unpublished observations, 1973.

BANGHAM, A. D.: Physical structure and behaviour of lipids and lipid enzymes. Adv. Lipid. Res. 1, 65-104 (1963).

BANGHAM, A. D., HILL, M. N., MILLER, N. G. A.: Preparation and use of liposomes as models of biological membranes. In: Methods in Membrane Biology (ed. E. D. KORN) Vol. I, pp. 1-68. New York: Plenum 1974.

BAYLOR, D. A., FUORTES, M. G. F.: Electrical responses of single cones in the retina of the turtle. J. Physiol. (London) 207, 77-92 (1970).

BEAN, R. C., SHEPHERD, W. C., CHAN, H., EICHNER, J. T.: Discrete conductance fluctuations in lipid bilayer protein membranes. J. Gen. Physiol. 53, 741-757 (1969).

BONTING, S. L., BANGHAM, A. D.: On the biochemical mechanism of the visual process. In: Biochemistry of the Eye (ed. M. M. DARDENNE, J. NORDMANN) pp. 493-513. Basel-New York: Karger 1968.

BOWNDS, D.: Site of attachment of retinal in rhodopsin. Nature (London) 216, 1178-1181 (1967).

BOWNDS, D., GORDON-WALKER, A., GAIDE-HUGUENIN, A. C., ROBINSON, W.: Characterization and analysis of frog photoreceptor membranes. J. Gen. Physiol. 58, 225-237 (1971).

BRANTON, D.: Fracture faces of frozen membranes. Proc. Nat. Acad. Sci. (Washington) 55, 1048-1056 (1966).

BROWN, H. M., HAGIWARA, S., KOIKE, H., MEECH, R. W.: Electrical characteristics of a barnacle photoreceptor. Fed. Proc. 30, 69-78 (1971).

BROWN, J. E., LISMAN, J. E.: An electrogenic sodium-pump in *Limulus* ventral photoreceptor cells. J. Gen. Physiol. 59, 720-733 (1972).

BROWN, J. E., PINTO, L. H.: Ionic mechanism for the photoreceptor potential of the retina of *Bufo Marinus*. J. Physiol. (London) 236, 575-591 (1974).

CELIS, H., ESTRADA-O., S., MONTAL, M.: Model translocators for divalent and monovalent ion transport in phospholipid membranes. I. The ion permeability induced in lipid bilayers by the antibiotic X537A. J. Membrane Biol. 18, 187-200 (1974).

CHEN, Y. S., HUBBELL, W. L.: Temperature and light-dependent structural changes in rhodopsin-lipid membranes. Exp. Eye Res. 17, 517-532 (1973).

CONE, R. A.: Rotational diffusion of rhodopsin in the visual receptor membrane. Nature (London) 236, 39-43 (1972).

CONE, R. A.: Rhodopsin, visual excitation and membrane viscosity. In: Perspectives in Membrane Biology (eds. S. ESTRADA-O., C. GITLER). New York: Academic Press, in press.

DAEMEN, F. J. M.: Vertebrate rod outer segment membranes. Biochim. Biophys. Acta 300, 255-288 (1973).

DAEMEN, F. J. M., BONTING, S. L.: Biochemical aspects of the visual process. IV. Aldehydes and cation permeability of artificial phospholipid micelles. Biochim. Biophys. Acta 183, 90-97 (1969).

DINGLE, J. T., LUCY, J. A.: Vitamin A, carotenoids and cell function. Biol. Rev. 40, 422-461 (1965).

EHRENSTEIN, G., BLUMENTHAL, R., LA TORRE, R., LECAR, H.: Kinetics of opening and closing of individual excitability inducing material channels in a lipid bilayer. J. Gen. Physiol. 63, 707-721 (1974).

EHRENSTEIN, G., LECAR, H., NOSSAL, R.: The nature of the negative resistance in bimolecular lipid membranes containing excitability-inducing material. J. Gen. Physiol. 55, 119-133 (1970).

EISENBERG, M., HALL, J. E., MEAD, C. A.: The nature of the voltage-dependent conductance induced by alamethicin in black lipid membranes. J. Membrane Biol. 14, 143-176 (1973).

EISENMAN, G., KRASNE, S.: The Ion-selectivity of carrier molecules, membranes and enzymes. In: MTP International Review of Science. Biochemistry Series (ed. C. F. FOX) Vol. II. London: Butterworths, in press.

FALK, G., FATT, P.: Rapid hydrogen ion uptake of rod outer segments and rhodopsin solutions on illumination. J. Physiol. (London) 183, 211-224 (1966).

FALK, G., FATT, P.: Conductance changes produced by light in rod outer segments. J. Physiol. (London) 198, 647-699 (1968).

FALK, G., FATT, P.: Physical changes induced by light in the rod outer segment of vertebrates. In: Photochemistry of Vision. Handbook of Sensory Physiology (ed. H. J. A. DARTNALL) Vol. VII, pt. 1, pp. 200-244. Berlin-Heidelberg-New York: Springer 1972.

FELDBERG, N. T.: The regeneration of rhodopsin following the removal of detergent. Invest. Ophthalmol. 13, 155-157 (1974).

FUORTES, M. G. F., O'BRYAN, P. M.: Generator potentials in invertebrate photoreceptors. In: Physiology of Photoreceptor Organs. Handbook of Sensory Physiology (ed. M. G. F. FUORTES) Vol. VII, pt. 2, pp. 279-319. Berlin-Heidelberg-New York: Springer 1972a.

FUORTES, M. G. F., O'BRYAN, P. M.: Responses to single photons. In: Physiology of Photoreceptor Organs. Handbook of Sensory Physiology (ed. M. G. F. FUORTES) Vol. VII, pt. 2, pp. 321-338. Berlin-Heidelberg-New York: Springer 1972b.

GITLER, C., MONTAL, M.: Thin-proteolipid films: A new approach to the reconstitution of biological membranes. Biochem. Biophys. Res. Commun. 47, 1486-1491 (1972a).

GITLER, C., MONTAL, M.: Formation of decane-soluble proteolipids; influence of monovalent and divalent cations. FEBS Letters 28, 329-332 (1972b).

GORDON, L. G. M., HAYDON, D. A.: The unit conductance channel of alamethicin. Biochim. Biophys. Acta 255, 1014-1018 (1972).

HAGINS, W. A.: Electrical signs of information flow in photoreceptors. Cold Spr. Harb. Symp. Quant. Biol. 30, 403-418 (1965).

HAGINS, W. A.: The visual process: Excitatory mechanisms in the primary receptor cells. Ann. Rev. Biophys. Bioengineer 1, 131-158 (1972).

HAGINS, W. A., ADAMS, R. G.: Movements of ^{24}Na and ^{42}K in the squid retina. Biol. Bull. 119, 318 (1960).

HAGINS, W. A., PENN, R. D., YOSHIKAMI, S.: Dark current and photocurrent in retinal rods. Biophys. J. 10, 380-412 (1970).

HAGINS, W. A., YOSHIKAMI, S.: A role for Ca^{2+} in excitation of retinal rods and cones. Exp. Eye Res. 18, 299-305 (1974).

HAGINS, W. A., ZONANA, H., ADAMS, R. G.: Local membrane current in the outer segments of squid photoreceptor. Nature (London) 194, 844-847 (1962).

HAYDON, D. A., HLADKY, S. B.: Ion-transport across thin-lipid membranes: A critical discussion of mechanisms in selected systems. Quart. Rev. Biophys. 5, 187-282 (1972).

HEITZMAN, H.: Rhodopsin is the predominant protein of rod outer segment membranes. Nature (London) 235, 114 (1972).

HELLER, J.: Structure of visual pigments. I. Purification, molecular weight and composition of bovine visual pigment 500. Biochemistry 7, 2906-2913 (1968).

HELLER, J., LAWRENCE, M. A.: Structure of the glycopeptide from bovine visual pigment 500. Biochemistry 9, 864-869 (1970).

HENDRICKS, Th., DAEMEN, F. J. M., BONTING, S. L.: Biochemical aspects of the visual process. XXV. Light-induced calcium movements in isolated frog rod outer segments. Biochim. Biophys. Acta 345, 468-473 (1974).

HILLE, B.: Ionic Channels in nerve membranes. Prog. Biophys. Mol. Biol. 21, 1-32 (1970).

HLADKY, S. B., HAYDON, D. A.: Discreteness of conductance change in bimolecular lipid membranes in the presence of certain antibiotics. Nature (London) 225, 451-453 (1970).

HODGKIN, A. L.: The Conduction of the Nervous Impulse. Liverpool: University of Liverpool 1963.

HOLLOWAY, P. W.: A simple procedure for removal of triton X-100 from protein samples. Anal. Biochem. 53, 304-308 (1973).

HONG, K., HUBBELL, W. L.: Preparation and properties of phospholipid bilayers containing rhodopsin. Proc. Nat. Acad. Sci. (Washington) 69, 2617-2621 (1972).

HONG, K., HUBBELL, W. L.: Lipid requirement for rhodopsin regenerability. Biochemistry 12, 4517-4523 (1973).

HUBBARD, R.: Geometrical isomerization of vitamin A, retinene and retinene oxime. J. Am. Chem. Soc. 78, 4662-4667 (1956).

HUBBARD, R., BOWNDS, D., YOSHIZAWA, T.: The chemistry of visual photoreception. Cold Spr. Harb. Symp. Quant. Biol. 30, 301-315 (1965).

HUBBARD, R., KROPF, A.: The action of light on rhodopsin. Proc. Nat. Acad. Sci. (Washington) 44, 130-139 (1958).

HUBBELL, W. L.: Characterization of rhodopsin in synthetic systems. Acc. Chem. Res. (in press).

INESI, G.: Active transport of calcium ions in sarcoplasmic reticulum membranes. Ann. Rev. Biophys. Bioengineer. 1, 191-210 (1972).

KAGAWA, Y.: Dissociation and reassembly of the inner mitochondrial membrane. In: Methods in Membrane Biology (ed. E. D. KORN) Vol. I, pp. 201-267. New York: Plenum 1974.

KAGAWA, Y., RACKER, E.: Partial resolution of the enzymes catalyzing oxidative phosphorylation. IX. Reconstruction of oligomycin-sensitive adenosine-triphosphatase. J. Biol. Chem. 241, 2467-2474 (1966).

KAGAWA, Y., RACKER, E.: Partial resolution of the enzymes catalyzing oxidative phosphorylation. XXV. Reconstitution of vesicles catalyzing ^{32}pi-adenosine-triphosphatase exchange. J. Biol. Chem. 246, 5477-5487 (1971).

KATZ, B., MILEDI, R.: The statistical nature of the acetylcholine potential and its molecular components. J. Physiol. (London) 224, 665-700 (1972).

KEYNES, R. D., RITCHIE, J. M., ROJAS, E.: The binding of tetrodotoxin to nerve membranes. J. Physiol. (London) 213, 235-254 (1971).

KOBAMOTO, N., TIEN, H. Ti.: Light-induced electrical effects in a retinal bilayer lipid membrane. Biochim. Biophys. Acta 241, 129-146 (1971).

KOLB, H. A., LÄUGER, P., BAMBERG, E.: Correlation analysis of electrical noise in lipid bilayer membranes: Kinetics of gramicidin A channels. J. Membrane Biol. (in press).

KORENBROT, J. I.: Ionic flux and membrane characteristics of isolated rod outer segments. Exp. Eye Res. 16, 343-355 (1973).

KORENBROT, J. I., CONE, R. A.: Dark ionic flux and the effects of light in isolated rod outer segments. J. Gen. Physiol. 60, 20-45 (1972).

KROPF, A.: The structure and reactions of visual pigments. In: Physiology of Photoreceptor Organs. Handbook of Sensory Physiology (ed. M. G. F. FUORTES) Vol. VII, pt. 2, pp. 239-278. Berlin-Heidelberg-New York: Springer 1972.
LASANSKY, A., MARCHIAFAVA, P. L.: Light-induced resistance changes in retinal rods and cones of the tiger salamander. J. Physiol. (London) 236, 171-191 (1974).
LA TORRE, R., EHRENSTEIN, G., LECAR, H.: Ion-transport through excitability-inducing material (EIM) channels in lipid bilayer membranes. J. Gen. Physiol. 60, 72-85 (1972).
LIEBMAN, P. A.: Light-dependent Ca^{++} content of rod outer segment disc membranes. Invest. Ophthalmol. 13, 700-701 (1974).
MARTONOSI, A.: Sarcoplasmic reticulum. IV. Solubilization of microsomal adenosine triphosphatase. J. Biol. Chem. 243, 71-81 (1968).
MATHEWS, R. G., HUBBARD, R., BROWN, P. K., WALD, G.: Tautomeric forms of metarhodopsin. J. Gen. Physiol. 47, 215-240 (1963).
MILLECCHIA, R., MAURO, A.: The ventral photoreceptor cells of *Limulus*. III. A voltage-clamp study. J. Gen. Physiol. 54, 331-351 (1969).
MONTAL, M.: Asymmetric lipid bilayers. Response to multivalent ions. Biochim. Biophys. Acta 298, 750-754 (1973).
MONTAL, M.: Formation of bimolecular membranes from lipid monolayers. In: Biomembranes, Cell Organelles and Membrane Components. Methods in Enzymology (eds. S. FLEISCHER, L. PACKER, R. W. ESTABROOK) Vol. 32. New York: Academic Press 1974.
MONTAL, M.: Lipid-protein assembly and the reconstitution of biological membranes. In: Perspectives in Membrane Biology (eds. S. ESTRADA-O., C. GITLER). New York: Academic Press, in press.
MONTAL, M., KORENBROT, J. I.: Incorporation of rhodopsin proteolipid into bilayer membranes. Nature (London) 246, 219-221 (1973).
MONTAL, M., MUELLER, P.: Formation of bimolecular membranes from lipid monolayers and a study of their electrical properties. Proc. Nat. Acad. Sci. (Washington) 69, 3561-3566 (1972).
MUELLER, P., RUDIN, D. O.: Translocators in bimolecular lipid membranes: Their role in dissipative and conservative bioenergy-transductions. In: Current Topics in Bioenergetics (ed. D. R. SANADI) Vol. III, pp. 157-249. New York: Academic Press 1969.
OHNISHI, T., EBASHI, S.: The velocity of Ca^{++}-binding of isolated sarcoplasmic reticulum. J. Biochem. (Tokyo) 55, 599-603 (1964).
OSTROY, S. E., ERHARDT, F., ABRAHAMSON, E. W.: Protein configuration changes in the photolysis of rhodopsin. II. The sequence of intermediates in the thermal decay of cattle metarhodopsin in vitro. Biochim. Biophys. Acta 112, 265-277 (1966).
OSTWALD, T. J., HELLER, J.: Properties of a magnesium- or calcium-dependent adenosine triphosphatase from frog rod photoreceptor outer segment discs and its inhibition by illumination. Biochemistry 11, 4679-4686 (1972).
PENN, R. D., HAGINS, W. A.: Kinetics of photocurrent of retinal rods. Biophys. J. 12, 1073-1094 (1972).
PINTO DA SILVA, P., BRANTON, D.: Membrane intercalated particles: The plasma membrane as a planar fluid domain. Chem. Phys. Lipids 8, 265-278 (1972).
POO, M. M., CONE, R. A.: Lateral diffusion of rhodopsin in *Necturus* rods. Exp. Eye Res. 17, 503-510 (1974).
PRESSMAN, B. C.: Carboxylic ionophores as mobile carriers for divalent ions. In: The Role of Membranes in Metabolic Regulation (ed. M. A. MEHLMAN, R. W. HANSON) pp. 149-164. New York: Academic Press 1972
RACKER, E.: A new procedure for the reconstitution of biologically active phospholipid vesicles. Biochem. Biophys. Res. Commun. 55, 224-230 (1973).

RACKER, E., HINKLE, P. C.: Effect of temperature on the function of a proton-pump. J. Membrane Biol. 17, 181-188 (1974).
RADDING, C. M., WALD, G.: Acid-base properties of rhodopsin and opsin. J. Gen. Physiol. 39, 909-922 (1956).
RAUBACH, R. A., NEMES, P. P., DRATZ, E. A.: Chemical labelling and freeze fracture studies on the localization of rhodopsin in the rod outer segment disk membrane. Exp. Eye Res. 18, 1-12 (1974).
RAZIN, S.: Reconstitution of biological membrane. Biochim. Biophys. Acta 265, 241-296 (1972).
ROBINSON, W. E., GORDON-WALKER, A., BOWNDS, D.: Molecular weight of frog rhodopsin. Nature (London) 235, 112-114 (1972).
ROBINSON, R. A., STOKES, R. H.: Electrolyte Solutions. London: Butterworth 1959.
SACHS, F., LECAR, H.: Acetylcholine noise in tissue culture muscle cells. Nature (London) 246, 214-216 (1973).
SCHADT, M.: Photoresponse of bimolecular lipid membranes pigmented with retinal and vitamin A acid. Biochim. Biophys. Acta 323, 351-366 (1973).
SHICHI, H.: Conformational aspects of rhodopsin associated with disc membranes. Exp. Eye Res. 17, 533-543 (1973).
SHICHI, H., LEWIS, M. S., IRREVERE, F., STONE, A. L.: Biochemistry of visual pigments. I. Purification and properties of bovine rhodopsin. J. Biol. Chem. 244, 529-536 (1969).
SNODDERLY, D. M.: Reversible and irreversible bleaching of rhodopsin in detergent solutions. Proc. Nat. Acad. Sci. (Washington) 57, 1356-1362 (1967).
SREBRO, R., BEHBEHANI, M.: A stochastic model for discrete waves in the *Limulus* photoreceptor. J. Gen. Physiol. 58, 267-286 (1971).
STARK, G., KETTERER, B., BENZ, R., LAUGER, P.: The rate constants of valinomycin-mediated ion-transport through thin-lipid membranes. Biophys. J. 11, 981-994 (1971).
STEINEMANN, A., STRYER, L.: Accessibility of the carbohydrate moiety of rhodopsin. Biochemistry 12, 1499-1502 (1973).
STEINEMANN, A., WU, C. W., STRYER, L.: Conformational aspects of rhodopsin and retinal disc membranes. J. Supramol. Struct. 2, 348-353 (1973).
SZUTS, E. Z., CONE, R. A.: Rhodopsin: Light-activated release of calcium. Fed. Am. Soc. Exp. Biol. Proc. 33, Abstr. 1403 (1974).
TOMITA, T.: Electrical activity of vertebrate photoreceptors. Quart. Rev. Biophys. 3, 179-222 (1970).
TOYODA, J., NOSAKI, H., TOMITA, T.: Light-induced resistance changes in single photoreceptors of *Necturus* and *Gekko*. Vision Res. 9, 453-463 (1969).
VAIL, W. J., PAPAHADJOPOULOS, D., MOSCARELLO, M. A.: Interaction of a hydrophobic protein with liposomes. Evidence for particles seen in freeze-fracture as being proteins. Biochim. Biophys. Acta 345, 463-467 (1974).
WALD, G.: The molecular basis of visual excitation. Nature (London) 219, 800-807 (1968).
WALD, G.: Visual pigments and photoreceptors--review and outlook. Exp. Eye Res. 18, 333-343 (1974).
WALD, G., BROWN, P. K.: The role of sulfhydryl groups in the bleaching and synthesis of rhodopsin. J. Gen. Physiol. 35, 797-821 (1952).
WALD, G., BROWN, P. K.: The molar extinction of rhodopsin. J. Gen. Physiol. 37, 189-200 (1953).
WALD, G., BROWN, P. K., GIBBONS, I. R.: The problem of visual excitation. J. Opt. Soc. Amer. 53, 20-35 (1963).
WRIGHT, W. E., BROWN, P. K., WALD, G.: Orientation of intermediates in the bleaching of shear-oriented rhodopsin. J. Gen. Physiol. 62, 509-522 (1973).

WU, C. W., STRYER, L.: Proximity relationships in rhodopsin. Proc. Nat. Acad. Sci. (Washington) 69, 1104-1108 (1972).
WULFF, V. J., ADAMS, R. G., LINSCHITZ, H., ABRAHAMSON, E. W.: Effects of flash illumination on rhodopsin in solution. Ann. N.Y. Acad. Sci. 74, 281-290 (1958).
YOSHIKAMI, S., HAGINS, W. A.: Ionic basis of dark current and photocurrent of retinal rods. Biophys. Soc. Ann. Meet. Abstr. 10, 60a (1971).
ZINGSHEIM, H. P., NEHER, E.: The equivalence of fluctuation-analysis and chemical relaxation measurements: A kinetic study of ion-pore formation in thin-lipid membranes. Biophys. Chem. 1, (in press).
ZUCKERMAN, R.: Ionic analysis of photoreceptor membrane currents. J. Physiol. (London) 235, 333-354 (1973).

Membrane Transport in Plants

Edited by U. Zimmermann, J. Dainty

252 figures, 49 tables. XIII, 473 pages. 1974
ISBN 3-540-06989-5 Cloth DM 73,–
ISBN 0-387-06989-5 (North America) Cloth $30.00

Contents: Thermodynamics and Electrochemistry of Membrane Transport. – Water Transport and Osmotic Processes. – Electrical Properties of Membranes. – Solute Transport in Algae and Cellsuspension Cultures. – Transport in Isolated Chloroplasts. – ATPases and Transport. – Kinetics of Transport. – Transport in Organs of Higher Plants. – Regulating Factors in Membrane Transport.

This book contains the proceedings of the 'International Workshop on Membrane Transport in Plants', held in February 1974 at the Nuclear Research Center, Jülich (FRG). The Papers cover a broad spectrum of topics in plant physiology, including the thermodynamics of transport processes, water relations, primary reactions of photosynthesis, hormonal regulation, phytochrome interaction with membranes, and the more conventional aspects of membrane transport. The aim was to bring advanced modern concepts of membrane transport to the attention of biologists and to give physical chemists an understanding of complex biological systems.

Springer-Verlag
Berlin
Heidelberg
New York

Prices are subject to change without notice